# An Introduction to Geosynthetic Engineering

# An Introduction to Geosynthetic Engineering

**Sanjay Kumar Shukla**
*School of Engineering, Edith Cowan University,*
*Joondalup, Perth, Australia*

**CRC Press**
Taylor & Francis Group
Boca Raton   London   New York   Leiden

CRC Press is an imprint of the
Taylor & Francis Group, an **informa** business

A BALKEMA BOOK

*CRC Press/Balkema is an imprint of the Taylor & Francis Group, an informa business*

© 2016 Taylor & Francis Group, London, UK

Typeset by V Publishing Solutions Pvt Ltd., Chennai, India
Printed and Bound by CPI Group (UK) Ltd, Croydon, CR0 4YY

Published by: CRC Press/Balkema
P.O. Box 11320, 2301 EH Leiden, The Netherlands
e-mail: Pub.NL@taylorandfrancis.com
www.crcpress.com – www.taylorandfrancis.com

*Library of Congress Cataloging-in-Publication Data*

Applied for

ISBN: 978-1-138-02774-9 (Hbk)
ISBN: 978-1-4987-7809-1 (eBook PDF)

Printed and bound in the United States of America by
Edwards Brothers Malloy on sustainably sourced paper

# Table of contents

# About the author

**Dr Sanjay Kumar Shukla** is the Founding Editor-in-Chief of the *International Journal of Geosynthetics and Ground Engineering* (Springer International Publishing). He is an Associate Professor and Program Leader of Civil and Environmental Engineering at the School of Engineering, Edith Cowan University, Perth, Australia. He is also a Distinguished Professor of Civil Engineering at the Institute of Engineering and Technology, Chitkara University, Himachal Pradesh, India. He graduated in 1988 with a first-class degree with distinction in Civil Engineering from BIT Sindri (Ranchi University, Ranchi), India, and earned his MTech in Civil Engineering (Engineering Geology) in 1992 and PhD in Civil Engineering (Geotechnical Engineering) in 1995 from the Indian Institute of Technology Kanpur, India. He has over twenty years of teaching, research and consultancy experience in the field of Geotechnical and Geosynthetic engineering. He is the author of ICE textbooks entitled *'Core Principles of Soil Mechanics'* and *'Core Concepts of Geotechnical Engineering'* published in 2014 and 2015, respectively, and has authored/co-authored/edited 5 other books and 12 book chapters. He is also the author/co-author of more than 150 research papers and technical articles, including over 100 refereed journal publications. He is a fellow of Engineers Australia, Institution of Engineers (India) and the Indian Geotechnical Society, and a member of the American Society of Civil Engineers (ASCE), the International Geosynthetics Society and the Indian Roads Congress. He serves on the editorial boards of *International Journal of Geotechnical Engineering, Ground Improvement, Geotechnical Research, Indian Geotechnical Journal* and *Cogent Engineering*, and he is the Scientific Editor of *Journal of Mountain Science*.

# Preface

Development and use of polymeric materials in the form of geosynthetics and fibres as a new class of construction materials have revolutionized the infrastructure and the environmental protection works in the construction industry during the past three to four decades. Geosynthetics are available in a wide range of compositions appropriate to different applications and environments. Engineers have grown increasingly interested in geosynthetics and their correct use because geosynthetics often provide efficient, cost-effective, environment-friendly and/or energy-efficient solutions to several problems. Simultaneously, significant advances have been made in the use of geosynthetics in civil (geotechnical, transportation, hydraulic and environmental) engineering applications as well as in some areas of mining, agricultural and aquacultural engineering. These developments have occurred because of the ongoing dialogue among leading engineers and researchers from industries and academic institutions.

Every four years since 1982, the engineering community holds the International Conference on Geosynthetics. There are currently three research journals, namely *Geotextiles and Geomembranes*, *Geosynthetics International*, and *International Journal of Geosynthetics and Ground Engineering*, dealing with different engineering aspects of geosynthetics and their applications, known as geosynthetic engineering. A few books dealing with geosynthetic engineering under different titles have been written in the past to cover the developments in this area. Each of them serves a purpose for students, researchers and practising engineers.

This textbook presents an introduction to geosynthetic engineering with up-to-date and in-depth coverage of the subject within nine chapters, which are primarily designed and developed for a one-semester course for senior undergraduate and postgraduate students as a part of a geotechnical/civil engineering program offered in the universities/institutes/colleges worldwide. The material in all the chapters of this book is presented clearly in simple and plain English, and includes the optimum amount of text and number of illustrations, tables, examples and questions/problems for practice. The book covers the basic concepts, and explains the fundamentals of analysis and design of most geosynthetic applications, including use of fibre-reinforced soils. Each chapter includes many useful references, quoted in the text and listed at the end of the chapter, for further study. As the practical solution to an engineering problem often requires the application of engineering judgment and experience, which can be acquired by regular professional practice and self-study, an attempt has been made to provide practical experience, including field application guidelines and case studies, throughout the book. The chapter summary presented at the end may help readers master some key learning points easily. Since I have written this introductory textbook based on over 20 years of personal teaching, research and consultancy experience in geotechnical and geosynthetic engineering as a student-centered learning resource, the students of civil, mining, agricultural and aquacultural engineering will find this textbook very useful for their needs in their study of geosynthetics and their applications and learning the fundamentals of the subject without any major assistance. In fact, some readers can learn the subject even by self-study. Apart from students, researchers and teachers, this textbook will be a valuable resource for the practising engineers working in the areas of civil, mining, agricultural and aquacultural engineering to refresh their learning of the basic concepts of geosynthetic engineering while dealing with construction with geosynthetics in their projects.

As already mentioned, for the applications of geosynthetics, the details are described focusing on their basic description, analysis and design concepts, application guidelines, and some case studies. *Chapter 1* provides a basic description of geosynthetics including their types, basic characteristics and manufacturing processes. The functions that can be performed by commercially available geosynthetics and several general aspects related to their selection are also described in this chapter. *Chapter 2* deals with the properties of geosynthetics and highlights the basic concepts of their determination along with their importance in the design process and the performance in field applications. This chapter also presents the basic concepts of design methods for the geosynthetic-related structures or systems. *Chapter 3* focuses on presenting the details of major geotechnical applications of geosynthetics, especially use of geosynthetics in retaining walls, embankments, shallow foundations and slopes. *Chapter 4* covers the use of geosynthetics in hydraulic and geoenvironmental engineering, especially in filters and drains, erosion control for slopes, containment facilities and tunnel linings. *Chapter 5* describes the applications of geosynthetics in transportation structures, especially in unpaved roads, paved roads and railway tracks. *Chapter 6* introduces some applications of geosynthetics, which are implemented specifically in mining, agricultural and aquacultural projects in order to benefit from either improved productivity or lower cost of operations in an environmentally friendly manner. For the success of the geosynthetic applications, the geosynthetics must be installed properly without any significant damages. In view of these requirements, to help the users of geosynthetics, *Chapter 7* describes the general application guidelines and installation survivability requirements. *Chapter 8* contains the details of quality evaluation, field performance monitoring, and the fundamentals of economic evaluation and related experiences from some completed geosynthetic-related projects and reported economic studies. *Chapter 9* presents the fundamental aspects of fibre-reinforced soils, including phase concepts, engineering behaviour, reinforcement models, and possible field application areas and guidelines.

I would like to thank Janjaap Blom, Senior Publisher, Alistair Bright, Acquisitions Editor, Lukas Goosen, Production Manager and other staff of CRC Press / Balkema, Taylor & Francis Group for their full support and cooperation at various stages of the preparation and production of this textbook.

I wish to extend sincere appreciation to my wife, Sharmila, for her encouragement and support throughout the preparation of the manuscript. I would also like to thank my daughter, Sakshi, and my son, Sarthak, for their patience during my work on this textbook at home.

Finally, I welcome suggestions from the readers and users of this textbook for improving its content in future editions.

Sanjay Kumar Shukla
*Perth, 2016*

# Chapter 1

# Basic description, functions and selection of geosynthetics

## 1.1  INTRODUCTION

The term 'geosynthetics' has two parts: the prefix 'geo', referring to an end use associated with improving the performance of civil engineering works involving earth/ground/soil/rock, and the suffix 'synthetics', referring to the fact that the materials are almost exclusively from man-made products. The materials used in the manufacture of geosynthetics are primarily synthetic polymers generally derived from the crude petroleum oils, although other materials such as rubber, fibreglass and bitumen are also sometimes used for manufacturing geosynthetics. Geosynthetics is, in fact, a generic name representing a broad range of planar products, such as *geotextiles*, *geogrids*, *geonets*, *geomembranes*, *geocells*, *geofoams*, *geocomposites*, etc., which are manufactured mainly from the polymeric materials and used in contact with soil, rock and/or any other civil engineering-related material as an integral part of a man-made project, structure, or system.

Over the past 35–40 years, civil engineers have shown an increasing interest in geosynthetics and their field applications, mainly because the use of geosynthetics offers technically efficient, cost-effective, environmentally friendly and/or energy-efficient alternative solutions to many civil engineering problems. Moreover, geosynthetics have now been added to the list of traditional construction materials such as soil, rock, brick, lime, cement, bitumen and steel in the relevant standards and codes of practice worldwide. Geosynthetics are also used in areas other than civil engineering, such as mining, agricultural and aquacultural engineering. Geosynthetics technology is basically a composite science involving the skills of polymer technologists, chemists, production engineers and application engineers. The applications of geosynthetics mainly fall within the discipline of civil engineering, and the analysis and design of these applications, due to the use of geosynthetics with soils, rocks and other similar materials, are closely associated with geotechnical engineering. For a given application, the knowledge of geotechnical engineering only serves to define and enumerate the functions and properties of a geosynthetic (Ingold, 1994). Note that for any application of a geosynthetic, there can be one or more functions that the geosynthetic will be expected to serve during its performance life. The selection of a geosynthetic for any field application is highly governed by the function(s) to be performed by the geosynthetic in that specific application as discussed in Section 1.6 and Section. 1.7.

The subject of geosynthetics and their applications is known as the 'Geosynthetic engineering', which is defined as follows (Shukla, 2013):

> Geosynthetic engineering deals with the application of scientific principles and methods to the acquisition, interpretation and use of knowledge of geosynthetic products for the solution to the problems in geotechnical, transportation, environmental and hydraulic engineering, and also in some areas of agriculture, aquaculture and mining engineering.

This book presents an introduction to geosynthetic engineering, focusing on its definition. This chapter provides a basic description of geosynthetics including their types, basic characteristics and manufacturing processes. The functions that can be performed by commercially available geosynthetics and several general aspects related to their selection are also described in this chapter.

## 1.2 TYPES OF GEOSYNTHETICS

Geosynthetics are commercially available in numerous types in the market with different product/brand names and descriptive numbers/codes, such as Tensar SS40, Secutex 301 GRK5, Terram W/20–4, Netlon CE131, etc. However, the most commonly available geosynthetics are classified into the following major types (Figs. 1.1–1.6):

- Geotextiles
- Geogrids
- Geonets
- Geomembranes
- Geocells
- Geofoams
- Geocomposites

*Geotextile*: It is a planar, permeable, polymeric textile product in the form of a flexible sheet (Fig. 1.1). Currently available geotextiles are classified into the following categories based on the manufacturing process:

- *Woven geotextile*: A geotextile produced by interlacing, usually at right angles, two or more sets of yarns (made of one or several fibres) or other elements using a conventional weaving process with a weaving loom (Fig. 1.1(a)).
- *Nonwoven geotextile*: A geotextile produced from directionally or randomly oriented fibres into a loose web by bonding with partial melting, needle-punching, or chemical binding agents (glue, rubber, latex, cellulose derivative, etc.) (Fig. 1.1(b)).
- *Knitted geotextile*: A geotextile produced by interlooping one or more yarns (or other elements) together with a knitting machine, instead of a weaving loom (Fig. 1.1(c)).
- *Stitched geotextile*: A geotextile in which fibres or yarns or both are interlocked/bonded by stitching or sewing (Fig. 1.1(d)).

*Figure 1.1* Typical geotextiles: (a) woven geotextile; (b) nonwoven geotextile; (c) knitted geotextile; (d) stitched geotextile.

*Geogrid*: This is a planar, polymeric product consisting of a mesh or net-like regular open network of intersecting tensile-resistant elements, called the *ribs*, integrally connected at the junctions (Fig. 1.2). The ribs can be linked by extrusion, bonding or interlacing; the resulting geogrids are called *extruded geogrid*, *bonded geogrid* and *woven geogrid*, respectively. The extruded geogrids are classified into the following three categories, principally based on the direction of stretching during their manufacture:

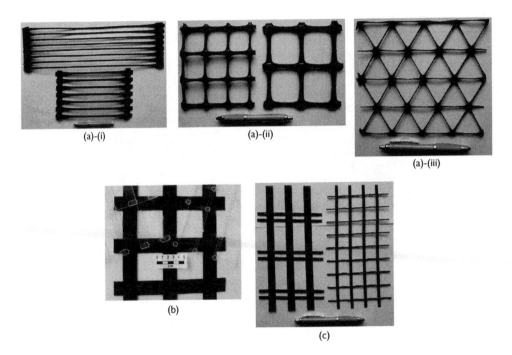

(a)-(i)  (a)-(ii)

(a)-(iii)

(b)

(c)

Figure 1.2 Typical geogrids: (a) extruded geogrids – (i) uniaxial, (ii) biaxial, (iii) triaxial; (b) bonded geogrids; (c) woven geogrids.

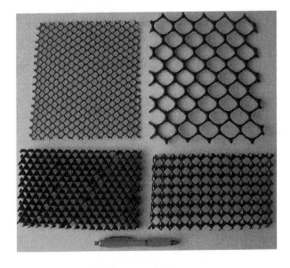

Figure 1.3 Some typical geonets.

*Figure 1.4* Some typical geomembranes.

- *Uniaxial geogrid*: A geogrid produced by the longitudinal stretching of a regularly punched polymer sheet, and therefore it possesses a much higher tensile strength in the longitudinal direction than the tensile strength in the transverse direction (Fig. 1.2(a)-(i)).
- *Biaxial geogrid*: A geogrid produced by stretching in both the longitudinal and the transverse directions of a regularly punched polymer sheet, and therefore it possesses an equal tensile strength in both the longitudinal and the transverse directions (Fig. 1.2(a)-(ii)).
- *Triaxial geogrid*: A geogrid produced to have an equal tensile strength in multi-directions, that is, almost 360° tensile properties (Fig. 1.2(a)-(iii)).

*Geonet*: This is a planar, polymeric product consisting of a regular dense network of integrally connected parallel sets of ribs overlying similar sets at various angles (Fig. 1.3). At first glance, the geonets appear similar to the geogrids; however, the geonets are different from the geogrids, not mainly in the material or their configuration, but in their functions to perform the in-plane drainage of liquids or gases, as described in Section 1.6.

*Geomembrane*: This is a planar, relatively impermeable, synthetic sheet manufactured from materials of low permeability to control the fluid migration in a project as a barrier or liner (Fig. 1.4). The materials may be polymeric or asphaltic or a combination thereof. The term *barrier* applies when the geomembrane is used within an earth mass. The term *liner* is usually reserved for the cases where the geomembrane is used as an interface or a surface revetment.

*Geocell*: This is a three-dimensional, permeable, polymeric honeycomb or web structure, produced in the factory using the strips of needle-punched polyester or solid

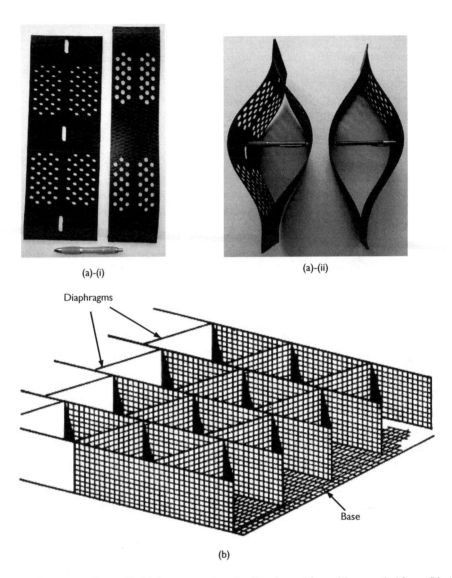

(a)-(i)

(a)-(ii)

Diaphragms

Base

(b)

*Figure 1.5* Some typical geocells: (a) factory produced – (i) collapsed form, (ii) expanded form; (b) site assembled.

high density polyethylene (Fig. 1.5(a), or assembled from geogrids and special bodkin couplings in triangular or square cells (Fig. 1.5(b)).

*Geofoam*: This is a lightweight product in slab or block form with a high void content, and has applications primarily as lightweight fills, thermal insulators and drainage channels. It is manufactured by the application of the polymer in semi-liquid form through the use of a foaming agent.

*Geocomposite*: This is a term applied to the product that is assembled or manu-factured in laminated or composite form from two or more materials, at least one of

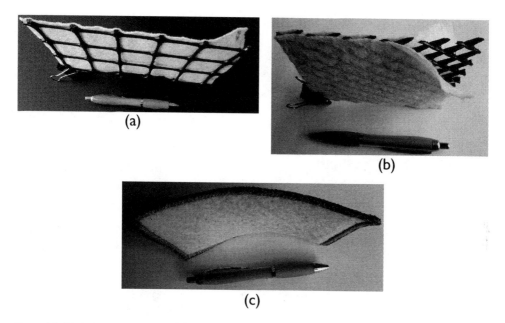

*Figure 1.6* Typical geocomposites: (a) reinforced drainage separator; (b) drainage composites; (c) geosynthetic clay liner.

which is a geosynthetic (geotextile, geogrid, geonet, geomembrane, or any other type), which, in combination, perform specific function (s) more effectively than when used separately. Figure 1.6 shows some typical geocomposites.

There are a large number of geosynthetics available today, including grids, nets, meshes, webs and composites, which are technically not textiles; however, they are used in combination with or in place of geotextiles. All such products are often called *geotextile-related products* (GTP). Some common GTP and other types of geosynthetic products are briefly described below.

- *Geofabric*: A planar flat sheet of geotextile or geotextile-related products.
- *Geomat*: A three-dimensional, permeable, polymeric structure made of coarse and rigid filaments bonded at their junctions, used to reinforce roots of vegetation such as grass and small plants and extend the erosion control limits of vegetation for permanent installation (Fig. 1.7(a)).
- *Geomesh*: A geosynthetic or geonatural generally with a planar woven structure having large pore sizes, which vary from several millimetres to several centimetres for use mainly in erosion control works (Fig. 1.7(b)).
- *Geopipe*: A plastic pipe (smooth or corrugated with or without perforations) placed beneath the ground surface and subsequently backfilled (Fig. 1.7(c)).
- *Geotube*: A factory assembled geosynthetic product, made from high strength woven geotextile or specifically designed and manufactured needle-punched

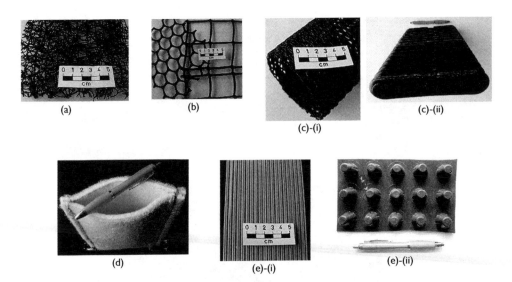

(a)

(b)

(c)-(i)

(c)-(ii)

(d)

(e)-(i)

(e)-(ii)

*Figure 1.7* Some typical geotextile-related products: (a) geomat; (b) geomeshes; (c) geopipes –
(i) circular cross-section, (ii) elliptical cross-section; (d) geotube; (e) geospacers.

nonwoven geotextile using staple fibres in tube form (Fig. 1.7(d)). This is hydraulically filled with sand, soil, recycled material, treated materials or a combination
thereof for coastal and dewatering applications such as shoreline/riverbank protection by constructing walls and revetments, construction of groynes and breakwaters,
etc. Geotubes can also be used to construct beach access stairs, raised garden beds,
artificial reefs, and emergency/temporary structures (bunds/cofferdams). In brief, the
geotubes are extremely robust geotextile containers used for several applications.

- *Geospacer*: A three-dimensional polymeric moulded structure consisting of cuspi-
  dated or corrugated plates with large void spaces (Figure 1.7(e)).
- *Geostrip*: A polymeric material in the form of a strip.
- *Electrokinetic geosynthetic (GEK)*: A mesh made from metal wire stringers coated
  in a conductive polymer; it resembles a reinforcing geomesh/geonet and is available in the form of sheets, strips or tubes.

   Products, based on natural fibres (jute, coir, cotton, wool, etc.), are also being used
in contact with soil, rock and/or other civil engineering-related material, especially in
temporary civil engineering applications. Such products, called the *geonaturals*, have
a short life span when used with earth materials due to their biodegradable characteristics, and therefore, they have not as many field applications as the geosynthetics
have (Shukla, 2003a). Though the geonaturals are significantly different from the
geosynthetics in material characteristics, they can be considered as a complementary
companion of geosynthetics, rather than a replacement, mainly because of some common field application areas. In fact, the geonaturals are also polymeric materials since
they contain a large proportion of naturally occurring polymers such as lignin and
cellulose. Figure 1.8 shows some geonaturals.

*Figure 1.8* Typical geonaturals: (a) geojutes – (i) woven, (ii) nonwoven; (b) geocoir (woven); (c) geonatural composites as erosion control mats.

*Table 1.1* Abbreviations and graphical symbols of geosynthetic products as recommended by the International Geosynthetics Society.

| Geosynthetics | Graphical symbols | Abbreviations |
|---|---|---|
| Geotextile | ━ ━ ━ ━ ━ ━ ━ ━ ━ ━ ━ ━ ━ | GTX |
| Geomembrane | ━━━━━━━━━━━━━━━━ | GMB |
| Geobar | ⁄⁄⁄⁄⁄⁄⁄⁄⁄⁄⁄⁄⁄⁄⁄ | GBA |
| Geoblanket | ∿∿∿∿∿∿∿∿ | GBL |
| Geocomposite drain with Geotextiles on both sides | ▽▽▽▽▽▽▽▽ | GCD |
| Geocell | IIIIIIIIIIIIIIIIIIIIIIIIIIIIIIIIIIIIIIIIIIII | GCE |
| Geocomposite clay layer | ⁊⁊⁊⁊⁊⁊⁊⁊⁊⁊⁊⁊⁊⁊⁊ | GCL |
| Surficial geosynthetic erosion control | ############################ | GEC |
| Electrokinetic geosynthetic | ƶƶƶƶƶƶƶƶƶƶƶƶƶƶƶƶƶƶƶƶƶƶƶƶƶƶƶ | GEK |
| Geogrid | ━●━●━●━●━●━ | GGR |
| Geomat | ∿∿∿∿∿∿∿ | GMA |
| Geomattress | ▨▨▨▨▨▨▨▨▨▨ | GMT |
| Geonet | ✕✕✕✕✕✕✕✕✕✕✕✕✕✕✕✕✕ | GNT |
| Geospacer | ⨆⨆⨆⨆⨆⨆⨆⨆⨆ | GSP |
| Geostrip | ━•━•━•━•━•━• | GST |

For convenience in making drawings or diagrams of geosynthetic applications with clarity, the geosynthetic products can be represented by abbreviations and/or graphical symbols (see Table 1.1) as recommended by the International Geosynthetics Society. The abbreviations GST and GCP can be used to represent geosynthetic and geocomposite, respectively.

## 1.3  BASIC CHARACTERISTICS OF GEOSYNTHETICS

Geosynthetics are versatile in use, adaptable to many field situations, and can be combined with several building materials. They are utilized in a range of applications in many areas of civil engineering, especially geotechnical, transportation, water resources, environmental, coastal, sediment and erosion control engineering, mining engineering, and agricultural and aquacultural engineering for achieving technical and/or economic benefits. The rapid growth in the past few decades all over the world is due mainly to the following favourable basic characteristics of geosynthetics:

- non-corrosiveness
- highly resistant to biological and chemical degradation
- long-term durability under soil cover up to 120 years
- high flexibility
- minimum volume
- lightness
- ease of storing and transportation
- simplicity of installation
- speeding the construction process
- making economical and environment-friendly solutions
- providing a good aesthetic look to structures.

The importance of geosynthetics can also be observed in their ability to partially or completely replace natural resources such as gravel, sand, clay, etc. In fact, the geosynthetics can be used for achieving better durability, aesthetics and environment of the civil engineering projects, and also of some projects in mining, agricultural and aquacultural engineering.

## 1.4 RAW MATERIALS FOR GEOSYNTHETICS

Almost exclusively, the raw materials from which the geosynthetics are produced are polymeric. Polymers are materials of very high molecular weight and are found to have multifarious applications in modern society. The polymers used to manufacture the geosynthetics are generally thermoplastics, which may be amorphous or semi-crystalline. Such materials melt on heating and solidify on cooling. The heating and cooling cycles can be applied several times without affecting the properties.

Any polymer, whether amorphous or semi-crystalline, consists of long chain molecules containing many identical chemical units bound together by covalent bonds. Each unit may be composed of one or more small molecular compounds called the *monomers*, which are most commonly hydrocarbon molecules. The process of joining monomers, end to end, to form long polymer chains is called the *polymerization*. In Figure 1.9, it is observed that that during the polymerization, the double carbon bond of the ethylene monomer forms a covalent bond with the carbon atoms of neighbouring monomers. The end result is a long polyethylene (PE) chain molecule in which one carbon atom is bonded to the next. If the chains are packed in a regular form and are highly ordered, the resulting configuration will have a crystalline structure, otherwise an amorphous structure. No polymers used for manufacturing the geosynthetics are completely crystalline, although the high density polyethylene (HDPE) can attain 90% or so crystallinity, but some

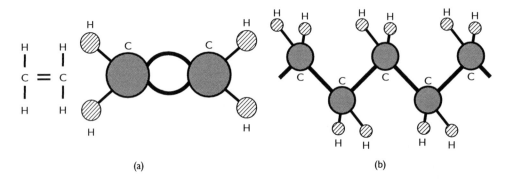

*Figure 1.9* The process of polymerization: (a) an ethylene monomer; (b) a polyethylene molecule.

are completely amorphous. Manufacture of polymers is generally carried out by chemical or petrochemical companies who produce polymers in the form of solid pellets, flakes or granules.

The number of monomers in a polymer chain determines the length of the polymeric chain and the resulting molecular weight. The molecular weight can affect physical and mechanical properties, heat resistance, and durability (resistance to chemical and biological attack) properties of geosynthetics. The physical and mechanical properties of the polymers are also influenced by the bonds within and between chains, the chain branching, and the degree of crystallinity. An increase in the degree of crystallinity leads directly to an increase in rigidity and tensile strength, hardness and softening point, and to a decrease in chemical permeability.

If the polymer is stretched in the melt, or in solid form above its final operating temperature, the molecular chains become aligned in the direction of stretch. This alignment, or *molecular orientation*, can be permanent if, still under the stress, the material is cooled to its operating temperature. The orientation of molecules of the polymers by mechanical drawing results in higher tensile properties and improved durability of the fibres. The properties of polymers can also be altered by including the introduction of side branches, or grafts, to the main molecular chain.

The polymers used for manufacturing geosynthetics are listed in Table 1.2 along with their commonly used abbreviations. The more commonly used types are polypropylene (PP), high density polyethylene (HDPE) and polyester (polyethylene terephthalate (PET)).

Most of the geotextiles are manufactured from PP or PET (see Table 1.3). Polypropylene is a semi-crystalline thermoplastic with a melting point of 165 °C and a density in the range of 0.90 to 0.91 g/cm³. Polyester is also a thermoplastic with a melting point of 260 °C and a density in the range of 1.22 to 1.38 g/cm³.

The primary reason for use of PP in geotextile manufacturing is its low cost. For non-critical structures, PP provides an excellent, cost-effective raw material. It exhibits a second advantage in that it has excellent chemical and pH range resistance

Table 1.2 Polymers commonly used for the manufacture of geosynthetics.

| Types of polymer | | Abbreviations |
|---|---|---|
| Polypropylene | | PP |
| Polyester (polyethylene terephthalate) | | PET |
| Polyethylene | Low density polyethylene | LDPE |
| | Very low density polyethylene | VLDPE |
| | Linear low density polyethylene | LLDPE |
| | Medium density polyethylene | MDPE |
| | High density polyethylene | HDPE |
| | Chlorinated polyethylene | CPE |
| | Chlorosulfonated polyethylene | CSPE |
| Polyvinyl chloride | | PVC |
| Polyamide | | PA |
| Polystyrene | | PS |

Note: The basic materials consist mainly of the elements carbon, hydrogen, and sometimes nitrogen and chlorine; they are produced from coal and petroleum oil.

Table 1.3 Major geosynthetics and the most commonly used polymers for their manufacture.

| Geosynthetics | Polymers used for manufacturing |
|---|---|
| Geotextiles | PP, PET, PE, PA |
| Geogrids | PET, PP, HDPE |
| Geonets | MDPE, HDPE |
| Geomembranes | HDPE, LLDPE, VLDPE, PVC, CPE, CSPE, PP |
| Geofoams | PS |
| Geopipes | HDPE, PVC, PP |

because of its semi-crystalline structure. Additives and stabilizers (such as carbon black) must be added to give PP ultraviolet light (UV) resistance during processing. As the critical nature of the structure increases, or the long-term anticipated loads go up, PP tends to lose its effectiveness. This is because of relatively poor creep deformation characteristics under long-term sustained load.

Polyester (PET) is increasingly being used for manufacturing the reinforcing geosynthetics such as geogrids because of high strength and high resistance to creep. The chemical resistance of polyester is generally excellent, with the exception of very high pH environments. It is inherently stable to UV light.

Polyethylene (PE) is one of the simplest organic polymers used extensively in the manufacture of geomembranes. It is used in its low density and less crystalline form (LDPE), which is known for its excellent pliability, ease of processing and good physical properties. It is also used as high density polyethylene (HDPE), which is more rigid and chemically more resistant.

Polyvinyl chloride (PVC) is the most significant commercial member of the family of vinyl-based resins. With plasticizers and other additives, it takes up a great variety of forms. Unless PVC has suitable stabilizers, it tends to become brittle and darken when exposed to UV light over time, and can undergo heat-induced degradation.

Polyamides (PA), better known as nylons, are melt processable thermoplastics that contain an amide group as a recurring part of the chain. Polyamide offers a combination of properties including high strength at elevated temperatures, ductility, wear and abrasion resistance, low frictional properties, low permeability for gases and hydrocarbons, and good chemical resistance. Its limitations include a tendency to absorb moisture, with resulting changes in physical and mechanical properties, and limited resistance to acids and weathering.

The raw material for the manufacture of geofoams is polystyrene (PS), which is known as a packaging material and insulating material in everyday life. Its genesis is from ethylene and is available in two forms: expanded polystyrene (EPS), and extruded polystyrene (XPS).

There are several environmental factors that affect the durability of polymers. The ultraviolet component of solar radiation, heat and oxygen, and humidity are the factors above ground that may lead to degradation. Below the ground, the main factors affecting the durability of polymers are soil particle size and angularity, acidity/alkalinity, heavy metal ions, presence of oxygen, moisture content, organic content, and temperature. The resistance of commonly used polymers to some environmental factors is compared in Table 1.4. It must be emphasized that the involved reactions are usually slow and can be retarded even more by the use of suitable additives. When the polymers are subjected to a higher temperature, they lose their weight. What remains above 500 °C is probably carbon black and ash (Fig. 1.10). Note that the ash content of a polymeric compound is the remains of inorganic ingredients used as fillers or cross-linking agents and the ash from the base polymer.

Table 1.4 A comparison of the resistance of polymers, commonly used in the production of geosynthetics (adapted from John, 1987; Shukla, 2002a).

| Influencing factors | Resistance of polymers | | | |
| --- | --- | --- | --- | --- |
| | PP | PET | PE | PA |
| Ultraviolet light (unstabilized) | Medium | High | Low | Medium |
| Ultraviolet light (stabilized) | High | High | High | Medium |
| Alkalis | High | Low | High | High |
| Acids | High | Low | High | Low |
| Salts | High | High | High | High |
| Detergents | High | High | High | High |
| Heat, dry (up to 100 °C) | Medium | High | Low | Medium |
| Steam (up to 100 °C) | Low | Low | Low | Medium |
| Hydrolysis (reaction with water) | High | High | High | High |
| Micro-organisms | High | High | High | Medium |
| Creep | Low | High | Low | Medium |

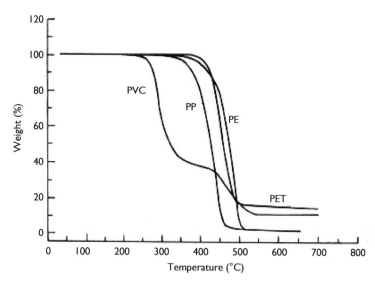

*Figure 1.10* Effect of temperature on some geosynthetic polymers (after Thomas and Verschoor, 1988).

The formulation of a polymeric material is a complex task. No geosynthetic material is 100% of the polymer resin associated with its name, because pure polymers are not suitable for production of geosynthetics. The primary resins are always formulated with additives, fillers, and/or other agents as UV light absorbers, antioxidants, thermal stabilizers, etc. to produce a plastic with the required properties. For example, PE, PET and PA have 97% resin, 2–3% carbon black (or pigment) and 0.5–1.0% other additives.

If the long molecular chains of the polymer are cross-linked to one another, the resulting material is called the *thermoset*, which, once cooled, remains solid upon the subsequent application of heat. Though a thermoset does not melt on reheating, it degrades. The thermoset materials (such as ethylene vinyl acetate (EVA), butyl, etc.), alone or in combination with thermoplastic materials, are sometimes used to manufacture the geomembranes.

Although most geosynthetics are made from synthetic polymers, a few specialist geosynthetics, especially geotextiles, may also incorporate steel wire or natural biodegradable fibres such as jute, coir, paper, cotton, wool, silk, etc. The biodegradable geotextiles are usually limited to erosion control applications where the natural vegetation will replace the role of geotextile as it degrades. The jute nets are marketed under various trade names, including *geojute, soil-saver, and anti-wash*. They are usually in the form of a woven net with a typical mesh open size of about 10 by 15 mm, a typical thickness of about 5 mm, and an open area of about 65%. Vegetation can easily grow through openings and use the fabric matrix as the support. The jute, which is about 80% natural cellulose, should completely degrade in about two years. An additional advantage of these biodegradable products is that the decomposed jute improves the quality of the soil for vegetation growth. Some

*Table 1.5* Chemical identification tests (based on Halse *et al.*, 1991).

| Method | Information obtained |
|---|---|
| Thermogravimetric analysis (TGA) | Polymer, additives and ash contents; carbon black content; decomposition temperatures |
| Differential scanning calorimetry (DSC) | Melting point, degree of crystallinity, oxidation time, glass transition |
| Thermo-mechanical analysis (TMA) | Coefficient of linear thermal expansion, softening point, glass transition |
| Infrared spectroscopy (IR) | Additives, fillers, plasticizers, and rate of oxidation reaction |
| Chromatography: gas chromatography (GC); and high pressure liquid chromatography (HPLC) | Additives and plasticizers |
| Density determination ($\rho$) | Density and degree of crystallinity |
| Melt index (MI) | Melt index and flow rate ratio |
| Gel permeation chromatography (GPC) | Molecular weight distribution |

non-polymeric materials like sodium or calcium bentonite are also used to make a few geosynthetic products.

It is sometimes important to know the polymer compound present in the geosynthetic being used. A quantitative assessment requires the use of a range of identification test techniques, as listed in Table 1.5 along with a brief description. More detailed descriptions can be found in the works of Halse *et al.* (1991), Landreth (1990), and Rigo and Cuzzuffi (1991).

## 1.5   MANUFACTURING PROCESSES FOR GEOSYNTHETICS

Geotextiles are manufactured in many different ways, partly using the traditional textile procedures, and partly using procedures not commonly recognized as textile procedures. The manufacturing process of a geotextile basically includes two steps (Giroud and Carroll, 1983): the first step consists in making linear elements such as *fibres or yarns* from the polymer pellets, under the agency of heat and pressure, and the second step consists in combining these linear elements to make a planar structure usually called the *fabric*.

The basic elements of a geotextile are its fibres. A *fibre* is a unit of matter characterized by flexibility, fineness, and high ratio of length to thickness. There are mainly four types of synthetic fibres: *filaments* (produced by extruding melted polymer through dies or spinnerets and subsequently drawing it longitudinally), *staple fibres* (obtained by cutting filaments to a short length, typically 2 to 10 cm), *slit films* (flat tape-like fibres, typically 1 to 3 mm wide, produced by slitting with blades an extruded plastic film and subsequently drawing it), and *strands* (a bundle of tape-like fibres that can be partially attached to each other). During the drawing process, the molecules become oriented in the same direction resulting in an increase of modulus of the fibres. A *yarn* consists of a number of fibres from the particular polymeric compound selected. Several types of yarn are used to construct woven geotextiles: *monofilament yarn* (made from a single filament), *multifilament yarn* (made from fine filaments aligned together), *spun yarn* (made from staple fibres interlaced or twisted together), *slit film yarn* (made from a single slit film fibre), and *fibrillated yarn* (made from strands). Note that the synthetic fibres are very efficient load-carrying elements, with tensile strengths equivalent to prestressing steel in some cases (e.g. in case of polyaramid fibres). Fibre technology in itself is a well-advanced science with an enormous database. It is in the fibre where control over physical and mechanical properties first takes place in a well-prescribed and fully automated manner.

As the name implies, the woven geotextiles are obtained by the conventional weaving processes. Although modern weaving looms are extremely versatile and sophisticated items, they operate on the basic principles embodied in a mechanical loom illustrated in Fig. 1.11. The weaving process gives these geotextiles their characteristic appearance of two sets of parallel yarns interlaced at right angles to each other as shown in Fig. 1.12. The terms '*warp and weft*' are used to distinguish between the two different directions of yarn. The longitudinal yarn, running along the length of the weaving machine or loom and hence running lengthwise in a woven geotextile roll, is called the *warp*. The transverse yarn, running across the width of the loom and hence running widthwise in a woven geotextile roll, is called the *weft*. Since the warp direction coincides with the direction in which the geotextile is manufactured on the mechanical loom, this is also called the *machine direction (MD) (i.e. production direction or roll length direction)*, whereas at right angles to the machine direction in the plane of the geotextile is the *cross-machine direction (CMD)*, which is basically the weft direction.

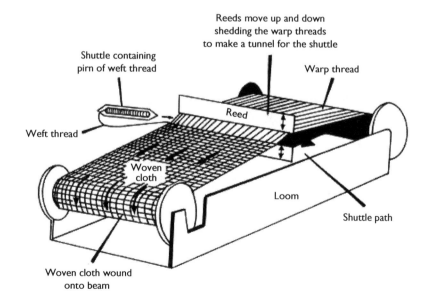

*Figure 1.11* Main components of a weaving loom (after Rankilor, 1981).

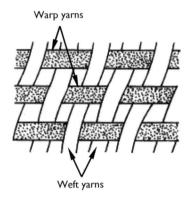

*Figure 1.12* A woven geotextile having a plain weave.

In Fig. 1.12, the type of weave described is a *plain weave*, of which there are many variations, such as *twill*, *satin* and *serge*; however, the plain weave is the one most commonly used in geotextiles.

Nonwoven geotextiles are obtained by processes other than weaving. The processing involves continuous laying of the fibres on a moving conveyor belt to form a loose web slightly wider than the finished product. This passes along the conveyor to be bonded by mechanical bonding (obtained by punching thousands of small barbed needles through the loose web), thermal bonding (obtained by partial melting of the fibres), or chemical bonding (obtained by fixing the fibres with a cementing medium such as glue, latex, cellulose derivative, or synthetic resin) resulting in the following three different types:

1   mechanically bonded nonwoven geotextile (or needle-punched nonwoven geotextiles)
2   thermally bonded nonwoven geotextile, and
3   chemically bonded nonwoven geotextile, respectively.

Figure 1.13 shows the diagrammatic representation of the production of needle-punched geotextiles. These geotextiles are usually relatively thick, with a typical thickness in the range of 0.5 to 5 mm.

Knitted geotextiles are manufactured using knitting process, which involves interlocking a series of loops of one or more yarns together to form a planar structure. There is a wide range of different types of knits used, one of which is illustrated in Fig. 1.14. These geotextiles are very extensible, and therefore they are used in very limited quantities.

Stitch bonded geotextiles are produced from multi-filaments by a stitching process. Even strong, heavyweight geotextiles can be produced rapidly. Geotextiles are sometimes manufactured in a tubular or cylindrical fashion without longitudinal seam. Such geotextiles are called the *tubular geotextiles*.

A geotextile can be saturated with bitumen, resulting in a *bitumen impregnated geotextile*. Impregnation aims at modifying the geotextile, to protect it against external forces, and, in some cases, to make it fluid impermeable.

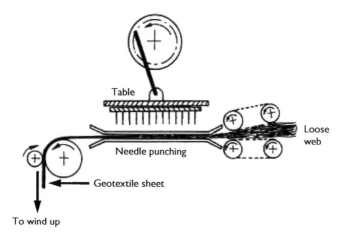

*Figure 1.13* The manufacturing process of needle-punched nonwoven geotextile.

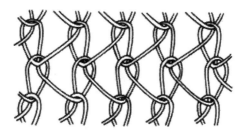

*Figure 1.14* A knitted geotextile.

All the geogrids share a common geometry comprising two sets of orthogonal load-carrying elements which enclose substantially rectangular or square patterns. Due to the requirement of high tensile properties and acceptable creep properties, all geogrids are produced from molecularly oriented plastic. The main difference between different grid structures lies in how the longitudinal and transverse elements are joined together.

Extruded geogrids are manufactured from polymer sheets in two or three stages of processing. The first stage involves feeding a sheet of polymer, several millimetres thick, into a punching machine, which punches out the holes on a regular grid pattern. Following this, the punched sheet is heated and stretched, or drawn, in the machine direction. This distends the holes to form an elongated grid opening known as the *aperture*. In addition to changing the initial geometry of the holes, the drawing process orients the randomly oriented long-chain polymer molecules in the direction of drawing. The degree of orientation will vary along the length of the grid; however, the overall effect is an enhancement of tensile strength and tensile stiffness. The process may be halted at this stage, in which case the end product is a *uniaxially oriented geogrid*. Alternatively, the uniaxially oriented

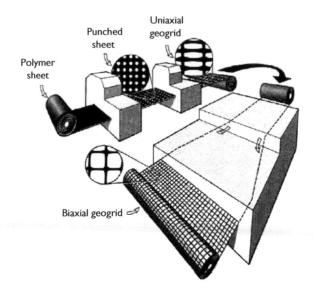

*Figure 1.15* 'Tensar' manufacturing process (courtesy of Netlon Limited, UK).

grid may proceed to a third stage of processing to be warm drawn in the trans-verse direction, in which case a *biaxially oriented geogrid* is obtained (Fig. 1.15). Although the temperatures used in the drawing process are above ambient, this is effectively a cold drawing process, as the temperatures are significantly below the melting point of the polymer. It should be noted that the ribs of geogrids are often quite stiff compared to the fibres of geotextiles.

Woven geogrids are manufactured by weaving or knitting processes from poly-ester multi-filaments. Where the warp and weft filaments cross they are interlaced at multiple levels to form a competent junction. The skeletal structure is generally coated with acrylic or PVC or bitumen to provide added protection against environmental attack and construction induced-damages.

Bonded geogrids are manufactured by bonding the mutually perpendicular PP or PET strips together at their crossover points using either laser or ultra-sonic welding. There are several bonded geogrids, which are extremely versatile because they can be used in isolated strip form, and as multiple strips for ground reinforcement.

Geonets are manufactured typically by an extrusion process in which a mini-mum of two sets of strands (bundles of tape-like fibres that can be attached to each other) are overlaid to yield a three dimensional structure. A counter-rotating die, with a simplified section as shown in Fig. 1.16, is fed with hot plastic by a screw extruder. The die consists of an inner mandrel mounted concentrically inside a heavy tubular sleeve. When both the inner and outer sections of the die are rotated then the two sets of spirals are produced simultaneously; however, at the instant that inner and outer slots align with one another there is only one,

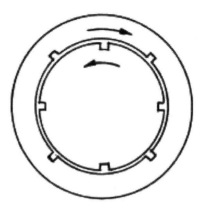

*Figure 1.16* Rotating die.

double thickness, set of strands extruded. It is at this instant that the crossover points of the two spirals are formed as extruded junctions. Consequently the extrudate takes the form of a tubular geonet. This continuously extruded tube is fed, coaxially over a tapering mandrel which stretches the tube to the required diameter. This stretching process results in inducing a degree of molecular orientation and it also controls the final size and geometry of the finished geonet. To convert the tubular geonet to flat sheet, the tube is cut and laid flat. If the tube is slit along its longitudinal axis, the resulting geonet appears to have a diamond shaped aperture. Alternatively, the tube may be cut on the bias, e.g. parallel to one of the strand arrays, in which case the apertures can appear to be almost square.

Unlike the geogrids, the intersecting ribs of geonets are generally not perpendicular to one another. In fact they intersect at typically 60° to 80° to form a diamond shaped aperture. It can be seen that one parallel array of elements sits on top of the underlying array so creating a structure with some depth. The geonets are typically 5 to 10 mm in thickness.

Most of the geomembranes are made in a plant using one of the following manufacturing processes: (i) extrusion, (ii) spread-coating, or (iii) calendering. The extrusion process is a method whereby a molten polymer is extruded into a non-reinforced sheet using an extruder. Immediately after extrusion, when the sheet is still warm, it can be laminated with a geotextile; the geomembrane thus produced is *reinforced*. The spread coating process usually consists in coating a geotextile (woven, nonwoven, knitted) by spreading a polymer or asphalt compound on it. The geomembranes thus produced are therefore also reinforced. *Non-reinforced geomembranes* can be made by spreading a polymer on a sheet of paper, which is removed and discarded at the end of the manufacturing process. Calendering is the most frequently used manufacturing process in which a heated polymeric compound is passed through a series of heated rollers of the calender, rotating under mechanical or hydraulic pressure (Fig. 1.17).

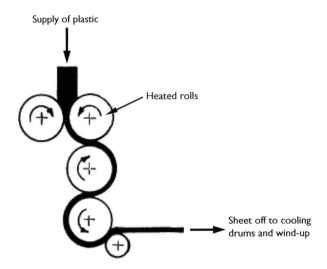

Supply of plastic

Heated rolls

Sheet off to cooling
drums and wind-up

*Figure 1.17* Calendering process of manufacturing geomembranes (after Ingold, 1994).

By utilizing auxiliary extruders, both HDPE and LLDPE geomembranes are sometimes coextruded, using flat dies or blown film methods. The coextruded sheet is manufactured such that a HDPE-LLDPE-HDPE geomembrane results. The HDPE on the upper and the lower surfaces is approximately 10% to 20% of the total sheet thickness. The objective is to retain the excellent chemical resistance of HDPE on the surfaces of geomembranes, with the flexibility of LLDPE in the core. The typical thickness of geomembranes ranges from 0.25 to 7.5 mm (10 to 300 mils; 1 mil = 0.001 inch ≈ 0.025 mm).

The textured surfaces (surfaces with projections or indentations) can be made on one or both sides of a geomembrane by blown film coextrusion, or impingement by hot PE particles or any other suitable method. A geomembrane with textured surfaces on one or both surfaces is called *textured geomembrane*. The textured surface greatly improves the stability, particularly on sloping ground, by increasing the interface friction between the geomembrane and the soils or the geosynthetics. The textured surfaces are generally produced with about a 6 inch (150 mm) nontextured border on both sides of the sheet. The smooth border provides a better surface for welding than a textured surface. The smooth edges also permit quick verification of the thickness and the strength before installation.

Geocomposites can be manufactured from two or more of the geosynthetic types described in Section 1.2. A geocomposite can therefore combine the properties of the constituent members in order to meet the needs of a specific application. Examples of geocomposites are band/strip/wick drains, and geosynthetic clay liners (GCLs).

*Band drains*, also called the *fin drains*, usually consist of a plastic fluted or nubbed water conducting core (drainage core) wrapped in a geotextile sleeve (Fig. 1.18). They are designed for an easy installation in either a slot or trench dug in the soil along the

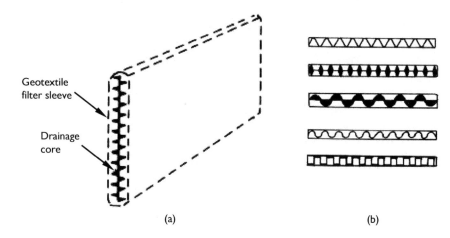

*Figure 1.18* Strip drain: (a) components; (b) various shapes of drainage cores.

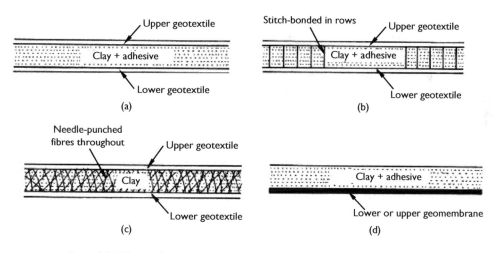

*Figure 1.19* Types of geosynthetic clay liners (after Koerner and Daniel, 1997).

edge of the highway/runway pavement or railway track or any other civil engineering structures that require drainage measures.

Geotextiles can be attached to geomembranes to form geocomposites. Geotextiles are commonly used in conjunction with geomembranes for puncture protection, drainage, and improved tensile strength.

A geosynthetic clay liner (GCL) is a manufactured hydraulic barrier used as alternative material to substitute a conventional compacted soil layer for the low-permeability soil component of various environmental and hydraulic projects

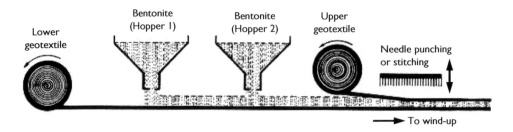

*Figure 1.20* Major steps of the manufacturing process for needle-punched and stitch-bonded geosynthetic clay liners.

including landfill and remediation projects. It consists of a thin layer of sodium or calcium bentonite (mass per unit area ≈ 5 kg/m²), which is either sandwiched between two sheets of woven or nonwoven geotextiles (Fig. 1.19), or mixed with an adhesive and attached to a geomembrane The sodium bentonite has a lower hydraulic conductivity. Figure 1.20 shows the major steps of the manufacturing process for a needle-punched or stitch bonded GCL. It should be noted that the GCLs are also known by several other names such as *clay blankets, bentonite blankets* and *bentonite mats*.

The combination of the geotextile (filtering action), geomembranes (waterproofing properties), and geonets (drainage and load distribution) offers a complete system of filter-drainage-protection, which is very compact and easy to install.

The geosynthetics manufactured in the factory environment have specific properties whose uniformity is far superior to soils, which are notoriously poor in homogeneity, as well as in isotropy. Based on many years' experience of manufacturing and the development of quality assurance procedures, the geosynthetics are made in such a way that better durability properties are obtained.

Most geosynthetics are supplied in rolls. Although there is no standard width for geosynthetics, most geotextiles are provided in widths of around 5 m while the geogrids are generally narrower and geomembranes may be wider. Geotextiles are supplied typically with an area of 500 m² per roll for a product of average mass per unit area. In fact, depending on the mass per unit area, thickness, and flexibility of the product, the roll lengths vary between a few tens of metres up to several hundreds of metres with the majority of roll lengths falling in the range 100 m to 200 m. The product of mass per unit area, roll width and length, gives the mass of the roll. Rolls with a mass not exceeding 100 kg can usually be handled manually. If allowed to become wet, the weight of a roll, particularly of geotextiles, can increase dramatically. Geonets are commercially available in rolls up to 4.5 m wide. The HDPE and VLDPE geomembranes are supplied in roll form with widths of approximately 4.6 to 10.5 m and lengths of 200 to 300 m. Geosynthetic clay liners are manufactured in panels that measure 4–5 m in width and 30–60 m in length, and are placed on rolls for shipment to the project site.

**EXAMPLE 1.1**

Consider a roll of geotextile with the following details:
  Mass of the roll, $M = 100$ kg
  Width of the roll, $B = 5$ m, and
  Mass per unit area of the roll, $m = 200$ g/m$^2$
Determine the roll length.

**SOLUTION**

Mass per unit area of the geotextile roll can be expressed as

$$m = \frac{M}{L \times B} \tag{1.1}$$

where $L$ is the roll length.
From Equation (1.1),

$$L = \frac{M}{m \times B} = \frac{100}{(0.2)(5)} = 100 \text{ m}$$

## 1.6  FUNCTIONS OF GEOSYNTHETICS

Geosynthetics have numerous application areas in civil engineering, and also in mining, agricultural and aquacultural engineering. They always perform one or more of the following basic functions when used in contact with soil, rock and/or any other civil engineering-related material:

- *Reinforcement*
- *Separation*
- *Filtration*
- *Drainage*
- *Fluid Barrier*
- *Protection*

### 1.6.1  Reinforcement

A geosynthetic performs the reinforcement function by improving the mechanical properties of a soil mass as a result of its inclusion. When soil and geosynthetic reinforcement are combined, a composite material, 'reinforced soil', possessing high compressive and tensile strengths (and similar, in principle, to the reinforced concrete) is produced. In fact, any geosynthetic applied as reinforcement has the main task of resisting the applied stresses and/or preventing inadmissible deformations in the geotechnical structures. In this process, the geosynthetic acts as a tensioned member coupled to the soil/fill material by friction, adhesion, interlocking and/or confinement and thus maintains the stability of the soil mass. Figure 1.21(a) explains the basic mechanism involved in the reinforcement function. Figure 1.21(b) shows a magnified view of the reinforced soil element as indicated in Figure 1.21(a). You may notice that the reinforcement is extended, resulting in a mobilized tensile force $T$, and the soil is compressed by a confining lateral stress $\sigma_R$ as a reinforcement restraint.

Different concepts have been advanced to define the basic mechanism of reinforced soils. The effect of inclusion of relatively inextensible reinforcements (such as metals, fibre-reinforced plastics, etc. having a high modulus of deformation) in the soil can be explained using either an induced stresses concept (Schlosser and Vidal, 1969) or an induced deformations concept (Basset and Last, 1978). According to the induced stresses concept, the tensile strength of the reinforcements and friction at the soil-reinforcement interfaces give an apparent cohesion to the reinforced soil system.

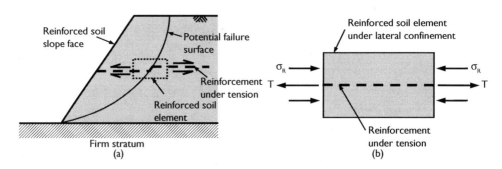

Figure 1.21 Basic mechanism involved in the reinforcement function: (a) a soil slope reinforced with a horizontal layer of reinforcement; (b) magnified view of the reinforced soil element (Shukla et al., 2009).

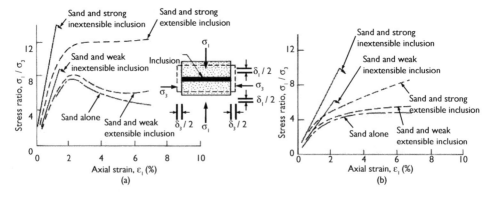

Figure 1.22 Postulated behaviour of a unit cell in plane strain conditions with and without inclusions: (a) dense sand with inclusions; (b) loose sand with inclusions (after McGown et al., 1978).

The induced deformations concept considers that the tensile reinforcements involve anisotropic restraint of the soil deformations. The behaviour of the soil reinforced with extensible reinforcements, such as geosynthetics, does not fall completely within these concepts. The difference, between the influences of inextensible and extensible reinforcements, is significant in terms of the load-settlement behaviour of the reinforced soil system (Fig. 1.22). The soil reinforced with an extensible reinforcement (termed *ply-soil* by McGown and Andrawes (1977)) has greater extensibility and smaller losses of post peak strength compared to soil alone or soil reinforced with inextensible reinforcement (termed *reinforced earth* by Vidal (1969)). However, some similarity between the ply-soil and the reinforced earth exists in that they inhibit the development of internal tensile strains in the soil and develop tensile strengths.

Fluet (1988) subdivided the reinforcement function into the following two categories:

1   A *tensile member*, which supports a planar load, as shown in Fig. 1.23(a).
2   A *tensioned member*, which supports not only a planar load but also a normal load, as shown in Fig. 1.23(b).

Jewell (1996) and Koerner (2005) consider not two but three mechanisms for soil reinforcement, because when the geosynthetic works as a tensile member, it might be due to two different mechanisms: shear or anchorage. Therefore, the three reinforcing mechanisms, concerned simply with the types of load that are supported by the geosynthetic, are:

1   *Shear*, also called the *sliding*: The geosynthetic supports a planar load due to slide of the soil over it.
2   *Anchorage*, also called the *pullout*: The geosynthetic supports a planar load due to its pullout from the soil.
3   *Membrane*: The geosynthetic supports both a planar and a normal load when placed on a deformable soil.

Shukla (2002b, 2004) describes the reinforcing mechanisms that take into account the reinforcement action of the geosynthetic, that is, how the geosynthetic reinforcement takes the stresses from the soil and which type of stresses are taken. This concept can be observed broadly in terms of the following roles of geosynthetics:

1   A geosynthetic layer reduces the outward horizontal/shear stresses transmitted from the overlying soil/fill to the top of the underlying foundation soil. This action of geosynthetics is known as the *shear stress reduction effect*, and it results in a general-shear, rather than a local-shear failure (Fig. 1.24(a)), thereby causing an increase in the load-bearing capacity of the foundation soil (Bourdeau *et al.*, 1982; Guido *et al.*, 1985; Love *et al.*, 1987; Espinoza, 1994; Espinoza and Bray, 1995; Adams and Collin, 1997). Through the shear interaction mechanism the geosynthetic can therefore improve the performance of the system with very little or no rutting. In fact, the reduction in shear stress and the change in the failure mode are the primary benefits of the geosynthetic layer at small deformations.

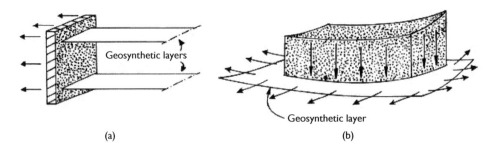

Figure 1.23 Reinforcement function: (a) tensile member; (b) tensioned member (Fluet, 1988).

2   A geosynthetic layer redistributes the applied surface load by providing a restraint of the granular fill if embedded in it, or by providing a restraint of the granular fill and the soft foundation soil, if placed at their interface, resulting in a reduction of applied normal stress on the underlying foundation soil (Fig. 1.24(b)). This is referred to as the *slab effect* or *confinement effect* of geosynthetics (Bourdeau, *et al.*, 1982; Giroud *et al.*, 1984; Madhav and Poorooshasb, 1989; Hausmann, 1990; Sellmeijer, 1990). The friction mobilized between the soil and the geosynthetic layer plays an important role in confining the soil.

3   The deformed geosynthetic, sustaining normal and shear stresses, has a membrane force with a vertical component that resists applied loads, that is, the deformed geosynthetic provides a vertical support to the overlying soil mass subjected to the loading. This action of geosynthetics is popularly known as its *membrane effect* (Fig. 1.24(c)) (Giroud and Noiray, 1981; Bourdeau *et al.*, 1982; Sellmeijer *et al.*, 1982; Love *et al.*, 1987; Madhav and Poorooshasb, 1988; Bourdeau, 1989; Sellmeijer, 1990; Shukla and Chandra, 1994, 1995). Depending upon the type of stresses - normal stress and shear stress, sustained by the geosynthetic during their action, the membrane support may be classified as 'normal stress membrane support', and 'interfacial shear stress membrane support', respectively (Espinoza and Bray, 1995). The edges of the geosynthetic layer are required to be anchored in order to develop the membrane support contribution resulting from the normal stresses, whereas the membrane support contribution resulting from the mobilized interfacial membrane shear stresses does not require any anchorage. The membrane effect of geosynthetics causes an increase in the load-bearing capacity of the foundation soil below the loaded area with a downward loading on its surface to either side of the loaded area, thus reducing its heave potential. Note that both the geotextile and the geogrid can be effective in membrane action in the case of high-deformation systems.

4   The use of geogrids has another benefit owing to the interlocking of the soil particles through the apertures (openings between the longitudinal and transverse ribs, generally greater than 6.35 mm (¼ inch)) of the grid known as the *interlocking effect* (Guido *et al.*, 1986) (Fig. 1.24(d)). The transfer of stress from the soil to the geogrid reinforcement is made through bearing (passive resistance) at the soil to the grid cross-bar interface. It is important to underline that owing to the small surface area and large apertures of the geogrids, the interaction is due mainly

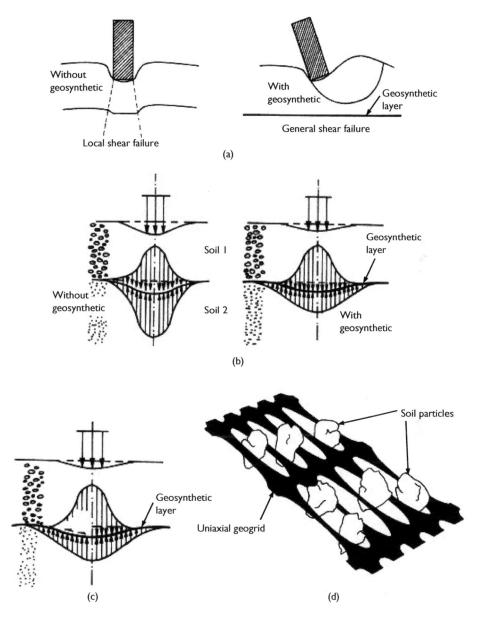

*Figure 1.24* Roles of the geosynthetic reinforcement: (a) causing change of failure mode (shear stress reduction effect); (b) redistribution of the applied surface load (confinement effect); (c) providing vertical support (membrane effect) (after Bourdeau *et al.* (1982) and Espinoza (1994)); (d) providing passive resistance through interlocking of the soil particles (interlocking effect).

to interlocking rather than to friction. However, an exception occurs when the soil particles are small, and in this situation, the interlocking effect is negligible because no passive strength is developed against the geogrid (Pinto, 2004).

## 1.6.2 Separation

If the geosynthetic has to prevent intermixing of adjacent dissimilar soils and/or fill materials during construction and over a projected service lifetime of the application under consideration, it is said to perform a separation function. Figure 1.25 shows that the geosynthetic layer prevents the intermixing of soft soil and granular fill, thereby keeping the structural integrity and functioning of both materials intact. This function can be observed if a geotextile layer is provided at the soil subgrade level in pavements or railway tracks to prevent pumping of soil fines into the granular subbase/base course and/or to prevent intrusion of granular particles into soil subgrade.

In many geosynthetic applications, especially in roads, rail tracks, shallow foundations, and embankments, a geosynthetic layer is placed at the interface of the soft foundation soil and the overlying granular layer (Fig. 1.26). In such a situation, it becomes a difficult task to identify the major function out of reinforcement and separation. Nishida and Nishigata (1994) have suggested that the separation can be a dominant function over the reinforcement function when the ratio $\sigma/c_u$ of the applied stress ($\sigma$) on the subgrade soil to the shear strength ($c_u$) of the subgrade soil has a low value (less than 8), and it is basically independent of the settlement of the reinforced soil system (Fig. 1.27).

It is important to note that the separation depends on the particle size of the soils involved. Most low-strength foundation soils are composed of small particles, whereas the placed layers (for roads, railways, foundations and embankments) are generally of coarser materials. In these situations, the separation is always needed, quite independently from the ratio of the applied stress to the strength of the subgrade soil, as Fig. 1.27 clearly shows. In general, the reinforcement will increase in importance as that the ratio $\sigma/c_u$ increases. Fortunately, separation and reinforce-

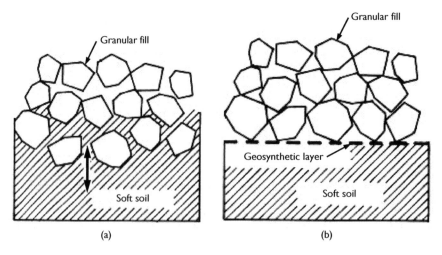

*Figure 1.25* Basic mechanism involved in the separation function: (a) granular fill – soft soil system without the geosynthetic separator; (b) granular fill – soft soil system with the geosynthetic separator.

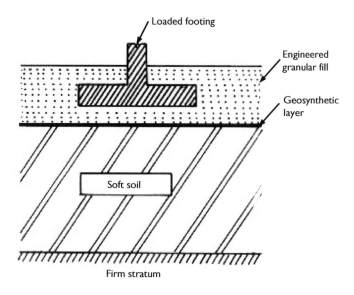

Figure 1.26 A loaded geosynthetic-reinforced granular fill – soft soil system.

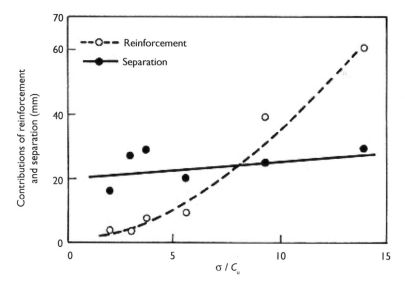

Figure 1.27 Relationship between reinforcement and separation functions (after Nishida and Nishigata, 1994).

ment are compatible functions. Furthermore, they work together, interacting: reinforcement reduces the deformation and therefore reduces the mixing of the particles (performing indirectly and to some extent the separation function); on the other hand, separation prevents mixing and consequently prevents the progressive loss of the strength of the subsequent layers. The ideal material to be used for roads, railways, foundations and embankments (i.e. when a coarse-grained soil is placed on top of a fine-grained soil with low strength) would be a continuous material such as a high-strength geotextile or a composite of a stiffer geogrid combined with a geotextile. In this way, the necessary separation and reinforcement functions can be performed simultaneously.

The selection of primary function out of reinforcement and separation can also be done on the basis of available empirical knowledge, if the California Bearing Ratio (CBR) of the subgrade soil is known. If the subgrade soil is soft, that is, the CBR of the subgrade soil is low, say its unsoaked value is less than 3 (or soaked value is less than 1), then the reinforcement can generally be taken as the primary function because of adequate tensile strength mobilization in the geosynthetic through large deformation, that is, deep ruts (say, greater than 75 mm) in the subgrade soil. Geosynthetics, used with subgrade soils with an unsoaked CBR higher than 8 (or soaked CBR higher than 3), will have a generally negligible amount of reinforcement role, and in such cases the primary function will uniquely be separation. For soils with intermediate unsoaked CBR values between 3 and 8 (or soaked CBR values between 1 and 3), the selection of the primary function is totally based on the site-specific situations.

### 1.6.3  Filtration

A geosynthetic may function as a filter that allows for adequate fluid flow with limited migration of soil particles across its plane over a projected service lifetime of the application under consideration. Figure 1.28 shows that a geosynthetic allows passage of water from a soil mass while preventing the uncontrolled migration of soil particles.

When a geosynthetic filter is placed adjacent to a base soil (the soil to be filtered), a discontinuity arises between the original soil structure and the structure of

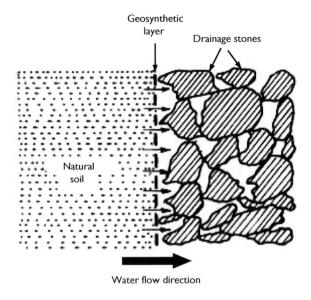

*Figure 1.28* Basic mechanism involved in the filtration function.

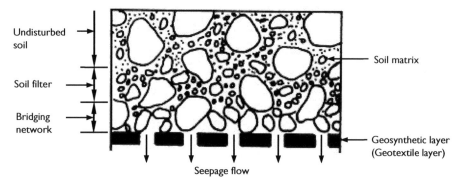

*Figure 1.29* An idealized interface conditions at equilibrium between the soil and the geosynthetic filter.

the geosynthetic. This discontinuity allows some soil particles, particularly particles closest to the geosynthetic filter and having diameters smaller than the *filter opening size* (see Chapter 2 for more explanation), to migrate through the geosynthetic under the influence of seepage flows. For a geosynthetic to act as a filter, it is essential that a condition of equilibrium is established at the soil-geosynthetic interface as soon as possible after installation to prevent soil particles from being piped indefinitely through the geosynthetic. At equilibrium, three zones may generally be identified: the undisturbed soil, a 'soil filter' layer which consists of progressively smaller particles as the distance from the geosynthetic increases, and a bridging layer which is a porous, open structure (Fig. 1.29). Once the stratification process is complete, it is actually the soil filter layer, which actively filters the soil.

Note that the filtration function also provides separation benefits. However, a distinction may be drawn between filtration function and separation function with respect to the quantity of fluid involved and the degree to which it influences the geosynthetic selection. In fact, if the water seepage across the geosynthetic is not a critical situation, then the separation becomes the major function. It is also a practice to use the separation function in conjunction with reinforcement or filtration; accordingly the separation is not specified alone in several applications.

## 1.6.4  Drainage

If a geosynthetic allows for adequate fluid flow with limited migration of soil particles within its plane from the surrounding soil mass to various outlets over a projected service lifetime of the application under consideration, it is said to perform the drainage (i.e. fluid transmission) function.

Figure 1.30 shows that the geosynthetic layer adjacent to the retaining wall collects water from the backfill and transports it to the weep holes constructed in the retaining wall.

Note that while performing the filtration and drainage functions, the geosynthetic dissipates the excess pore water pressure by allowing flow of water in plane and across its plane.

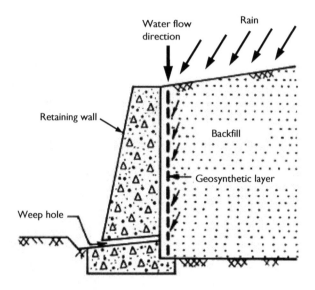

*Figure 1.30* Basic mechanism involved in the drainage function.

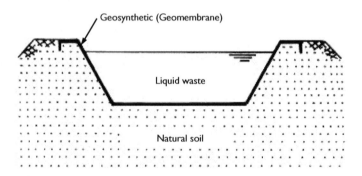

*Figure 1.31* Basic mechanism involved in the fluid barrier function.

## 1.6.5  Fluid barrier

A geosynthetic performs the fluid barrier function, if it acts like an almost impermeable membrane to prevent the migration of liquids or gases over a projected service lifetime of the application under consideration.

Figure 1.31 shows that a geosynthetic layer, installed at the base of a pond, prevents the infiltration of liquid waste into the natural soil.

## 1.6.6  Protection

A geosynthetic, placed between two materials, performs the protection function when it alleviates or distributes stresses and strains transmitted to the material to be protected against any damage (Fig. 1.32). In some applications, a geosynthetic layer is needed as a localized stress reduction layer to prevent or reduce the local damage to a geotechnical/structural system.

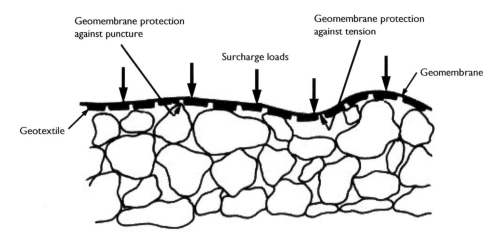

*Figure 1.32* Basic mechanism involved in the protection function.

## 1.6.7   Other functions

The basic functions of geosynthetics presented in Section 1.6.1 to Section 1.6.6 can be quantitatively described by standard tests or design techniques, or both. Geosynthetics can also perform some other functions that are, in fact, qualitative descriptions, mostly dependent on the basic functions, and are not yet supported by standard tests or generally accepted design techniques. Such functions of geosynthetics, basically describing their performance characteristics, are the following:

- *Absorption*: A geosynthetic provides absorption if it is used to assimilate or incorporate a fluid. This may be considered for two specific environmental aspects: water absorption in erosion control applications and the recovery of floating oil from surface waters following ecological disasters.
- *Containment*: A geosynthetic provides containment when it is used to encapsulate/contain a civil-engineering related material such as soil, rock, or fresh concrete to a specific geometry and prevent its loss.
- *Cushioning*: A geosynthetic provides cushioning when it is used to control and eventually to damp the dynamic mechanical actions. This function has to be emphasized particularly for the geosynthetic applications in canal revetments, shore protections, pavement overlay protection from reflective cracking and seismic base isolation of earth structures.
- *Insulation*: A geosynthetic provides insulation when it is used to reduce the passage of electricity, heat, or sound.
- *Screening*: A geosynthetic provides screening when it is placed across the path of a flowing fluid carrying fine particles in suspension to retain some or all particles while allowing the fluid to pass through. After some period of time, the particles accumulate against the geosynthetic, and hence it is required that the geosynthetic should be able to withstand the pressures generated by the accumulated particles and the increasing fluid pressure.

- *Surface stabilization (Surficial erosion control)*: A geosynthetic provides the surface stabilization when it is placed on a soil surface to restrict movement and prevent dispersion of surface soil particles subjected to erosion actions of rain and wind, often while allowing or promoting growth of vegetation.
- *Vegetative reinforcement*: A geosynthetic provides vegetative reinforcement when it extends the erosion control limits and performance of vegetation.

The relative importance of each function is governed by the site conditions, especially soil type and groundwater drainage, and the construction application. In many cases/applications, two or more basic functions of the geosynthetic are required.

All the functions of geosynthetics are listed along with their symbols/abbreviations in Table 1.6. The suggested symbols may help in making the drawing or diagram of a geosynthetic application with clarity. For example, if a geotextile is required to be represented for reinforcement function or filtration function, then this can be done as shown in Fig. 1.33.

## 1.7 SELECTION

Geosynthetics are available with a variety of geometric and polymer compositions to meet a wide range of functions and applications. Depending on the type of application, the geosynthetics may have specific requirements.

*Table 1.6* Functions of geosynthetics and their symbols.

| Functions | Symbols |
| --- | --- |
| Reinforcement | R |
| Separation | S |
| Filtration | F |
| Drainage (i.e. fluid transmission) | D (or FT) |
| Fluid barrier | FB |
| Protection | P |
| Absorption | A |
| Containment | C |
| Cushioning | Cus |
| Insulation | I |
| Screening | Scr |
| Surface stabilization | SS |
| Vegetative reinforcement | VR |

*Figure 1.33* Symbolic representations: (a) a reinforcement geotextile; (b) a filtration geotextile.

When installed, a geosynthetic may perform more than one of the listed functions (see Table 1.6) simultaneously, but generally one of them will result in the lower factor of safety; thus it becomes a primary function. The use of a geosynthetic in a specific application needs the classification of its functions as primary or secondary. Table 1.7 shows such a classification, which is useful while selecting the appropriate type of geosynthetic to solve the problem in hand. Each function uses one or more properties of the geosynthetic (see Chapter 2), such as tensile strength or water permeability, referred to as the *functional properties*. The function concept is generally used in the design with the formulation of a factor of safety, *FS*, in the traditional manner as:

$$FS = \frac{\text{Allowable (or test) functional property}}{\text{Required (or design) functional property}} \tag{1.2}$$

The allowable functional property is the available property, measured by the *performance test* or the *index test* as explained in Chapter 2, possibly factored down to account for uncertainties in its determination or in other site-specific conditions during the design life of the soil-geosynthetic system; whereas the required value of functional property is established by the designer/specifier using the accepted methods of analysis and design or empirical guidelines for the actual field conditions. The entire process, generally called the 'design-by-function' is widespread in its use. The actual magnitude of the factor of safety depends upon the implication of failure, which is always site-specific. If the factor of safety is sufficiently larger than one, the geosynthetic is acceptable for utilization because it ensures the stability and serviceability of the structure. However, as might be anticipated with a new technology, the universally accepted values of minimum factor of safety have not yet been established, and a conservative approach in this regard is still warranted.

*Table 1.7* Selection of geosynthetics based on their functions.

| Functions to be performed by the geosynthetics | | Geosynthetics that can be used |
|---|---|---|
| Separation | Primary | GTX, GCP, GFM |
| | Secondary | GTX, GGR, GNT, GMB, GCP, GFM |
| Reinforcement | Primary | GTX, GGR, GCP |
| | Secondary | GTX, GCP |
| Filtration | Primary | GTX, GCP |
| | Secondary | GTX, GCP |
| Drainage | Primary | GTX, GNT, GCP, GPP |
| | Secondary | GTX, GCP, GFM |
| Fluid Barrier | Primary | GMB, GCP |
| | Secondary | GCP |
| Protection | Primary | GTX, GCP |
| | Secondary | GTX, GCP |

Notes: GTX = Geotextile, GGR = Geogrid, GNT = Geonet, GMB = Geomembrane, Geofoam = GFM, Geopipe = GPP, GCP = Geocomposite.

---

**EXAMPLE 1.2**

A geogrid-reinforced soil structure is designed considering the tensile strength of the geogrid equal to 80 kN/m. If the factor of safety against the tensile failure is 1.5, what will be the allowable tensile strength of the geogrid?

**SOLUTION**

Given: Design tensile strength, $\sigma_{design}$ = 80 kN/m and factor of safety, $FS = 1.5$
If $\sigma_{all}$ is the allowable tensile strength of the geogrid, then from Equation (1.2),

$$FS = \frac{\sigma_{all}}{\sigma_{design}}$$

or

$$\sigma_{all} = (FS)(\sigma_{design}) = (1.5)(80) = \textbf{120 kN/m}$$

Note: In Chapter 2, you will learn that the tensile strength of the geosynthetic is expressed in kN/m, not in kN/m² as generally done for most other construction materials.

---

It is observed that only geotextiles and geocomposites perform most of the functions, and hence they are used in many applications. Geotextiles are porous to water flow across their manufactured planes and also within their planes. Thick, nonwoven needle-punched geotextiles have considerable void volume in their structure, and thus they can transmit fluid within the structure to a very high degree. The degree of porosity, which may vary widely, is used to determine the selection of specific geotextiles. Geotextiles can also be used as a fluid barrier on impregnation with materials like bitumen. The geotextiles vary with the type of polymer used, the type of fibre and the fabric style.

Geogrids are used to primarily function as reinforcement, and separation may be an occasional function, especially when soils having very large particle sizes are involved. The performance of the geogrid as reinforcement relies on its rigidity or high tensile modulus and on its open geometry, which accounts for its high capacity for interlocking with soil particles.

It has been observed that for the geotextiles to function properly as reinforcement, friction must develop between the soil and the reinforcement to prevent sliding, whereas for the geogrids, it is the interlocking of the soil particles through the apertures of the geogrid that achieves an efficient interlocking effect. In this respect, the geotextiles are frictional resistance-dependent reinforcement, whereas the geogrids are passive resistance-dependent reinforcement. The laboratory studies have shown that the geogrids are a superior form of reinforcement owing to the interlocking of the soil particles within the grid apertures. Note that the use of triaxial geogrids in some applications such as pavements reduces the amount of aggregates, thus requiring less excavation below the ground level and hence providing cost savings.

Geonets, unlike geotextiles, are relatively stiff, netlike materials with large open spaces between structural ribs. They are used exclusively as the fluid conducting cores in the prefabricated drainage geocomposites. Geonets play a major role in landfill leachate collection and leak detection systems in association with geomembranes or geotextiles. For a fair drainage function, the geonets should not be laid in contact with soils or waste materials, but used as the drainage cores with geotextile, geomembrane or other material on their upper and lower surfaces, thus avoiding the soil particles from obstructing the drainage net channels. Geonets as a drainage material generally fall intermediate in their flow capability between thick needle-punched nonwoven geotextiles and numerous drainage geocomposites.

Geomembranes are mostly used as a fluid barrier or liner. Sometimes, a geomembrane is also known as the *flexible membrane liner* (FML), especially in landfill applications. The permeability of typical geomembranes ranges from $0.5 \times 10^{-12}$ to $0.5 \times 10^{-15}$ m/s. Thus, the geomembranes are from $10^3$ to $10^6$ times lower in permeability than the compacted clay. In this context, we speak of geomembranes as being essentially impermeable. The recommended minimum thickness for all geomembranes is 30 mil (0.75 mm), with the exception of HDPE (high-density polyethylene), which should be at least 60 mil (1.5 mm) to allow for the extrusion seaming (Qian *et al.*, 2002). The most widely used geomembrane in the waste management industry is HDPE because this offers an excellent performance for landfill liners and covers, lagoon liners, wastewater treatment facilities, canal linings, floating covers, tank linings and so on. If a greater flexibility than that of HDPE is required, then a linear low-density polyethylene (LLDPE) is used because it has a lower molecular weight resin that allows LLDPE to conform to non-uniform surfaces, making it suitable for landfill caps, pond liners, lagoon liners, potable water containment, tunnels and tank linings.

Geofoams are available in slab or block form. Since these geosynthetic products are very lightweight material with a unit weight ranging from 0.11 to 0.48 kN/m$^3$, they can be selected as a highway fill over compressible subgrade soils and frost-sensitive soils, and as a backfill material for retaining walls to reduce the lateral earth pressure, thereby functioning basically as a separator. They can also function as a thermal insulator beneath buildings and as drainage channels beneath building slabs.

Geopipes are available in a wide range of diameters and wall dimensions for carrying liquids and gases. For applications like subdrainage systems and leachate collection systems, the geopipes should have perforations through the wall section to allow for the inflow of water and gas. The *standard dimension ratio*, defined as the ratio of outside pipe diameter to minimum pipe wall thickness, varies from a minimum of about 10 to a maximum of 40. This ratio can be related to the external strength and the internal pressure capability. Compared to steel pipes, they are cheap, light, and easy to install and join together along with a better durability.

It may happen that the geotextile, geogrid, geonet, geomembrane, geocell, geofoam or geopipe chosen to meet the requirements of a particular function does not match any other function, which has to be served simultaneously in an application. In such a situation, geocomposites can be used. In fact, the geocomposites can be manufactured to perform a combination of the functions described earlier. For example, a geomembrane-geonet-geomembrane composite can be made where the interior net acts as a drain to the leak detection system. Similarly a geotextile-geonet composite improves the separation, filtration and drainage. A geocomposite consisting of a geotextile cover and a drainage core, called the *band drain* or *wick drain* or *vertical strip drain* or *prefabricated vertical drain* (PVD) (Fig. 1.6(b)), provides the drainage for accelerating the consolidation of soil when installed vertically into the consolidating soil.

Geocomposites are generally, but certainly not always, completely polymeric. Other options include using fibreglass or steel for tensile reinforcement, sand in compression or as a filler, dried clay for subsequent expansion as a liner, or bitumen as a waterproofing agent. Geomembrane-clay composites are used as the liners, where the geomembrane decreases the leakage rate while the clay layer increases the breakthrough time. In addition, the clay layer reduces the leakage rate from any holes that might develop in the geomembrane, while the geomembrane will prevent cracks in the clay layer due to changes in moisture content.

The selection of a geosynthetic for a particular application is governed by several other factors such as specification, durability, availability, cost, and construction technique. The durability and other properties including the cost of geosynthetics are dependent on the type of polymers used as raw materials for their manufacturing. To be able to accurately specify a geosynthetic, which will provide the required properties, it is essentially to have at least a basic understanding of how polymers and production processes affect the properties of the finished geosynthetic products, as described in Section 1.5. Table 1.3 lists major types of geosynthetics and the most commonly used polymers for their manufacture. Table 1.8 provides the basic properties of some of these polymers, helping in the selection of geosynthetics.

For example, geotextiles can perform several basic functions as reinforcement, separation, filtration, drainage, and protection (see Table 1.7). They are manufactured using polypropylene, polyester, polyethylene or polyamide (see Table 1.3). Geotextiles as a reinforcement require a strong, relatively stiff and preferably a

*Table 1.8* Typical properties of polymers used for the manufacture of geosynthetics.

| Polymers | Specific gravity | Melting temperature (°C) | Tensile strength at 20 °C (MPa) | Modulus of elasticity (GPa) | Strain at break (%) |
|---|---|---|---|---|---|
| PP | 0.90–0.91 | 160–165 | 400–600 | 1.3–1.8 | 10–40 |
| PET | 1.22–1.38 | 260 | 800–1200 | 12–18 | 8–15 |
| PE | 0.91–0.96 | 100–135 | 80–600 | 0.2–1.4 | 10–80 |
| PVC | 1.38–1.55 | 160 | 20–50 | 2.7–3 | 50–150 |
| PA | 1.05–1.15 | 220–250 | 700–900 | 3–4 | 15–30 |

water-permeable material. An inspection of Table 1.8 indicates that the polyester has a very high tensile strength at a relatively low strain. Thus a woven geotextile of polyester is a logical choice for the reinforcement applications. For separation/filtration applications, the geotextile has to be flexible, water-permeable, and soil-tight. A nonwoven geotextile or a lightweight woven geotextile of polyethylene can be a logical choice for separation/filtration applications. Note that the environmental factors and the site conditions also greatly govern the selection of geosynthetics for a particular application (Shukla, 2003b).

Sometimes, during the selection process, you may find that several geosynthetics satisfy the minimum requirements for the particular application. In such a situation, the geosynthetic should be selected on the basis of cost-benefit ratio, including the value of field experience and product documentation.

The properties of geosynthetics can change unfavourably in several ways such as ageing, mechanical damage (particularly by installation stresses), creep, hydrolysis (reaction with water), chemical and biological attack, ultraviolet light exposure, etc., which have been discussed in detail in Chapter 2. These factors have to be taken into account when the geosynthetics are selected. In permanent installations, there must be proper care for maintaining the long-term satisfactory performance of the geosynthetics, that is, the *durability*.

Considering the risks and consequences of failure, especially for critical projects, great care is required in the selection of the appropriate geosynthetic. One should not try to economize by eliminating the soil-geosynthetic performance testing when such testing is required for the selection.

For some applications, the geosynthetics are selected on the basis of empirical guidelines. In these cases proper care should be taken to clearly define the required properties of the geosynthetic, in physical as well as in statistical terms.

## 1.8    HISTORICAL DEVELOPMENTS OF GEOSYNTHETIC ENGINEERING

The present form of geosynthetic engineering has been achieved because of ongoing efforts by researchers, practising engineers and polymer technologists during the past several decades. The following are some key stages of the historical developments:

1    The first use of fabrics in reinforcing roads was attempted by the South Carolina Highway Department in 1926 (Beckham and Mills, 1935).
2    The first geotextile used in a dam, in 1970, was a needle-punched nonwoven geotextile used as a filter for the aggregate downstream drain in the Valcross Dam (17 m high), France (Giroud, 1992).
3    The first sample of Tensar grid was made in the Blackburn laboratories of Netlon Ltd, UK in July 1978.
4    Geofoam was originally applied as a lightweight fill in Norway in 1972.
5    Soil confinement systems based on cellular geotextile nets were first developed and evaluated in France during 1980.

6  The first known environmental application of geonet was in 1984 for leak detection in a double-lined hazardous liquid-waste impoundment in Hopewell, Virginia.

7  The first conference on geosynthetics was held in Paris in 1977.

8  Koerner and Welsh wrote the first book on geosynthetics in 1980.

9  The International Geosynthetics Society was established in 1983.

10  There are currently three geosynthetic-related international journals, namely *Geotextiles and Geomembranes, Geosynthetics International,* and *International Journal of Geosynthetics and Ground Engineering,* which were first published in 1984, 1995 and 2015, respectively.

Note that the rapid and significant development of geosynthetic engineering worldwide is mainly based on the fact that the geosynthetics often provide cost-effective, environmentally friendly, energy-efficient and sustainable solutions in the construction industry. Details of the sustainability aspects of geosynthetic applications can be found in Bouazza and Heerten (2012).

## Chapter summary

1  Geosynthetics are planar products manufactured mainly from polymers (PP, PET, PE, PA, etc.) with some additives (e.g. carbon black), and their major types are geotextiles, geogrids, geonets, geomembranes, geocells, geofoams and geocomposites. Geosynthetic clay liners and prefabricated vertical drains are most common examples of geocomposites.

2  Geonaturals (geojutes, geocoirs, etc.) are similar to geosynthetics in physical structure and some applications, but they are manufactured from natural fibres (jute, coir, etc.), which have a short life span due to their biodegradable characteristics.

3  Geotextiles are manufactured in many different ways, including the conventional weaving processes, resulting in the woven geotextiles. The manufacture of non-woven geotextiles from suitable fibres involves needle-punching, thermal bonding and chemical bonding processes. The geogrids share a common geometry comprising load-carrying elements which enclose substantially rectangular, square or triangular patterns. Most geomembranes are made by extrusion, spread-coating, or calendaring process. Geocells are factory-manufactured products or they are assembled at the construction site. Geocomposites are generally, but certainly not always, completely polymeric.

4  Most geosynthetics are supplied in rolls having widths from 4 m to 10 m, and lengths from 30 to 300 m. Rolls with a mass not exceeding 100 kg can usually be handled manually.

5  In civil, mining, agricultural and aquacultural applications, the geosynthetics perform several functions, including separation, reinforcement, filtration, drainage, fluid barrier and protection as the major functions. Geosynthetics are used mainly to achieve technical and/or economic benefits, and they also provide environmental-friendly and energy-efficient solutions to several construction problems in a sustainable manner.

6    The geosynthetic reinforcement within a soil mass improves its strength characteristics through the following actions: shear stress reduction effect, slab/confinement
effect, membrane effect and interlocking effect. The geosynthetic as a separator at
the interface of two dissimilar soil layers prevents mixing and consequently prevents the progressive loss of the strength and functions of the soil layers.

7    When a geosynthetic filter is placed adjacent to a base soil, at the equilibrium,
three zones may generally be identified: the undisturbed soil, a 'soil filter' layer
which consists of progressively smaller particles as the distance from the geosynthetic filter increases, and a bridging layer which is a porous, open structure.
Once this stratification process is complete, it is actually the soil filter layer, which
actively filters the soil. The geosynthetic layer as a drainage medium collects the
water from the adjacent soil mass and conveys it to the drain pipe network or the
disposal area.

8    A geosynthetic may act like an almost impermeable membrane to prevent the
migration of fluids (liquids and gases) from one area to the other. This fluid
barrier function is generally achieved through the use of geomembranes, which
require a suitable protection in most field applications.

9    The use of a geosynthetic in a specific application needs the classification of its
function as primary or secondary because the function concept is commonly used
in the design with the formulation of a factor of safety, defined as the ratio of
allowable/test functional property to the required or design functional property.
If the factor of safety is sufficiently larger than one, the geosynthetic is acceptable
for utilization because it ensures the stability and serviceability of the structure.

10   The selection of a geosynthetic for a particular application is governed by several factors such as specification, durability, availability, cost, and construction
technique. Woven geotextiles, geogrids and some geocomposites are commonly
used for reinforcement applications. Nonwoven geotextiles are commonly used
in many separation, filtration, drainage and protection applications alone or in
combination with others as geocomposites. Geonets are used exclusively as the
fluid conducting cores in prefabricated drainage geocomposites. Geofoams are
used as a very lightweight fill material, thermal insulator or drainage channel.
Geopipes are used for carrying liquids and gases.


## Questions for practice

(Select the most appropriate answer to the multiple-choice questions from Q 1.1 to
Q 1.15.)

1.1   A planar, polymeric product consisting of a mesh or net-like regular open network of intersecting tensile-resistant elements, integrally connected at the junctions is called
(a) geotextile
(b) geogrid
(c) geonet
(d) geocell.

1.2   The materials used in the manufacture of geosynthetics are primarily synthetic polymers generally derived from
(a)  rubber
(b)  fibre glass
(c)  crude petroleum oils
(d)  jute.

1.3   The most widely used polymers for manufacturing the reinforcing geosynthetics are
(a)  PP and PET
(b)  PP and PA
(c)  PP and PE
(d)  PET and PE.

1.4   The resistance to creep is high for
(a)  PP
(b)  PET
(c)  PE
(d)  PA.

1.5   The term 'weft' refers to
(a)  the longitudinal yarn of the geotextile
(b)  the yarn running in the roll length direction
(c)  the yarn running in the machine direction
(d)  all of the above.

1.6   Which one of the following geosynthetics is a geocomposite?
(a)  geofoam
(b)  geonet
(c)  geosynthetic clay liner
(d)  geocell

1.7   If geosynthetic allows for adequate fluid flow with limited migration of soil particles across its plane over a projected service lifetime of the application under consideration, then this function of geosynthetic is called
(a)  filtration
(b)  separation
(c)  drainage
(d)  fluid barrier.

1.8   The deformed geosynthetic provides a vertical support to the overlying soil mass subjected to a loading. This action of the geosynthetic is known as
(a)  shear stress reduction effect
(b)  membrane effect
(c)  confinement effect
(d)  interlocking effect.

1.9   For a geotextile, the separation can be a dominant function over the reinforcement function when the ratio of the applied normal stress on the subgrade soil to its undrained shear strength is generally

(a) equal to 8
(b) greater than 8
(c) lower than 8
(d) none of the above.

1.10   In geosynthetic engineering, most of the functions are served by
(a) GTX and GGR
(b) GTX and GCP
(c) GNT and GXT
(d) GGR and GNT.

1.11   Which one of the following geosynthetics can serve the *protection* function?
(a) GTX
(b) GGR
(c) GMB
(d) GNT

1.12   The following geosynthetics are used as a drainage medium:
(A) thick needle-punched nonwoven geotextiles
(B) geonets
(C) drainage geocomposites

The correct decreasing order of flow capability is generally
(a) (A), (B), (C)
(b) (B), (A), (C)
(c) (C), (A), (B)
(d) (C), (B), (A).

1.13   The specific gravity is smaller than one for
(a) PP and PET
(b) PET and PE
(c) PE and PP
(d) PA and PET.

1.14   The melting temperature of the polyester is
(a) 130 °C
(b) 165 °C
(c) 220 °C
(d) 260 °C

1.15   Which one of the following polymers has the highest modulus of elasticity?
(a) PP
(b) PET
(c) PE
(d) PVC.

1.16   What do you mean by geosynthetics and geonaturals? Explain these two terms making a point-wise comparison.
1.17   Explain the process of polymerization and its role in improving the characteristics of polymer fibres.

1.18   What are the additives that are used to avoid ultraviolet light degradation of polymers?

1.19   What is the effect of temperature variation on the geosynthetic polymers?

1.20   What is the difference between thermoplastic and thermoset polymers? Why are thermoset polymers rarely used?

1.21   Describe the major steps of manufacturing process for the following types of geosynthetics:
  (a)  woven geotextiles
  (b)  nonwoven geotextile
  (c)  extruded geogrids
  (d)  geonets
  (e)  geomembranes

1.22   Describe the major steps of the manufacturing process for the needle-punched and stitch-bonded geosynthetic clay liners with the help of a neat sketch.

1.23   What are the advantages of a textured geomembrane? Where should a textured geomembrane be used in field applications?

1.24   What would be the benefits of having a geotextile bonded directly to the geomembrane on its lower side?

1.25   What are the components of a strip drain? Draw a neat sketch in support of your answer.

1.26   Consider a roll of geotextile with the following details:
  Mass of the roll, $M = 90$ kg
  Length of the roll, $L = 120$ m, and
  Mass per unit area of the roll, $m = 150$ g/m$^2$
  Determine the roll width.

1.27   List the basic functions of geosynthetics.

1.28   Explain the basic mechanisms involved in the separation and filtration functions with the help of neat sketches.

1.29   Explain the mechanism involved in drainage function comparing with the mechanism involved in filtration function?

1.30   What are the performance characteristics of geosynthetics other than their basic functions? How do they differ from the basic functions?

1.31   What are the characteristics of a soil reinforced with an extensible reinforcement? Are these characteristics similar for the soil reinforced with an inextensible reinforcement? Can you list the differences, if any?

1.32   What are the different mechanisms for soil reinforcement? Explain briefly.

1.33   How will you decide the primary function out of reinforcement and separation in any field application?

1.34   Describe idealized interface conditions at equilibrium between the soil and geosynthetic filter.

1.35   What do you mean by functional properties? Explain with some examples.

1.36   Define the factor of safety required for the acceptance of a geosynthetic for a specific application.

1.37   A geotextile-reinforced soil structure is designed considering the tensile strength of the geotextile equal to 60 kN/m. If the allowable tensile strength of the geogrid is 80 kN/m, what will be the factor of safety against the tensile failure of the geotextile?

1.38 Which manufactured style of a geotextile is best suited to its application as a drainage medium?

1.39 How does a geonet differ from a geogrid in terms of functions?

1.40 If a geotextile is placed adjacent to a geonet, what function(s) does the geotextile provide? How does the combination of geotextile and geonet accommodate the flow?

1.41 What are the advantages of geomembrane-clay composite liner?

1.42 List the major factors to be considered in the selection of a geosynthetic for field applications.

1.43 What is the role of cost-benefit ratio in the selection of geosynthetics?

1.44 Give the symbolic representation for the reinforcement geotextile.

## References

Adams, M.T. and Collin, J.G. (1997). Large model spread footing load tests on geosynthetic reinforced soil foundations. *Journal of Geotechnical and Geoenvironmental Engineering, ASCE*, **123**, 1, pp. 66–72.

Bassett, R.H. and Last, N.C. (1978). Reinforcing earth below footings and embankments. *Proceedings of the Symposium on Earth Reinforcement*, ASCE, New York, pp. 202–231.

Beckham, W.K. and Mills, W.H. (1935). Cotton-fabric reinforced roads. *Engineering News Record*, pp. 453–455.

Bourdeau, P.L. (1989). Modeling of membrane action in a two-layer reinforced soil system. *Computers and Geotechnics*, 7, 1–2, pp. 19–36.

Bourdeau, P.L., Harr, M.E., and Holtz, R.D. (1982). Soil-fabric interaction – an analytical model. *Proceedings of the 2nd International Conference on Geotextiles*. Las Vegas, USA, pp. 387–391.

Bouazza, A. and Heerten, G. (2012). Geosynthetic applications – sustainability aspects. Chapter 18, *Handbook of Geosynthetic Engineering*, Shukla, S.K., Editor, ICE Publishing, London, pp. 387–396.

Espinoza, R.D. (1994). Soil-geotextile interaction: evaluation of membrane support. *Geotextiles and Geomembranes*, **13**, 5, pp. 281–293.

Espinoza, R.D. and Bray, J.D. (1995). An integrated approach to evaluating single-layer reinforced soils. *Geosynthetics International*, 2, 4, 723–739.

Fluet, J.E. (1988). Geosynthetics for soil improvement: a general report and keynote address. *Proceedings of the Symposium on Geosynthetics for Soil Improvement*, Tennessee, USA, pp. 1–21.

Giroud, J.P. (1992). Geosynthetics in dams: Two decades of experience. *Geotechnical Fabrics Report*, **10**, 5, pp. 6–9.

Giroud, J. P. and Noiray, L. (1981). Geotextile-reinforced unpaved road design. *Journal of the Geotechnical Engineering Division*, ASCE, **107**, 9, pp. 1233–1254.

Giroud, J.P. and Carroll, R.G. (1983). Geotextile products. *Geotechnical Fabrics Report*, pp. 12–15.

Giroud, J.P., Ah-Line, A. and Bonaparte, R. (1984). Design of unpaved roads and trafficked areas with geogrids. *Proceedings of the Symposium on Polymer Grid Reinforcement*, London, pp. 116–127.

Guido, V.A., Biesiadecki, G.L. and Sullivan, M.J. (1985). Bearing capacity of a geotextile-reinforced foundation. *Proceedings of the 11th International Conference on Soil Mechanics and Foundation Engineering*. San Francisco, Calif. pp. 1777–1780.

Guido, V.A., Dong, K.G. and Sweeny, A. (1986). Comparison of geogrid and geotextile reinforced earth slabs. *Canadian Geotechnical Journal*, **23**, 1, pp. 435–440.

Halse, Y.H., Wiertz, J. and Rigo, J.M. (1991). Chemical identification methods used to characterize polymeric geomembranes. Chapter 15, *Geomembranes Identification and Performance Testing*, Chapman & Hall, London, pp. 316 – 336.

Hausmann, M.R. (1990). *Engineering Principles of Ground Modification*. McGraw-Hill, New York.

Ingold, T.S. (1994). *The Geotextiles and Geomembranes Manual*. Elsevier Advanced Technology, UK.

Jewell, R.A. (1996). *Soil Reinforcement with Geotextiles*. Construction Industry Research and Information Association, London, CIRIA, Special Publication 123.

John, N.W.M. (1987). *Geotextiles*. Blackie, London.

Koerner, R. M. (2005). *Designing with Geosynthetics*. Fifth Edition, Prentice Hall, New Jersey, USA.

Koerner, R.M. and Daniel, D.E. (1997). Final covers for solid waste landfills and abandoned dumps. ASCE Press, Reston, VA, and Thomas Telford, London, UK.

Landreth, R.E. (1990). Chemical resistance evaluation of geosynthetics used in the waste management applications. *Geosynthetics Testing for Waste Containment Applications*, Special Technical Publication STP 1081, ASTM, Philadelphia, pp. 3–11.

Love, J.P., Burd, H.J., Milligan, G.W.E. and Houlsby (1987). Analytical and model studies of reinforcement of a layer of granular fill on soft clay subgrade. *Canadian Geotechnical Journal*, 24, 4, pp. 611–622.

Madhav, M.R. and Poorooshasb, H.B. (1988). A new model for geosynthetic-reinforced soil. *Computers and Geotechnics*, 6, 4, pp. 277–290.

Madhav, M.R. and Poorooshasb, H.B. (1989). Modified Paternak model for reinforced soil. *Mathematical and Computational Modelling, an International Journal*, 12, 12, pp. 1505–1509

McGown, A. and Andrawes, K.Z. (1977). The influence of nonwoven fabric inclusions on the stress-strain behaviour of a soil mass. *Proceedings of the International Conference on the Use of Fabrics in Geotechnics*, Paris, pp. 161–166.

McGown, A., Andrawes, K.Z. and Al-Hasani, M.M. (1978). Effect of inclusion properties on the behaviour of sand. *Geotechnique*, 28, 3, pp. 327–346.

Nishida, K. and Nishigata, T. (1994). The evaluation of separation function for geotextiles. *Proceedings of the 5th Int. Conference on Geotextiles, Geomembranes and Related Products*. Singapore, 1994, pp. 139–142.

Pinto, M.I.M. (2004). Reply to the discussion of "Applications of geosynthetics for soil reinforcement' by M.I.M. Pinto, Ground Improvement, 7, 2, 2003, pp. 61–72 by S.K. Shukla'. *Ground Improvement*, UK, 8, 4, pp. 181–182.

Qian, X, Koerner, R.M. and Gray, D.H. (2002). *Geotechnical Aspects of Landfill Design and Construction*. Prentice Hall, New Jersey.

Rankilor, P.R. (1981). *Membranes in Ground Engineering*. John Wiley & Sons, Chichester, England, 1981.

Rigo, J.M. and Cuzzuffi, D.A. (1991). Test standards and their classification. *Geomembrane Identification and Performance Testing*, Chapman and Hall, London, pp. 22–58.

Schlosser, F. and Vidal, H. (1969). Reinforced earth. *Bulletin de Liaison des Laboratoires des Ponts et Chaussées*, No. 41, France.

Sellmeijer, J.B. (1990). Design of geotextile reinforced unpaved roads and parking areas. *Proceedings of the 4th International Conference on Geotextiles, Geomembranes and Related Products*. The Hague, Netherlands, pp. 177–182.

Sellmeijer, J.B., Kenter, C.J. and Van den Berg, C. (1982). Calculation method for fabric reinforced road. *Proceedings of the 2nd International Conference on Geotextiles*. Las Vegas, USA, pp. 393–398.

Shukla, S.K. (2002a). Fundamentals of geosynthetics. Chapter 1, *Geosynthetics and Their Applications*, Shukla, S.K., Editor, Thomas Telford, London, pp. 1–54.

Shukla, S.K. (2002b). Shallow foundations. Chapter 5, *Geosynthetics and Their Applications*, Shukla, S.K., Editor, Thomas Telford, London, pp. 123–163.

Shukla, S.K. (2003a). Geosynthetics in civil engineering constructions. *Employment News*, New Delhi, March 1–7, pp. 1–3.

Shukla (2003b). How to select geosynthetics for field applications. *Geosynthetic Pulse*, India, **1**, 5, pp. 2–2.

Shukla, S.K. (2004). Discussion of "Applications of geosynthetics for soil reinforcement" by M.I.M. Pinto' *Ground Improvement*, 8, 4, pp. 179–181.

Shukla, S.K. (2013). Articles of professional interest: editorial of the special issue on geosynthetic engineering, *Indian Geotechnical Journal*, **43**, 4, pp. 281–282.

Shukla, S.K. and Chandra, S. (1994). A generalized mechanical model for geosynthetic-reinforced foundation soil. *Geotextiles and Geomembranes*, **13**, 12, pp. 813–825.

Shukla, S.K. and Chandra, S. (1995). Modelling of geosynthetic-reinforced engineered granular fill on soft soil". *Geosynthetics International*, USA, **2**, 3, pp. 603–618.

Shukla, S.K., Sivakugan, N. and Das, B.M. (2009). Fundamental concepts of soil reinforcement – an overview. *International Journal of Geotechnical Engineering*, **3**, 3, pp. 329–342.

Thomas, R.W. and Verschoor, K.L. (1988). Thermal analysis of geosynthetics. *Geotechnical Fabrics Report*, **6**, 3, pp. 24–30.

Vidal, H. (1969). The principle of reinforced earth. *Highway Research Record*, No. 282, pp. 1–16.

## Answers to selected questions

| 1.1 | (b) |
| 1.3 | (a) |
| 1.5 | (d) |
| 1.7 | (a) |
| 1.9 | (c) |
| 1.11 | (a) |
| 1.13 | (c) |
| 1.15 | (b) |
| 1.26 | 5 m |
| 1.37 | 1.33 |

# Chapter 2

# Geosynthetics – properties, applications and design concepts

## 2.1 INTRODUCTION

Geosynthetics cover a wide range of materials, applications and environments. The evaluation of the properties of a geosynthetic is important in ensuring that it will adequately perform the intended function when used in the man-made project, structure or system as an integral part. Not all the properties of a geosynthetic may be important for every application. The required properties and characteristics of geosynthetics depend on their purpose and the desired function in a given application.

This chapter deals with the properties of geosynthetics and highlights the basic concepts of their determination along with their importance in the design process and the performance in field applications. The detailed description of standard procedures and standardized test equipment can be obtained from the test standards to be followed at the place of work. Geosynthetics, being polymer-based products, are viscoelastic, and under working conditions, their performance is dependent on several factors such as the ambient temperature, the level of stress, the duration of the applied stress, and the rate at which the stress is applied. For evaluating the properties by testing, geosynthetics are generally permitted to come to hygroscopic and thermal equilibrium with the surrounding or standard atmosphere; this process is called the *conditioning*. The properties of geosynthetics should therefore be determined keeping these factors in view.

The design of a structure incorporating geosynthetics aims to ensure its strength, stability and serviceability over its intended life span. This chapter also presents the basic concepts of design methods for the geosynthetic-related structures or systems.

## 2.2 PHYSICAL PROPERTIES

The physical properties of geosynthetics that are of prime interest are *specific gravity*, *unit mass (weight)*, *thickness* and *stiffness*. They are all considered to be *index properties* of geosynthetics. There are some more physical properties, which are important only in the case of geogrids and geonets and they are *type of structure, junction type, aperture size and shape, rib dimensions, planar angles made by intersecting ribs* and *vertical angles made at the junction point*. The physical properties of geosynthetics are more dependent on temperature and humidity than those of soils and rocks. In order to achieve consistent results in the laboratory, a good environmental control during the testing is therefore important.

## 2.2.1   Specific gravity

The specific gravity of a polymer, from which the geosynthetic is manufactured, is expressed as a ratio of its unit volume weight (without any voids) to that of pure water at 4°C. It can be determined by the displacement method. In case of geomembranes, a known mass is weighed in air and then in water. The specific gravity of the geomembrane specimen is the ratio of its weight in air to its weight difference in air and in water.

The specific gravity of a base polymer is an important property as it can assist in identifying the base polymer of the geosynthetics. Specific gravity is widely used for geomembrane identification and quality control. In case of polyethylene (PE), specific gravity, or more correctly density, is an important property as it forms the basis upon which PE is classified as very low, low, medium or high density. The typical values of specific gravity of commonly used polymeric materials are given in Table 1.8 (see Chapter 1). When the additives are added, the specific gravity of the resulting polymer may be higher or lower than that of the base polymer depending on the specific gravity and proportion of additive used. Note that the specific gravity of some of the polymers (e.g., PP and PE) is smaller than 1.0, which is a drawback when working with geosynthetics in underwater applications as some of them may float.

## 2.2.2   Unit mass

The *unit mass (or weight)* of a geosynthetic is measured in terms of *mass (or weight) per unit area* as opposed to mass (or weight) per unit volume due to variations in thickness under applied compressive stresses. It is usually given in units of gram per square metre (g/m$^2$). It is determined by weighing square or circular test specimens of known dimensions (generally of area not less than 100 cm$^2$), cut from locations distributed over the full width and length of the laboratory sample. The linear dimensions should be measured without any tension in the specimen. The calculated values are then averaged to obtain the mean mass per unit area of the laboratory sample.

Mass (weight) per unit area, with knowledge of the structure of the geosynthetic, can be a good indicator of cost and several other properties, such as tensile strength, tear strength, puncture strength etc., which are defined in Section 2.3. It can be used for the quality control of delivered geosynthetics to determine the specimen conformance. For commonly used geosynthetics, the mass per unit area varies in order of magnitude from typically 100 g/m$^2$ to 1000 g/m$^2$. For 'Tensar' SR2 and SS2 geogrids, the values of mass per unit area are 930 g/m$^2$ and 345 g/m$^2$, respectively. In comparison to the geotextiles, the geomembranes may have substantially larger values of mass per unit area, even up to several thousands of grams per square metre.

## 2.2.3   Thickness

The *thickness* of a geosynthetic is the distance between its upper and lower surfaces, measured normal to the surfaces at a specified normal compressive stress (generally

2.0 kPa for geotextiles, and 20 kPa for geogrids and geomembranes) applied for 5 s. It should be measured by using a thickness-testing instrument to an accuracy of at least 1 mil ( = 0.001 inch ≈ 0.025 mm). The thickness-testing instrument is basically a thickness gauge that consists of a base (or anvil) and a free-moving pressure foot-plate with parallel planar faces having an area of more than 2000 mm². Normally the thickness of geotextiles should be determined by measuring one layer only. In cases when two or more layers are used in contact with each other in an application, a test may be made with a specific number of layers instead of one, keeping in view the relevance of such findings. Thickness is not normally quoted for geotextiles, except for thicker nonwovens, but thickness is invariably quoted for geomembranes. The thickness of commonly used geosynthetics ranges from 10 to 300 mils. Most geomembranes used today are 20 mils thick or greater.

Thickness is one of the basic physical properties used to control the quality of many geosynthetics. The thickness values are required in the calculation of some geo-synthetic parameters such as the permittivity and transmissivity (see Section 2.4). Since many geosynthetics, particularly geotextiles and some drainage geocomposites, are highly compressible, the thickness measure will greatly depend upon the applied normal compressive stress. For this reason, it may be desirable to measure the thickness at various normal compressive stresses and to study the general relationship between the thickness and the stress. The thickness of a geosynthetic decreases when the applied normal compressive stress is increased as seen in Fig. 2.1. This decrease in thickness may result in the partial closing or opening of the voids of geotextile, depending on its initial structure and the boundary conditions. Care should be exercised to minimize the effects of cutting and handling the test specimens for their thickness measurement.

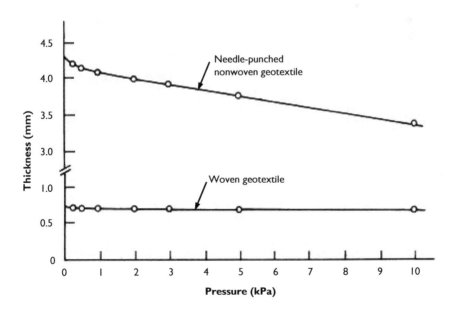

*Figure 2.1* Variation of the thickness of geotextiles with the applied normal pressure (after Shamsher, 1992).

The mass per unit area of a geosynthetic can be related to its mass density (simply called the density) $\rho$ or weight density (also called the unit weight) $\gamma$ as

$$m = \frac{M}{A} = \frac{M \times \Delta x}{A \times \Delta x} = \left(\frac{M}{V}\right)\Delta x = \rho \Delta x = \frac{\gamma \Delta x}{g} \tag{2.1}$$

where $M$ is the mass, $A$ is the surface area, $V$ is the total volume, $\Delta x$ is the thickness of the geosynthetic specimen, and $g\ (= 9.81\ \mathrm{m/s^2})$ is the acceleration due to gravity.

If it is assumed that the density of the geosynthetic $\rho$ is equal to the density of the polymer solid $\rho_s$, then Equation (2.1) reduces to

$$m = \rho_s \Delta x = G\rho_w \Delta x = \frac{G\gamma_w \Delta x}{g} \tag{2.2}$$

where $G$ is the specific gravity of the polymer solid, $\rho_w\ (= 1000\ \mathrm{kg/m^3})$ is the density of water, and $\gamma_w\ (= 9810\ \mathrm{kN/m^3})$ is the unit weight of water. Note that $\gamma = \rho g$ and $\gamma_w = \rho_w g$. Note that for the geomembranes, $\rho \approx \rho_s$, and hence Equations (2.1) and (2.2) are more correctly applicable.

### EXAMPLE 2.1

For a geomembrane specimen, consider the following:
  Thickness, $\Delta x = 3$ mm, and
  Mass per unit area, $m = 2826\ \mathrm{g/m^2}$.

Determine the specific gravity of the polymeric material of the geomembrane.

### SOLUTION

Given: $\Delta x = 3\ \mathrm{mm} = 0.003\ \mathrm{m}$ and $m = 2826\ \mathrm{g/m^2} = 2.826\ \mathrm{kg/m^2}$
From Equation (2.2), the specific gravity of the polymeric material is obtained as

$$G = \frac{m}{\rho_w \Delta x} = \frac{2.826}{(1000)(0.003)} = 0.942$$

## 2.2.4  Stiffness

The *stiffness* (also known as the *flexural rigidity*) of a geosynthetic is its ability to resist flexure/bending under its own weight. It is measured by its capacity to form a cantilever beam without exceeding a certain amount of downward bending under its own weight. In the commonly used test, known as the *single cantilever test*, the geosynthetic specimen is placed on a horizontal platform with a weight on it. Holding the weight, the specimen along with weight is slid slowly and steadily in a direction parallel to its long dimension until the leading edge projects beyond the edge of the platform. The length of overhang is measured when the tip of the test specimen is depressed under its own weight to the point where the line joining the tip to the edge of the platform makes an angle of 41.5° with the horizontal. One half of this length is the bending length. The cube of this quantity multiplied by the weight per unit area of the geosynthetic is the flexural rigidity.

The stiffness of a geosynthetic indicates the feasibility of providing a suitable working surface for installation. The survivability (also known as the workability or constructability) of a geosynthetic, defined as its ability to support work-personnel in an uncovered state and construction equipment during the initial stages of cover fill placement, depends on the geosynthetic stiffness as well as on some other factors such as the water absorption and the buoyancy. When placing a geotextile or geogrid on extremely soft soils, a high stiffness is desirable. The stiffness of geosynthetics can also have some effects on their performance when they are used in the mitigation of soil erosion of hill slopes. If the geosynthetic (geotextile or geomat) does not have a low stiffness to conform to the contours of the ground, then a gap may be left between the ground and the geosynthetic through which the water can flow and thereby erode.

Properties like aperture size and shape, rib dimensions, etc. can be measured directly and are relatively easily determined.

## 2.3    MECHANICAL PROPERTIES

Mechanical properties are important in those applications where a geosynthetic is required to perform a structural role under the applied loads, or where it is required to survive the installation damages and the localized stresses. There are several mechanical properties, but only some of them are important in the case of a particular geosynthetic.

### 2.3.1    Compressibility

The *compressibility* of a geosynthetic is measured by the decrease in its thickness at the increasing applied normal pressure. This mechanical property is very important for nonwoven geotextiles, because they are often used to convey liquid within the plane of their structure. Figure 2.1 shows the changes in thickness under the applied pressure for typical woven and needle-punched nonwoven geotextiles. For most geotextiles, except needle-punched nonwoven geotextiles, the compressibility is relatively very low.

The compression behaviour of geosynthetics, particularly geocomposites, can be studied by applying compressive loads at a constant rate of deformation to specimens mounted between parallel plates in a loading frame. The deformations are recorded as a function of load and plotted as shown in Fig. 2.2(a). Being an artifact caused by the alignment or seating of the specimen, the toe region OA may not represent a compressive property of the material. The yield point and the strain should be calculated considering the zero deformation point as shown in Fig. 2.2(a). Many geosynthetics exhibit compressive deformation, but not all may exhibit a well-defined compressive

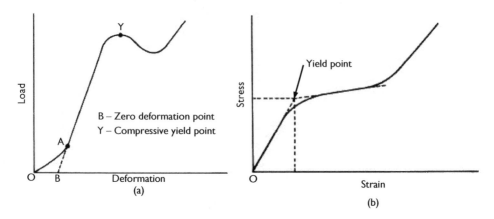

*Figure 2.2* Compression behaviour of geosynthetics: (a) typical load-deformation curve; and (b) typical stress-strain curve.

yield point; however, the significant change in the slope of the stress-strain curve can be used to determine yield point for comparative purposes (Fig. 2.2(b)). Variable inclined plates or set angled blocks, as described in ASTM D6364-06 (ASTM, 2011a), may be used to evaluate the deformation of the geosynthetic(s) under loading at various angles. The compressive loading test is generally used for the quality control to evaluate uniformity and consistency within a *lot* (a unit of production) or between the lots where the test specimen geometry factors such as thickness or materials may change.

## 2.3.2 Tensile properties

The determination of tensile properties, mainly *tensile strength* and *tensile modulus*, of geosynthetics is important, especially when they need to resist tensile stresses transferred from the soil in reinforcement applications, such as reinforced embankments over soft subgrades, reinforced soil retaining walls and reinforced slopes. The *tensile strength* of a geosynthetic is its maximum resistance to deformation developed when it is subjected to tension by an external force. Due to the specific geometry and irregular cross-sectional area that cannot be easily defined, the tensile strength of a geosynthetic cannot be expressed conveniently in terms of stress as we usually define it. It is, therefore, defined as the peak (or maximum) load that can be applied per unit length along the edge of the geosynthetic in its plane. The tensile properties of geosynthetics are studied by conducting the tensile strength test in which the geosynthetic specimen is loaded and the corresponding force-elongation curve is obtained.

The tensile strength is usually determined by the wide-width strip tensile test on a 200-mm wide geosynthetic strip with a gauge length of 100 mm (Fig. 2.3) (ASTM, 2011b). The entire width of a 200-mm wide geosynthetic specimen is gripped in the jaws of a tensile strength testing machine and it is stretched in one direction at a prescribed constant rate of extension until the specimen ruptures/breaks. During the

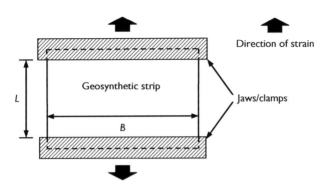

*Figure 2.3* Wide-width strip tensile test.
*(Note B = 200 mm, L = 100 mm).*

extension process, both load and deformations are measured. The width of the specimen is kept greater than its length, as some geosynthetics have a tendency to contract (or 'neck down') under load in the gauge length area. The greater width reduces the contraction effect of such geosynthetics, and by approximating the plane-strain conditions, it more closely simulates the deformation experienced by a geosynthetic when embedded in soil under field conditions. The test provides parameters such as peak strength, elongation and tensile modulus. The tensile properties depend on the geosynthetic polymer and manufacturing process leading to the structure of the finished product. The measured strength and the rupture strain are functions of many test variables, including sample geometry, gripping method, strain rate, temperature, initial preload, conditioning, and the amount of any normal confinement applied to the geosynthetic.

Figure 2.4 shows the influence of the specimen width on the tensile strength of the geotextile. To minimize the effects, the test specimen should have a width-to-gauge length ratio (i.e. the *aspect ratio*) of at least two, and the test should be carried out at a standard temperature. The actual temperature has a great influence on the strength properties of many polymers (Fig. 2.5). The tensile strength of geosynthetics is closely related to the mass per unit area (Fig. 2.6). A heavyweight geotextile, with a higher mass per unit area, will usually be stronger than a lightweight geotextile. For a given geosynthetic, the tensile strength is also a function of the rate of strain at which the specimen is tested. At a low strain rate, the measured strength tends to be lower and occurs at a higher failure strain. Conversely, at a high strain rate, the measured strength tends to be higher and occurs at a lower failure strain.

Other forms of the tensile strength tests such as grab test, biaxial test, plain-strain test, and multi-axial test are shown schematically in Fig. 2.7. The grab tensile test is used to determine the strength of the geosynthetic in a specific width, together with the additional strength contributed by adjacent geosynthetic or other material. This test is basically a uniaxial tensile test in which only the central portion of the geosynthetic specimen is gripped in the jaws (Fig. 2.7(a)). The test normally uses 25.4-mm (1-in.) wide jaws to grip a 101.6-mm (4-in.) wide geosynthetic specimen (ASTM, 2013). A continually increasing load is applied longitudinally to the specimen and the

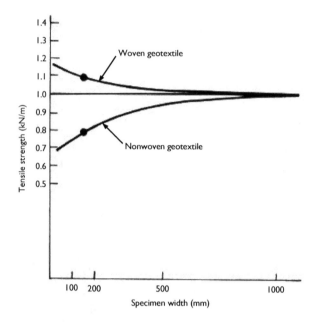

*Figure 2.4* Influence of geotextile specimen width on its tensile strength (after Myles and Carswell, 1986).

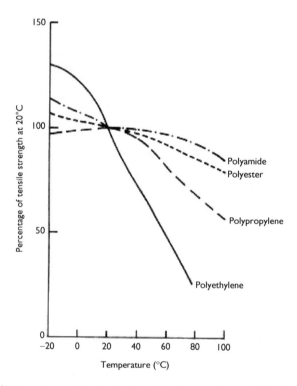

*Figure 2.5* Influence of temperature on the tensile strength of some polymers (after Van Santvoort, 1995).

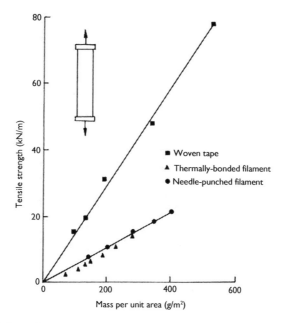

*Figure 2.6* Variation of tensile strength with mass per unit area for PP geotextiles (after Ingold and Miller, 1988).

test is carried to rupture. It is not clear how the force is distributed across the width of the specimen. This test simulates the field situation as shown in Fig. 2.8. It is difficult to relate the grab tensile strength to the wide-width strip tensile strength in a simple manner without direct correlation tests. The grab tensile test is often useful as a quality control or acceptance test for geotextiles. The typical range of the grab tensile strength of geotextiles is 300–3000 N.

The plain-strain tensile test is a uniaxial tensile test in which the entire width of the specimen is gripped in the jaws with the specimen being restrained from necking during the tensile load application (Fig. 2.7(c)). This test can be carried out to assess the strength of the geotextile when installed within the soil mass.

A geosynthetic layer installed within the soil mass is subjected to forces from more than one direction, including the forces perpendicular to the surfaces of the geosynthetic causing out-of-plane deformation. The multi-axial tensile test can be carried out to measure the out-of-plane response of a geosynthetic to a force that is applied perpendicular to the initial plane of the geosynthetic specimen. In this test, the geosynthetic specimen is clamped at the edges of a large diameter, generally 0.6 m, pressure vessel (Fig. 2.7(d)) (ASTM, 2010). The pressure is applied to the specimen to cause out-of-plane deformation and failure. This deformation with pressure information is then analyzed to evaluate the geosynthetic strength. When a geosynthetic deforms to a simplified geometric shape, such as an arc of a sphere or an ellipsoid, the data obtained from the test can be converted to the biaxial tensile stress-strain values. In geosynthetic applications where local subsidence is expected, the multi-axial tensile test can be considered a performance test.

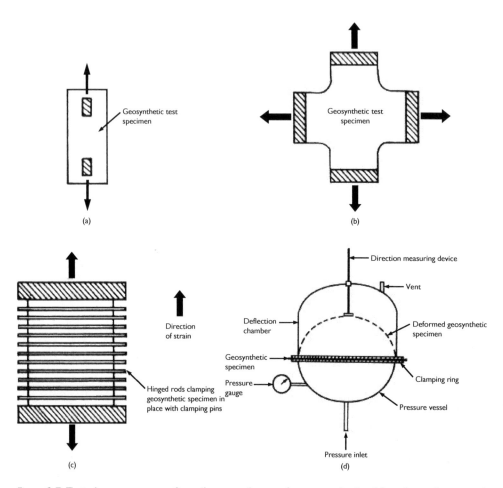

*Figure 2.7* Typical arrangements of tensile strength tests for geosynthetics: (a) grab tensile strength test; (b) biaxial tensile strength test; (c) plain-strain tensile strength test; (d) multi-axial tensile strength test.

During the manufacturing process, the variability in the geosynthetic properties may occur as happens with other civil engineering construction materials. Based on the quality control tests, a manufacturer of geosynthetics can represent the properties statistically in a normal distribution curve as shown in Fig. 2.9. The project specifications tend to include several qualifiers such as minimum, average/mean/typical, maximum, and minimum average roll value (MARV). If $X_1$, $X_2$, $X_3$, ...., $X_N$ are individual property values in a sample of size $N$, then these qualifiers as well as the standard deviation can be determined using the following expressions (Narejo *et al.*, 2001):

$$\text{Average, } \overline{X} = \frac{X_1 + X_2 + X_3 + ...... + X_N}{N} \tag{2.3}$$

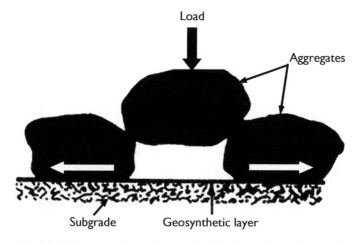

*Figure 2.8* A field situation that can be simulated by the grab tensile strength test.

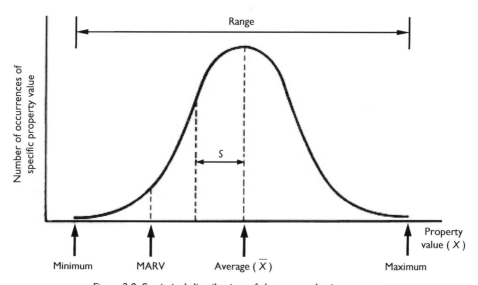

*Figure 2.9* Statistical distribution of the geosynthetic property.

$$\text{Standard deviation, } S = \sqrt{\frac{(X_1 - \bar{X})^2 + (X_2 - \bar{X})^2 + \ldots + (X_N - \bar{X})^2}{N-1}} \qquad (2.4)$$

$$\text{MARV} = \bar{X} - 2 \times S \qquad (2.5)$$

$$\text{Minimum} = \bar{X} - 3 \times S \qquad (2.6)$$

$$\text{Maximum} = \bar{X} + 3 \times S \tag{2.7}$$

$$\text{Range} = \text{Maximum} - \text{Minimum} \tag{2.8}$$

The significance of standard deviation lies in the variation in material properties and testing values of the particular property under investigation. The current trend is to report the strength value as a MARV in the weakest direction. For normally distributed data, the MARV is calculated statistically as the average/mean/typical value minus two times the standard deviation (see Eq. (2.5)). A specification based on the MARV means that 97.5% of the geosynthetic samples from each tested roll are required to meet or exceed the designer's specified value for the geosynthetic product to be acceptable. MARV has now become a manufacturing quality control tool used to allow manufacturers to establish the published values such that the users/purchasers of geosynthetics will have a 97.5% confidence that the property in question will meet or exceed the published values. MARV is applicable to the intrinsic physical properties of geosynthetics, such as weight, thickness, and strength, but it may not be appropriate for some hydraulic, degradation or endurance properties. It has been observed that for the design engineers, the use of MARV results in a better communication with manufacturers, lower number of change requests, and simpler and economical designs, thus resulting in cost savings for everyone involved in the process.

As already mentioned, the tensile strength of most geosynthetics, including woven geotextiles, is generally not the same in all directions in their plane, that is, they behave as anisotropic materials. Particularly for the woven geotextiles, the tensile strength is governed by the weaving structure. The strength in the warp direction (or machine direction, MD), called the *warp strength*, may not be equal to the strength in the weft direction (or cross-machine direction, CMD), called the *weft strength*. A uniaxial tensile strength of 100 kN/m measured in the machine direction would be written as 100 kN/m MD. Similarly, a uniaxial tensile strength of 40 kN/m measured in the cross-machine direction would be written as 40 kN/m CMD. Where the warp and weft strengths are usually found to be different, the strengths may be written as 100/40 kN/m in which case the first figure is taken as the warp strength and the second as the weft strength. For the woven geotextiles, the warp strength is generally greater than the weft strength.

The *tensile modulus* of a geosynthetic is the slope of its stress-strain or load-strain curve, as determined from the wide-width strip tensile test. This is basically a ratio of the change in tensile force per unit width of the geosynthetic to a corresponding change in strain, and is equivalent to Young's modulus for other construction materials, such as concrete, steel, timber, structural plastic, etc. The *tensile modulus* needs to be considered in designs, as the geosynthetic needs to resist tensile stresses under deformations compatible with those allowable for the soil.

Figure 2.10 shows the typical load-strain curves for geotextiles with an explanation of tensile moduli. Note that the typical S-shaped load-strain curve (Fig. 2.10(a)) generally results from a change in the orientation of 'tie' molecules, which run from one crystallite to another, linking them together. The tensile strengths corresponding to the breaking point and the highest peak point on the load-strain curve are called the *breaking tensile strength* and the *ultimate tensile strength*, respectively. At the com-

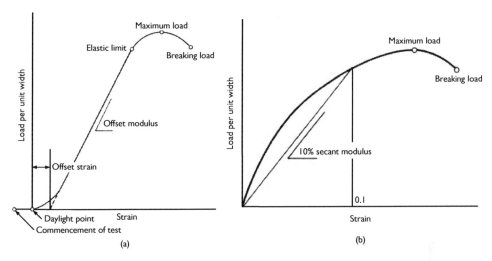

*Figure 2.10* Load – strain curves for geotextiles exhibiting: (a) linear behaviour; (b) non-linear behaviour (after Myles and Carswell, 1986).

mencement of the tensile test, the load will be zero unless a preload is used. As the test is begun, the geotextile strains without loading until it reaches the *daylight point* (a point where the load extension curve parts from the strain). The slope of the load per unit width–strain curve at any strain is the *tangent modulus*. The *offset modulus (i.e. the working modulus)* is the maximum value of the tangent modulus and is obtained from the slope of the linear portion of the load per unit width–strain curve. An offset strain is then defined by extending the linear portion of the curve back to the zero load line. It is important to understand that the (unknown) strain from the indicated start of the test to the daylight point is eliminated by preloading and that the amount of offset strain is influenced by the amount of preloading. For the geotextiles that do not have a linear range, the modulus is typically defined as the *secant modulus* at a specified strain, usually 5 or 10% strain (Fig. 2.10(b). The designer and specifier must have a clear understanding of the interpretation of these moduli.

Note that the property of a geosynthetic by virtue of which it can absorb energy is called the *toughness*. It is expressed as the actual work-to-break per unit surface area and is proportional to the area under the load per unit width – strain curve from the origin to the rupture point.

Figure 2.11 shows the typical strength properties of some geosynthetics. It is noticed that the woven geotextiles display generally the lowest extensibility and the highest strength of all the geotextiles. Geogrids have relatively high dimensional stability, high tensile strength and high tensile modulus at low strain levels. They develop reinforcing strength even at strain equal to 2%. The high tensile modulus results from the prestressing during the manufacture, which also creates integrally formed structures without weak points either in ribs or junctions. In the case of geonets, there is a preferential direction in strength between the MD and the CMD. Geonets have the greatest strength in the MD.

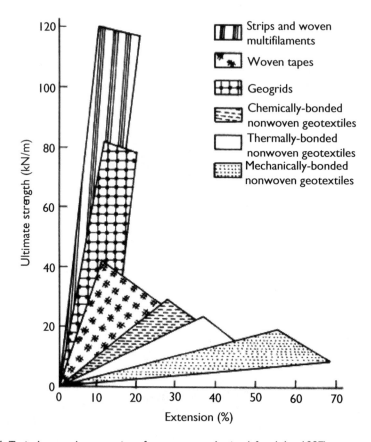

*Figure 2.11* Typical strength properties of some geosynthetics (after John, 1987).
Note: Overlapping zones have not been shown for clarity, some non-typical geosynthetics may lie
outside the zones indicated, and some geosynthetics are more sensitive to the test method.

The viscoelastic behaviour of geosynthetics can produce misleading results for both short-term and rapid rate tensile tests. The tests conducted to provide the design data should also consider the long-term conditions and account for the effect of the surrounding soil. The *geosynthetic confinement* within soil in the field and the resultant interlocking of soil particles with the geosynthetic structure are found to have a significant effect on the stress-strain properties. It is generally found that the modulus of a geosynthetic confined in soil is likely to be higher than when tested in isolation. The mechanism of this enhancement is simply the frictional force development. The deformation of a geosynthetic structure is, therefore, likely to be overestimated if the in-isolation modulus is used in the calculations. This fact tends to support the use of a working modulus as an appropriate modulus type. The details of confined tensile test methods have been presented by McGown *et al.* (1982) and El-Fermaoui and Nowatzki (1982). Due to the high costs involved, the confined tensile testing is not carried out on a routine basis. Keeping these facts in view, note that the wide-width strip tensile test is essentially an *index test*.

Geosynthetics in field applications are subjected to loadings which may cause longitudinal and lateral normal strains simultaneously. The ratio of lateral normal strain to longitudinal normal strain is known as the *Poisson's ratio*, which varies in the range of 0 to 0.5 for most linear elastic engineering materials, but the theoretical range is −1.0 to 0.5. Although the Poisson's ratio is often assumed as a material constant, it is strain dependent. It is a critical parameter in some engineering applications where the geosynthetics are being used as reinforcements (e.g. capping of contaminated high water content geomaterials or hydraulic filling of geosynthetic containers). The uniaxial tensile strength test data reported by Wesseloo *et al.* (2004) indicate that the Poisson's ratio for 0.2 mm thick HDPE geomembrane decreases from 0.50 to 0.24 when the axial strain increases from 0 to 200% at strain rates ranging from 50%/min to 0.05%/min at 22 ± 1°C. Based on the wide-width tensile strength tests, Kutay *et al.* (2006) reported that the lateral strain was greater than the axial strain at failure for one needle-punched polypropylene nonwoven geotextile (mass per unit area = 278 g/m$^2$) and two PET (polyester) woven geotextiles (mass per unit area = 290/1907 g/m$^2$, ultimate tensile strength = 70/632 kN/m, tensile strength at 5% strain = 16/316 kN/m), where the Poisson's ratio at failure ranged from 1.0 to 2.1; whereas the Poisson's ratio was in a range of 0.3 to 0.81 for four PET geogrids (mass per unit area = 305/387/617/1284 g/m$^2$, ultimate tensile strength = 63/102/137/371 kN/m, tensile strength at 5% strain = 25/37/65/146 kN/m) depending on the clamping technique used. Giroud (2004) presented analytical expressions for Poisson's ratio ($v$) as a function of axial strain ($\varepsilon$) for both incompressible and compressible materials. The expression for compressible material is based on an assumption where the small strain theory is extended to large strains. Shukla *et al.* (2009) presented an improved understanding of the theoretical expressions for the Poisson's ratio of geosynthetics, and proposed a simple method by presenting a graphical chart for estimating the Poisson's ratio of geosynthetics at zero strain ($v_0$), which is required to calculate the realistic values of Poisson's ratio of geosynthetics at other non-zero strains using the analytical expression.

At this stage, it is worthwhile mentioning index and performance tests. The *index test* (i.e. *in-isolation test* or *identification test*) is a test procedure which may contain a known bias but which may be used to establish an order for a set of geosynthetic specimens with respect to the property of interest. Index tests do not take into account the interaction which may occur between the geosynthetic and the soil. In fact, the index tests are carried out to compare the basic properties (e.g., wide-width tensile strength, creep under load, friction properties, etc.) of geosynthetic products. They are generally used routinely for *quality control* and *quality assurance* (see Section 8.2) of the manufactured geosynthetics. Index tests are also used to monitor changes that may occur after a geosynthetic has had some sort of exposure. Index tests generally do not reflect design features of applications. Geosynthetics, when correctly processed and stabilized, are resistant to chemical and microbiological attacks encountered in normal soil environments. In such situations and with well-understood properties of geosynthetics, only a minimum number of index tests are necessary. Index tests are generally simple tests which can be carried out quickly and cheaply.

*Performance test*, on the other hand, is carried out by placing the geosynthetic in contact with a soil/fill under standardized conditions in the laboratory to provide as closely as practicable simulation of the selected field conditions, which can be used in the design. Performance testing, if possible, should also be carried out at a full-scale at the site. Since the geosynthetics vary randomly in thickness and weight in a given sample roll due to normal manufacturing techniques, the tests must be conducted on the representative samples collected as per the guidelines of available standards, which ensure that all areas of the sample roll and a full variation of the product are represented within each sample group. For applications in more severe environments (soil treated with lime or cement, landfills or industrial wastes, or highly acidic volcanic soils, high temperature conditions, etc.), and with indeterminate design lives, the performance tests with site-specific parameters may be required.

### 2.3.3 Survivability properties

There are some mechanical properties of geosynthetics, which are related to geosynthetic survivability/constructability and separation function. The tests to determine such properties are generally treated as integrity/index tests. These properties are as follows:

- *Tearing strength*: ability of a geosynthetic to withstand stresses causing to continue or propagate a tear in it, often generated during the installation.
- *Static puncture strength*: ability of a geosynthetic to withstand localized stresses generated by penetrating or puncturing objects such as aggregates or roots, under quasi-static conditions (Fig. 2.12).
- *Impact strength (dynamic puncture strength)*: ability of a geosynthetic to withstand stresses generated by the sudden impact and penetration of falling objects, such as coarse aggregates, tools, and other construction items during the installation.
- *Bursting strength*: ability of a geosynthetic to withstand a pressure applied normal to its plane while constrained in all directions in that plane (Fig. 2.12).
- *Fatigue strength*: ability of a geosynthetic to withstand repetitive loading before undergoing failure.

The *tearing strength test* aims to measure the propensity of a geosynthetic to the tearing force once a tear has been initiated. The tearing strength of geotextiles under in-plane loading is determined by the *trapezoid tearing strength test*. In this test, a

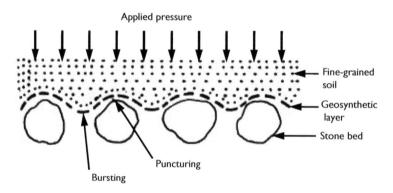

*Figure 2.12* Field situations showing puncturing and bursting of the geosynthetic (after Giroud, 1984).

trapezoidal outline is marked centrally on a rectangular test specimen (Fig. 2.13). Note that an initial 15-mm cut is made to start the tearing process. The specimen is gripped along the two non-parallel sides of the trapezoid in the jaws of a tensile test-ing machine. A continuously increasing force is applied in such a way that the tear propagates across the width of the specimen. The load actually stresses the individual fibres gripped in the clamps rather than stressing the geosynthetic structure. The value of tearing strength of the specimen is obtained from the force/extension curve, and is taken as the maximum force thus recorded (Fig. 2.14). Note that a geotextile may exhibit a single maximum or several maxima (Standards Australia, 2012a). Also, the failure pattern in tear is different in nonwoven geotextiles from that in woven geotex-tiles. The failure of a woven geotextile occurs essentially through the sequential rup-ture of yarns in tension, whereas the failure of a nonwoven geotextile is significantly affected by the inter-fibre friction forces. A typical range of the trapezoid tearing strength of geotextiles is 90–1300 N.

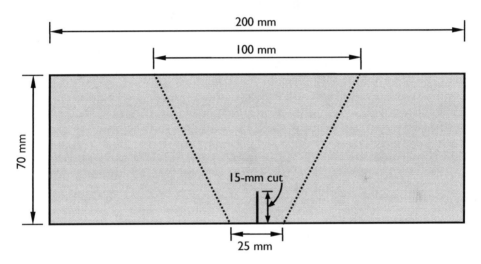

*Figure 2.13* Trapezoidal template for the trapezoid tearing strength test.

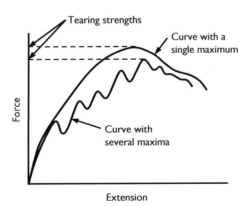

*Figure 2.14* Typical tearing force – extension curves for two different types of geotextiles.

In the *static puncture strength test*, a circular geosynthetic specimen is gripped without tension around its entire circumference between two steel clamping rings in a loading frame. A flat-ended cylindrical steel plunger attached to the load indicator is forced through the centre of the test specimen and perpendicularly to it at a constant rate of displacement (generally 50 mm/min.) until rupture of the specimen occurs (Fig. 2.15). The diameter of the plunger is generally 50 mm and the internal diameter of the ring is 150 mm. The relatively large size of the plunger provides a multidirectional force on the geosynthetic. The clamping system should prevent pretensioning of the specimen before and slippage during the test. Since this test utilizes the concept of the California bearing ratio (CBR) test for soils to determine the puncture resistance and an approximate indication of the resulting strain, it is known as the *CBR plunger test*. The force applied by the plunger and the corresponding displacement are measured.

Fig. 2.16 shows a typical graph of plunger force versus plunger displacement. The maximum force as shown on the curve, where available, or the highest recorded force is the value of puncture strength of the specimen. A typical range of puncture strength of geotextiles is 45–450 N. Note that the CBR plunger test is generally not recommended for geosynthetics having aperture greater than 10 mm. It is generally applicable to isotropic geotextiles, and may also be used for geomembranes. Because of clamping and equipment limitations, this test may not be suited for some woven geotextiles with high tensile strengths exceeding approximately 90 kN/m. This test has been shown to be practically independent of speed in the range of 5–100 mm/minute for relatively low-strength geotextiles (Standards Australia, 2012b).

The *impact strength* (i.e. *dynamic puncture strength* or *dynamic perforation strength*) of a geosynthetic is evaluated by the cone drop test method. This test involves

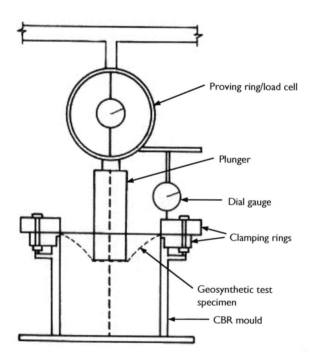

Proving ring/load cell

Plunger

Dial gauge

Clamping rings

Geosynthetic test specimen

CBR mould

*Figure 2.15* A typical test arrangement for static puncture test (CBR plunger test).

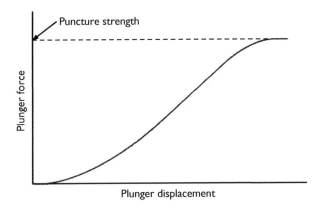

*Figure 2.16* A typical plunger force – displacement curve.

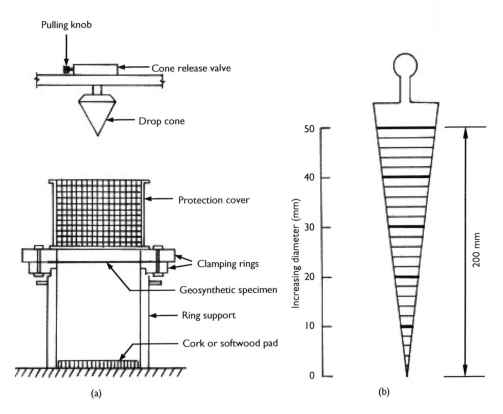

*Figure 2.17* Impact strength (dynamic puncture strength) test: (a) typical test arrangement; (b) penetration measuring cone.

the determination of the diameter of the punctured hole made by dropping a standard brass or stainless steel cone weighing 1 kg from a specified height onto the surface of a circular geosynthetic specimen gripped between clamping rings (Fig. 2.17(a)). The geosynthetic may be supported by water or soil to simulate the field conditions.

The diameter of the punctured hole, measured using a penetration-measuring cone (Fig. 2.17(b)), in combination with the drop height, gives a measure of the impact resistance/strength. The smaller the diameter of the hole, the greater is the impact resistance of the geosynthetic to damage during installation. The impact resistance/strength can be expressed as either diameter of the hole at a standard drop height of 500 mm, or drop height that will produce a hole of diameter 50 mm. The relationship between the drop height and the diameter of the hole has been developed from testing a wide range of geotextiles without providing any support during the test as (Standards Australia, 2014):

$$d_2 = d_1 \left( \frac{h_2}{h_1} \right)^{0.68} \tag{2.9}$$

$$h_2 = h_1 \left( \frac{d_2}{d_1} \right)^{1.47} \tag{2.10}$$

where $h_1$ is the drop height (first value), in mm; $h_2$ is the drop height (second value), in mm; $d_1$ is the diameter of hole corresponding to a drop height $h_1$, in mm; and $d_2$ is the diameter of hole corresponding to a drop height $h_2$, in mm. Note that Equations (2.9) and (2.10) are valid only where the diameter of the hole produced experimentally exceeds 15 mm.

The *Bursting strength* is measured by the bursting test (multi-axial tensile test) using the apparatus shown in Fig. 2.7(d). This test is performed by applying a normal pressure, usually by air pressure against a geosynthetic specimen clamped in a ring, as mentioned earlier. The normal stress against the geosynthetic at failure gives the value of the bursting strength. A typical range of bursting strength of geotextiles is 350–5200 kPa.

The *fatigue strength* of a geosynthetic can be assessed by measuring the change of its physical or mechanical properties under the repeated application of a cyclic force, usually leading to failure. It may be influenced by the following three factors: (a) range of force, (b) mean force, and (c) number of cycles of force applied.

### 2.3.4 Soil-geosynthetic interface properties

When a geosynthetic is used in reinforcing a soil mass, it is important that the bond developed between the soil and the geosynthetic is sufficient to stop the soil from sliding over the geosynthetic or the geosynthetic from pulling out of the soil when the tensile load is mobilized in the geosynthetic. The bond between the geosynthetic and the soil depends on the interaction of their contact surfaces. The *soil-geosynthetic interaction* through *interface friction* and/or *interlocking characteristics* is thus the key element in the performance of the geosynthetic-reinforced soil structures, such as retaining walls, slopes, foundations and embankments, and other applications where the resistance of a geosynthetic to sliding or pullout under simulated field conditions is important. The interaction is mainly responsible for the transference of stresses from the soil to the geosynthetic. In many applications, it is used to determine the bond length of the geosynthetic needed beyond the critical zone. The two test proce-dures, currently used to evaluate the soil-geosynthetic interaction, are the *direct shear test*, using a shear box, and the *pullout/anchorage test*. The basic principle of these tests is that to move a solid object, of weight $W$, along a horizontal plane, requires the application of a horizontal force of $\mu W$, where $\mu$ is the coefficient of friction between the material of the object and the material of the plane.

In the *direct shear test*, the shear resistance between a geosynthetic and a soil is determined by placing the geosynthetic and soil within a direct shear box, about 300 mm square in plan, divided into upper and lower halves (Fig. 2.18). The geosyn-thetic specimen is anchored along the edge of the box where the shear force is applied. A constant normal force representative of the design stresses is applied to the box, and keeping the lower half of the box fixed, the upper half is subjected to a shear force, under a constant rate of deformation. The shear force is recorded as a function of the horizontal displacement of the upper half of the shear box. The test is performed at a minimum of three different normal compressive stresses, selected to model appropri-ate field conditions. The limiting values of shear stresses, typically peak and residual shear stresses, are plotted against their corresponding values of the applied normal stresses. The test data are generally plotted by a best-fit straight line whose slope (tan $\phi$, $\phi$ being the peak/residual interface friction angle between the soil and the geosynthetic or the friction angle of soil as the case may be) is the peak/residual coefficient of inter-face friction between the soil and the geosynthetic (Fig. 2.19). Any intercept of the best-fit straight line with the shear stress axis defines an apparent adhesion. The shear stress and the normal stress axes must be drawn to the same scale. The test value may be a function of the applied normal stress, geosynthetic material characteristics, soil gradation, soil plasticity, soil density and water content, size of specimen, drainage conditions, displacement rate, magnitude of displacement, etc.

Note that the direct shear test is not suited for the development of exact stress-strain relationships for the test specimen due to the non-uniform distribution of shear-ing force and displacement. The total resistance may be a combination of sliding, rolling, interlocking of soil particles and geosynthetic surfaces, and shear strain within the geosynthetic specimen. Shearing resistance may be different on the two faces of a geosynthetic and may vary with direction of shearing relative to orientation of the geosynthetic. The direct shear test data can be used in the design of geosynthetic

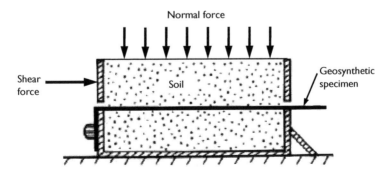

*Figure 2.18* Details of the direct shear test.

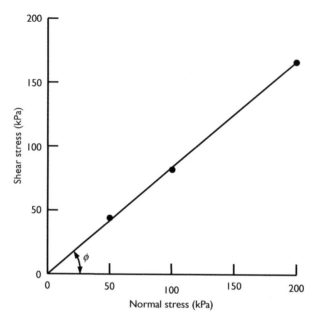

Figure 2.19 A typical plot of the results from the direct shear test.

Note: $\phi$ = 43°, 38°, 36°, 35°, 41° for sand-sand, sand-needle-punched nonwoven geotextile, sand-heat-bonded nonwoven geotextile, sand-lightweight woven geotextile and sand-geogrid, respectively (based on BS EN ISO 13738 (BSI, 2004a)).

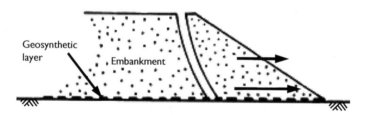

Figure 2.20 A reinforcing geosynthetic application with the sliding failure mode.

applications in which the sliding may occur between the soil and the geosynthetic (Fig. 2.20).

In the *pullout test*, a geosynthetic specimen, embedded between two layers of soil in a rigid box, is subjected to a horizontal force, keeping the normal stress applied to the upper layer of soil constant and uniform. Figure 2.21 depicts the general test arrangement of the pullout test. The force required to pull the geosynthetic out of the soil is recorded. The pullout resistance is calculated by dividing the maximum load by the test specimen width.

The ultimate pullout resistance $P$ of the geosynthetic reinforcement is given by:

$$P = 2 \times L_e \times W \times \sigma'_n \times C_i \times F \tag{2.11}$$

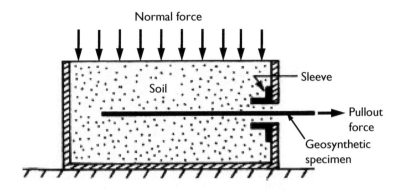

*Figure 2.21* Details of the pullout test.

where $L_e$ is the embedment length of the test specimen; $W$ is the width of the test specimen; $\sigma'_n$ is the effective normal stress at the soil-test specimen interfaces; $C_i$ is the coefficient of interaction (a scale effect correction factor) depending on the geosynthetic type, soil type and normal load applied; and $F$ is the pullout resistance (or friction bearing interaction) factor. For the preliminary design or in the absence of specific geosynthetic test data, $F$ may be conservatively taken as $(2/3) \tan \phi$ for the geotextiles, and $0.8 \tan \phi$ for the geogrids. Equation (2.11) is known as the *pullout capacity formula*.

The pullout resistance versus normal stress plot is a function of soil gradation, plasticity, as-placed dry unit weight, water content, embedment length and surface characteristics of the geosynthetic, displacement rate, normal stress and other test parameters. Therefore, the results should be expressed in terms of the actual test conditions. Figure 2.22 shows the effect of specimen embedment length on the pullout behaviour of a geogrid. A typical plot of maximum pullout resistance versus normal stress is shown in Fig. 2.23. The pullout test data can be used in the design of geosynthetic applications in which the pullout may occur between the soil and the geosynthetic as shown in Fig. 2.24.

The following points regarding the mechanical properties of geosynthetics are worth mentioning:

1   The strength of a woven geotextile is higher at 45° to the warp and weft directions, but is lower parallel to the warp/weft direction. Compared with the tensile strength of woven geotextiles, the nonwoven geotextiles tend to have a lower but generally more uniform strength in all directions.
2   The direct shear test can be conducted to study the geosynthetic-geosynthetic interface frictional behaviour by placing the lower geosynthetic specimen flat over a rigid medium in the lower half of the direct shear box and the upper geosynthetic specimen over the previously placed lower specimen.
3   A designer of the geosynthetic-reinforced soil structures must consider the potential failure mode, and then the appropriate test procedure should be used to evaluate the soil-geosynthetic interaction properties.

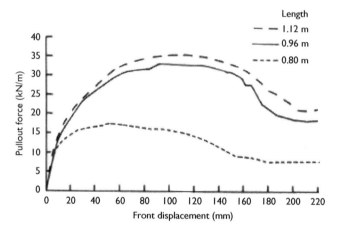

*Figure 2.22* Influence of the specimen embedment length on the pullout behaviour of a geogrid (after Lopes and Ladeira, 1996).

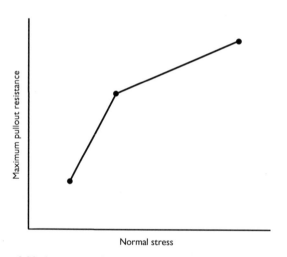

*Figure 2.23* A typical pullout resistance versus normal stress plot.

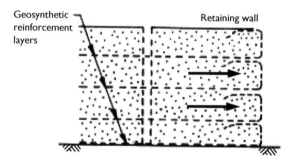

*Figure 2.24* A reinforcing geosynthetic application with the pullout failure mode.

4    In the case of an unpaved road with a geosynthetic layer at the subgrade level, the recommended test should be a combination of direct shear and pullout tests conducted simultaneously (Giroud, 1980).

5    It is common practice to assume a soil-geotextile friction angle between $(2/3)\phi$ and $\phi$, where $\phi$ is the angle of shearing resistance of soil.

## 2.4   HYDRAULIC PROPERTIES

The hydraulic properties of geosynthetics influence their ability to function as filters and drains. Hydraulic testing of geosynthetics is completely based on new and original concepts, methods, devices, interpretation and databases unlike physical and mechanical testing, as discussed in the previous sections. The reason behind this is that the traditional textile tests rarely have hydraulic applications. *Porosity, permittivity* and *transmissivity* are the most important hydraulic properties of geosynthetics, mainly of geotextiles, geonets, and many drainage geocomposites, which are commonly used in filtration and drainage applications.

### 2.4.1   Geosynthetic pore (or opening) characteristics

The voids (or holes) in a geosynthetic are called the *pores* or *openings*. The measurement of sizes of pores and the study of their distribution is known as *porometry*.

The *porosity* of a geosynthetic is related to its ability to allow liquid to flow through it and is defined as the ratio of the void volume to the total volume of the geosynthetic, usually expressed as a percentage. It may be calculated for geotextiles using the relationship derived below:

$$n = \frac{V_v}{V} = \frac{V - V_s}{V} = 1 - \frac{V_s}{V} = 1 - \frac{\dfrac{mA}{\rho_s}}{A\Delta x} = 1 - \frac{m}{\rho_s \Delta x} \tag{2.12}$$

where $n$ is the porosity; $V_v$ is the void volume; $V_s$ is the volume of solid polymer; $V(= V_v + V_s)$ is the total volume; $A$ is the surface area of geotextile; $m$ is the mass per unit area; $\rho_s$ is the density of solid polymer; and $\Delta x$ is the thickness of geotextile.

## EXAMPLE 2.2

Calculate the porosity of the geotextile with the following properties:
    Thickness, $\Delta x = 2.7$ mm
    Mass per unit area, $m = 300$ g/m²
    Density of polymer solid, $\rho_s = 900$ kg/m³

## SOLUTION

Given: $\Delta x = 2.7$ mm $= 0.0027$ m, $m = 300$ g/m² $= 0.3$ kg/m² and $\rho_s = 900$ kg/m³. From Equation (2.12), the porosity $n$ of the geotextile is obtained as

$$n = 1 - \frac{m}{\rho_s \Delta x} = 1 - \frac{0.3}{(900)(0.0027)} = 0.876 \text{ or } 87.6\%$$

## 2.4.2 Percentage open area

The percentage open area (POA) of a geosynthetic is the ratio of the total area of its openings to the total area, expressed as a percentage. This characteristic is considered to be a design parameter only for woven geotextiles, which have area of openings as the void spaces between adjacent filaments and yarns. Note that a higher POA generally indicates a greater number of openings per unit area in the geotextile. For the filter applications of a geotextile, its POA should be higher to avoid the *clogging* phenomenon, as described in Section 4.2, to occur throughout the design life of the particular application.

The pores in a geotextile are not of one size but are of a range of sizes. The pore-size distribution can be represented in much the same way as the particle-size distribution for a soil. In fact, a geotextile is similar to a soil in that it has voids (pores) and particles (filaments and fibres). However, because of the shape and arrangement of filaments and the compressibility of the structure with geotextiles, the geometric relationships between filaments and voids are more complex than in soils. Therefore, in geotextiles the pore size is measured directly, rather than using the particle size as an estimate of pore size, as is done with soils. In the determination of the particle-size distribution of soil, the soil, which initially has particles of unknown sizes, is passed through a series of sieves of different known sizes to determine the percentages of soil particles of the various sizes present. In determining the pore-size distribution of a geotextile the process is reversed. The geotextile is used as a sieve, of unknown sizes, and the particles of different known sizes are passed through the geotextile as a sieve. From the measured weights of particles of various known sizes which either pass through the geotextile or are retained on the geotextile, the pore-size distribution of the geotextile can be obtained. Due to the importance of pore-size distribution in the design of geotextiles for use as filters and separators, various test methods have been developed for measuring the size of openings in the geotextiles. Bhatia *et al.* (1994) made a comparison of six methods as presented in Table 2.1.

In the *dry sieving test method*, the known-sized spherical solid glass beads (or calibrated quartz sand particles) are sieved in a dry condition through a screen made of the geotextile specimen, being tested, in a sieve frame (Fig. 2.25) for a constant period of time, generally 10 min. Sieving is done by allowing beads of successively coarser size until they are 5% or less in weight, to pass through the geotextile. A mechanical sieve shaker, which imparts lateral and vertical motions to the sieve, thus causing the particles thereon to bounce and turn so as to present different orientations to the sieving surface, should be used for carrying out the sieving operations. Note that for the measurement of fine pores, difficulties are encountered in the dry sieving of sand particles through the geotextiles, particularly through thick nonwovens, due to the particles being trapped in the geotextile. On the other hand with the use of glass

*Table 2.1* Comparison of methods for determining pore-size distribution of geotextiles.

| Test method | Test mechanism | Test material | Sample size (cm²) | Time for one test |
|---|---|---|---|---|
| Dry sieving | Sieving-dry | Glass beads fraction | 434 | 2 h |
| Hydrodynamic sieving | Alternating water flow | Glass beads mixture | 257 | 24 h |
| Wet Sieving | Sieving-wet | Glass beads mixture | 434 | 2 h |
| Bubble point | Comparison of air flow, dry vs. saturated | Pore wick | 22.9 | 20 min |
| Mercury intrusion | Intrusion of a liquid in a pore | Mercury | 1.77 | 35 min |
| Image analysis | Direct measurement of pore spaces in cross-section of the geotextile | None | 1.5 | 2–3 days |

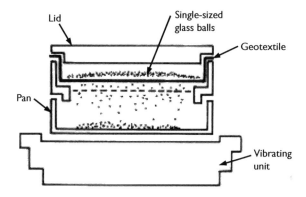

*Figure 2.25* Basic details of the dry sieving method.

beads, the electrostatic forces can affect the sieving, but no practical alternative dry methods of determining pore sizes for these types of geotextiles are available.

The *hydrodynamic test method* is based on hydrodynamic filtration, where a glass bead mixture is sieved with a basket with a geotextile bottom by alternating water flow that occurs as a result of the immersion and emersion of the basket several times in water. In the *wet sieving method*, a glass bead mixture is sieved through a screen made of geotextile while a continuous water spray is applied. The *bubble point method* is based on a process in which (i) a dry porous material will pass air through all of its pores when any amount of air pressure is applied to one side of the material; and (ii) a saturated porous material will only allow a fluid to pass when the pressure applied exceeds the capillary attraction of the fluid in the largest pore. The *mercury intrusion method* is based on the relationship between the pressure required to force a non-wetting fluid (mercury) into the pores of a geotextile and the radius of the pores intruded. The *image analysis* is a technique used for the direct measurement of pore spaces within a cross-sectional plane of the geotextile with the help of a microscope. The pore openings, which are obtained experimentally, are dependent on the technique used for their determination. It is believed that, despite some limitations, both wet and hydrodynamic sieving methods are better techniques than the dry sieving.

Note that the pore sizes measured by all these methods are not actual dimensions of the openings through the geotextile.

In the case of most of the geogrids, the open areas of the grids are greater than 50% of the total area. In this respect, a geogrid may be looked on as a highly permeable polymeric structure.

Figure 2.26 shows the pore-size distribution curves for typical woven and nonwoven geotextiles. The pore size (or opening size), at which 95% of the pores in the geotextile are finer, is originally termed the *equivalent opening size (EOS)* designated as $O_{95}$. In the USA, this pore size is determined by the dry sieving method and is termed the *apparent opening size (AOS)*, whereas in Europe and Canada, this is determined by the wet and the hydrodynamic sieving methods and is termed the *filtration opening size (FOS)*. If a geotextile has an $O_{95}$ value of 300 μm, then 95% of geotextile pores are 300 μm or smaller. In other words, 95% of particles with a diameter of 300 μm are retained on the geotextile during sieving. This notation is similar to that used for the soil particle-size distributions where, for instance, $D_{10}$ is the sieve size through which 10%, by weight, of the soil passes. *AOS* or *FOS* is, in fact, considered as the property that indicates the approximately largest particle that would effectively pass through the geotextile and thus reflects the approximately largest opening dimension available in the geotextile for soil to pass through. The opening size is also quoted for other percentages retained, such as $O_{50}$ or $O_{90}$ to determine the pore-size distribution of the geotextile. Note that the meaning of opening size values and their determination in the laboratory are still not uniform throughout the engineering profession and hence the filter criteria developed in different countries may not be directly comparable.

In Fig. 2.26, it is noted that the pores in a woven geotextile tend to be fairly uniform in size and regularly distributed. In general, the nonwoven geotextiles exhibit smaller $O_{90}$ pore sizes than the woven geotextiles; however, there is a degree of overlap in the commonly employed $O_{90}$ sizes, which vary from approximately 50 μm to 350 μm for the nonwoven geotextiles and from 150 μm to 600 μm for the woven geo-

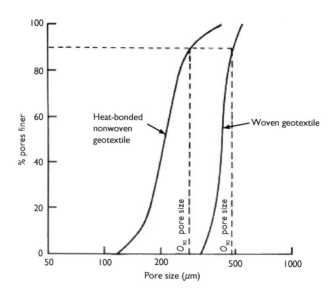

*Figure 2.26* Pore-size distributions of geotextiles (after Ingold and Miller, 1988).

textiles (Ingold and Miller, 1988). For the filtration applications, a geotextile high in *POA* should be selected, with a controlled opening size to suit the soil being filtered. Most nonwoven geotextiles and some woven geotextiles suit this application.

### 2.4.3 Permeability

The ability of a geosynthetic to transmit a fluid (liquid or gas) is called the *permeability*. The permeability (i.e. hydraulic conductivity) of a geosynthetic to fluid flow may be expressed by *Darcy's coefficient, permittivity* (as defined below) or *volume flow rate*. The Darcy's coefficient is the volume rate of flow of fluid under laminar flow conditions through a unit cross-sectional area of the geosynthetic under a unit hydraulic gradient and standard temperature conditions (generally $22 \pm 3°C$). The advantage in expressing the geosynthetic permeability in terms of Darcy's coefficient is that it is easy to relate the geosynthetic permeability directly with the soil permeability. A major disadvantage is that Darcy's law assumes a laminar flow, whereas geosynthetics, especially geotextiles, are often characterized as exhibiting semi-turbulent, or turbulent flows.

The simplest method of describing the permeability characteristics of geosynthetics is in terms of volume flow rate at a specific constant water head (generally 10 cm) (Fig. 2.27). The advantage of this method is that it is a simplest test to carry out, it

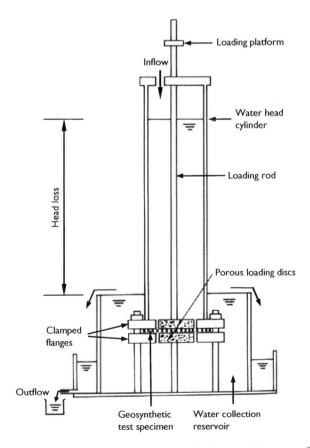

*Figure 2.27* A typical test arrangement of the constant head cross-plane water flow apparatus.

does not rely on Darcy's law for its authenticity, and it can easily be used to compare different geosynthetics used for drainage and filtration applications. In this method, due to the high hydraulic gradient, the turbulent flow can occur in many geotextiles. Thus, the measured permeability value cannot be compared with the actual permeability value measured for the laminar flow conditions.

The measurement of in-plane water permeability of a geosynthetic, particularly of a geotextile or a band/fin drain is important when it is used to carry water within itself and parallel to its plane. The in-plane water permeability, normally described in terms of the *transmissivity* (as defined in this section later on), is determined by measuring the volume of water that passes along the test specimen in a known time and under specified normal stress and hydraulic gradient. The test used to measure the in-plane drainage characteristics of geosynthetics is essentially the same as that used to measure water permeability normal to the plane of the geosynthetic (Fig. 2.27), except that the hydraulic gradient is applied along the length of the geosynthetic (Fig. 2.28) rather than across the thickness of the geosynthetic. The test can be conducted to model the particular field conditions, for example, by employing specific contact surfaces, and varying compressive stresses and hydraulic gradients.

*Permittivity* of a geosynthetic (generally a geotextile) is simply the coefficient of permeability for water flow normal to its plane (Fig. 2.29(a)) divided by its thickness. This property is a preferred measure of the water flow capacity across the geosynthetic plane and is quite useful in filter applications. Darcy's law in terms of permittivity can be expressed as:

$$Q_n = k_n \frac{\Delta h}{\Delta x}(LB) = \psi \Delta h \, A_n \qquad (2.13)$$

where $Q_n$ is the cross-plane volumetric flow rate of water, that is, the volumetric flow rate of water for flow across the plane of the geosynthetic, in m³/s; $k_n$ is the coefficient of cross-plane permeability, in m/s; $\Delta h$ is the hydraulic head causing the flow, in m; $\Delta x$ is the thickness of the strip of geosynthetic measured along the flow direction under a specified normal stress, in m; $L$ is the length of the strip of geosynthetic, in m; $B$ is the width of the strip of geosynthetic, in m; $\psi = k_n/\Delta x$, which is the permittivity of

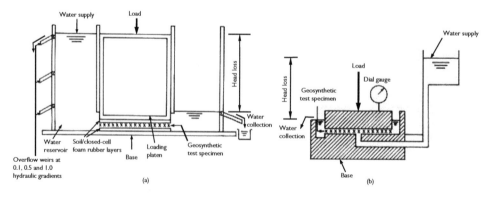

*Figure 2.28* Typical test arrangements of constant head in-plane water flow apparatus: (a) full width flow; (b) radial flow.

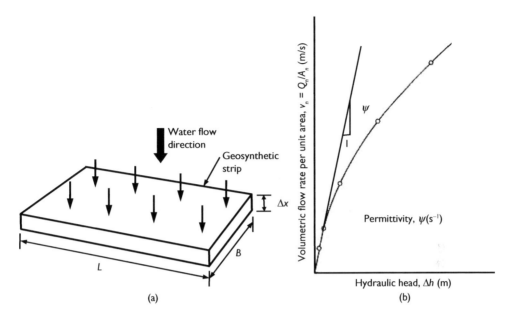

*Figure 2.29* Geosynthetic permittivity: (a) normal flow of water through a geosynthetic strip; (b) definition.

Note: In the laminar flow region, the volumetric flow rate per unit area versus the hydraulic head curve is linear and intersects the origin.

the geosynthetic, in s$^{-1}$; and $A_n = LB$ is the area of cross-section of geosynthetic for the cross-plane flow, in m$^2$. Permittivity may thus be defined as the volumetric flow rate of water per unit cross-sectional area of the geosynthetic per unit head, under the laminar conditions of flow in a direction normal to the plane of the geosynthetic (Fig. 2.29 (b)).

*Transmissivity* of a geosynthetic (generally the thick nonwoven geotextile, geonet or geocomposite) is simply the product of the coefficient of permeability for in-plane water flow (Fig. 2.30(a)) and its thickness. This property is a preferred measure of the in-plane water flow capacity of a geosynthetic and is widely used in drainage applications. Darcy's law in terms of transmissivity can be expressed as:

$$Q_p = k_p \frac{\Delta h}{L} A_p = k_p \frac{\Delta h}{L}(B\Delta x) = \theta i B \qquad (2.14)$$

where $Q_p$ is the in-plane volumetric flow rate of water, that is, the volumetric flow rate of water for flow within the plane of the geosynthetic, in m$^3$/s; $k_p$ is the coefficient of in-plane permeability, in m$^2$/s; $\theta = k_p \Delta x$, which is the transmissivity of the geosynthetic, in m$^2$/s; $i = \Delta h/L$ is the hydraulic head causing the flow; and $A_p = B\Delta x$ is the area of cross-section of geosynthetic for in-plane flow, in m$^2$. Transmissivity may thus be defined as the volumetric flow rate of water per unit width of the geosynthetic, per unit hydraulic gradient, under the laminar conditions of flow within the plane of

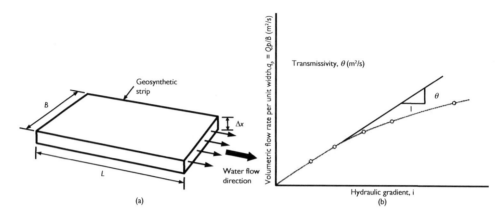

*Figure 2.30* Geosynthetic transmissivity: (a) in-plane flow of water through a geosynthetic strip; (b) definition.

Note: In the laminar flow region, the volumetric flow rate per unit width versus the hydraulic gradient curve is linear and intersects the origin.

## EXAMPLE 2.3

In a laboratory constant head in-plane permeability test on a 300-mm length (flow direction) by 200 mm-width geotextile specimen, the following parameters are measured:

  Thickness, $\Delta x = 2.0$ mm

  Flow rate of water in the plane of the geotextile, $Q_p = 52$ cm³/min

  Head loss in the plane of the geotextile, $\Delta h = 200$ mm

Determine the transmissivity ($\theta$) and the in-plane coefficient of permeability ($k_p$) of the geotextile.

## SOLUTION

Given: $L = 300$ mm $= 0.3$ m, $B = 200$ mm $= 0.2$ m, $\Delta x = 2.0$ mm $= 0.002$ m,

$Q_p = 52$ cm³/min $= \dfrac{52 \times 10^{-6}}{60} = 8.67 \times 10^{-7}$ m³/s, and $\Delta h = 200$ mm $= 0.2$ m

From Equation (2.14), the transmissivity is

$$\theta = \frac{Q_p}{iB} = \frac{Q_p}{(\Delta h/L)B} = \frac{8.67 \times 10^{-7}}{(0.2/0.3)(0.2)} = 6.5 \times 10^{-6} \; m^2/s$$

From the definition, the in-plane coefficient of permeability is

$$k_p = \frac{\theta}{\Delta x} = \frac{6.5 \times 10^{-6}}{0.002} = 3.25 \times 10^{-3} \; m/s$$

the geosynthetic (Fig. 2.30(b)). To exhibit a large transmissivity, a geotextile must be thick and/or have the large permeability in its plane.

Equations (2.13) and (2.14) indicate that once permittivity ($\psi$) and transmissivity ($\theta$) are successfully determined, the flow rates $Q_n$ and $Q_p$ do not depend on the thickness of the geosynthetic strip $\Delta x$, which is highly dependent on the applied pressure and therefore, is difficult to get measured accurately in the case of some types of geotextiles. Thus, it is preferable to determine and report the permittivity and transmissivity values of geotextiles rather than their coefficients of in-plane and cross-plane permeability, respectively.

If the transmissivity of a geotextile is determined by the radial transmissivity method (see Fig. 2.28(b) for the schematic diagram of the test) using a circular specimen, Darcy's law in terms of transmissivity can be expressed as:

$$Q_p = k_p \frac{dh}{dr}(2\pi r \Delta x) \tag{2.15}$$

where $r$ is any radius between the outer radius $r_0$ and the inner radius $r_i$ of the geotextile specimen; and $dh$ is the head loss across the radial distance $dr$.

On rearranging the parameters and integrating within proper limits, Equation (2.15) reduces to

$$\theta \int_{h_i}^{h_0} dh = \frac{Q_p}{2\pi}\int_{r_i}^{r_o} \frac{dr}{r}$$

or

$$\theta(h_o - h_i) = \frac{Q_p \ln(r_o/r_i)}{2\pi}$$

or

$$\theta = \frac{Q_p \ln(r_o/r_i)}{2\pi(h_o - h_i)} = \frac{Q_p \ln(r_o/r_i)}{2\pi\Delta h} \tag{2.16}$$

where $h_i$ and $h_0$ are the hydraulic heads at the inner and outer edges of the geotextile specimen, respectively, and $\Delta h$ ($= h_0 - h_i$) is the head loss across the radial distance $\Delta r$ ($= r_0 - r_i$). Equation (2.16) can be directly used to determine the transmissivity by the radial method.

Note that the determination of permittivity and transmissivity of geotextiles is based on Darcy's law of water flow. This means that the permittivity and the transmissivity of a geotextile are constants for a given thickness of the geotextile at a given confining pressure only when the laminar flow conditions exist, which is likely in a typical soil environment where geotextiles are used. It appears that for most geotextiles, Darcy's law holds if the approach velocity, that is, the velocity of the water approaching the geotextile, is kept at or below 0.035 m/s (Standards Australia, 2012c). The permittivity and the transmissivity, when determined for the region of transient or turbulent flow conditions, are called the permittivity and the transmissivity under nonlinear flow conditions.

It is further stressed that Darcy's law is valid only for the laminar flow. This means that permeability, permittivity and transmissivity are constants, that is, independent of the hydraulic gradient only when the water flow is not turbulent. These properties are governed by several other factors, such as fibre type, size and orientation; porosity or void ratio; confining pressure; repeated loading; contamination; and aging. When dry, some geotextiles exhibit resistance to wetting. In such cases, the initial permeability is low but rises until the geotextile reaches saturation. Permeability may also be reduced through air bubbles trapped in the geosynthetic. This is the reason why the testing standards usually require a careful saturation of the geosynthetic specimens before they are subjected to water flow. In addition, the permeability measurements will be more consistent with the use of the deaired water rather than tap water. Woven geotextiles are much less affected by the stress level, but their permeability is dramatically controlled by the structure of the fabric. The common and generally less expensive, tape-on-tape geotextiles have a low open area ratio and, in consequence, exhibit water permeabilities typically in the range 10–30 l/s/m$^2$ for a 10-cm head. In contrast, the woven monofilament-on-monofilament geotextiles have much larger open area ratios, giving water permeabilities in the range 100–1000 l/s/m$^2$ for a 10-cm head (Ingold and Miller, 1988). The tests performed at the University of Grenoble (France) have shown that the thickness and the permeability of the needle-punched nonwoven geotextiles are significantly affected by confining pressure as shown in Fig. 2.31. In this Figure, the values of $k_n$ and $k_p$ were close for the

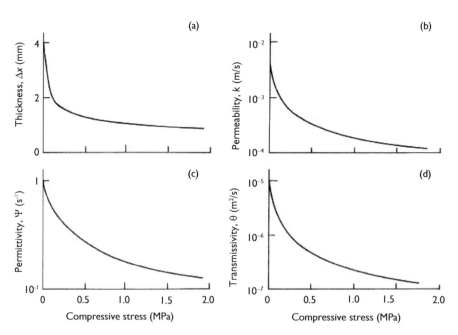

Figure 2.31 Influence of compressive stress on (a) thickness; (b) permeability; (c) permittivity; (d) transmissivity of a needle-punched nonwoven geotextile (after Giroud, 1980).

considered geotextile; therefore only an average value $k$ is presented. Note that the flow in the plane of the geotextile is more affected by the confining pressure than a normal flow. Permittivity, transmissivity and apparent opening size of certain geosynthetics (such as geotextiles) may also change if they are subjected to tension or creep deformation. Thus, for example, the ability of geotextiles to drain water, retain soil particles, and resist clogging may get altered under such deformations.

Geomembranes are nonporous homogeneous materials that are permeable in varying degrees to gases, vapours and liquids on a molecular scale in a three-step process: (1) absorption of the permeant, (2) diffusion of the dissolved species, and (3) desorption (evaporation) of the permeant, controlled mainly by the chemical potential gradient (or concentration gradient) that decreases continuously in the direction of the permeation. Geomembranes are mostly used as a fluid barrier or liner. Note that there are other liner materials that are porous, such as soils and concretes, in which the driving force for permeation is the hydraulic gradient. Sometimes, a geomembrane is also known as a flexible membrane liner (FML), especially in landfill applications. The wide range of uses of geomembranes under different service conditions to many different permeating species requires determination of permeability by test methods that relate to and simulate as closely as possible the actual environmental conditions in which the geomembrane will be in service. Various test methods for the measurement of permeation and transmission through geomembranes of individual constituents in complex mixtures, such as waste liquids, are recommended by ASTM D5886-95 (ASTM, 2011c).

It is found that the inorganic salts do not permeate geomembranes, but some organic species do. The rate of diffusion of an organic within a geomembrane is governed by several factors, including solubility of the permeant in the geomembrane, microstructure of the polymer, size and shape of the diffusing molecules, temperature at which the diffusion is taking place, the thickness of the geomembrane, and the chemical potential gradient across the geomembrane. A steady state of the flow of constituents will be established when, at every point within the geomembrane, the flow can be defined by *Fick's first law of diffusion* as:

$$Q_i = -D_i \times \frac{dc_i}{dx} \tag{2.17}$$

where $Q_i$ is the mass flow of constituent '$i$', in $g/cm^2/s$; $D_i$ is the diffusivity of constituent '$i$', in $cm^2/s$; $c_i$ is the concentration of constituent '$i$' within the mass of the geomembrane in $g/cm^3$; and $x$ is the thickness of the geomembrane in cm.

Note that typical values of permeability are $10^{-5} - 1$ m/s for geotextiles and $10^{-13}$ m/s or less for geomembranes. The permeability of geotextiles is of the same order of magnitude as the permeability of highly permeable soils, such as sand and gravel. The woven geotextiles and the thermally-bonded nonwoven geotextiles have almost no transmissivity and cannot be used as the drains. The permeability of geomembranes is much smaller than the permeability of clay, which is the least permeable soil. The needle-punched geotextiles have permeability values of the order of $10^{-4}$ or $10^{-3}$ m/s and the geonets have permeability values of the order of $10^{-2}$ or $10^{-1}$ m/s. A maximum saturated hydraulic conductivity ranging from $5 \times 10^{-11} - 1 \times 10^{-12}$ m/s is typical of geosynthetic clay liners (GCLs) over the range of confining pressures typically encountered in practice.

## 2.5   ENDURANCE AND DEGRADATION PROPERTIES

The endurance and degradation properties (e.g. creep behaviour, abrasion resistance, long-term flow capability, durability – construction survivability and longevity, etc.) of geosynthetics are related to their behaviour during service conditions, including the time in use.

### 2.5.1   Creep

*Creep* is the time-dependent increase in accumulative strain or elongation in a geosynthetic resulting from an applied constant load. Depending on the type of polymer and ambient temperature, the creep may be significant at stress levels as low as 20% of the ultimate tensile strength. In the test for determining the creep behaviour of a geosynthetic, the specimen of wide-width variety (say, 200 mm wide) is subjected to a sustained load using weights, or mechanical, hydraulic or pneumatic systems in one step while maintaining the constant ambient conditions of temperature and humidity. The longitudinal extensions/strains are recorded continuously or are measured at specified time intervals. Unless otherwise specified, the duration of testing is generally not less than 10,000 hours or to failure if this occurs in a shorter time. The test duration of 100 hours is useful for monitoring of products, but for a full analysis of creep properties, durations of up to 10,000 hours will be necessary. The percent strain versus log of time is plotted for each stress increment to calculate the creep rate, defined as the slope of the creep-time curve at a given time. Figure 2.32 compares strain versus the time behaviour of various yarns of different polymers. As shown, both the total strain and the rate of strain differ markedly.

More recent developments allow for the accelerated determination of creep characteristics via stepped isothermal methods (SIM). Curves are developed such that a prediction of the total likely creep effect over significant time intervals, perhaps 100 years as a design life, can be extrapolated.

Creep is an important factor in the design and performance of some geosynthetic-reinforced structures, such as retaining walls, steep-sided slopes, embankments over weak foundations, etc. In all these applications, the geosynthetic reinforcements may be required to endure exposure to high tensile stresses for long periods of time – typically 75-plus years. Creep should also be carefully considered as a relevant design

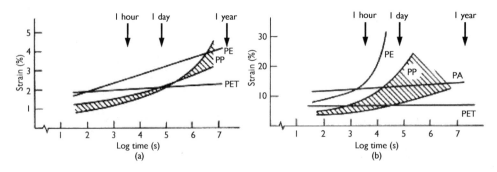

*Figure 3.32* Results of creep tests on various yarns of different polymers: (a) creep at 20% load; (b) creep at 60% load (after den Hoedt, 1986).

criterion in some drainage applications and some containment applications when the geosynthetic is under the load and is expected to perform in a specified manner for a defined time period. At higher loads, the creep leads ultimately to *stress rupture*, also known as the *creep rupture* or *static fatigue*. The higher the applied load, the shorter is the time to rupture. Thus the design load will itself limit the lifetime of the geosynthetic. An understanding of the geosynthetic creep thus helps the design engineer in selection of allowable load to be used in designs. In practical design, it is generally accepted that the creep data should not be extrapolated beyond one order of magnitude. The two approaches to evaluate the allowable load are given below:

a   *Allowable load based on limiting creep strains*: This requires the analysis of creep strains versus time plots for various stress levels. Details of this procedure were described by Jewell (1986), and Bonaparte and Berg (1987).
b   *Allowable load using factor of safety*: It is required to reduce the geosynthetic strength by a factor of safety corresponding to the specific polymer type to obtain the allowable load. The typical values of factor of safety are given in Table 2.2. Although not as technically accurate as the previous method, this approach is sometimes the only one available to the designer.

As the polymers are viscoelastic materials, strain rate and temperature are important while testing the geosynthetics (Andrawes *et al.*, 1986). When a low strain is applied in a wide-width tensile test, the geosynthetic specimen takes longer to come to failure and, therefore, the creep strain is greater. High rates of strain, which can be as much as 100% per minute, tend to produce lower failure strains and sometimes yield higher strengths than the strengths caused by low rates of strain. The creep rate of geosynthetics depends on the temperature. Higher creep rates are associated with higher temperatures, resulting in larger strains of geosynthetics to rupture. The rate of creep is also related to the level of load to which the polymer is subjected (Greenwood and Myles, 1986; Mikki *et al.*, 1990). Chang *et al.* (1996) have reported that under the same confining pressure on geotextiles, the amount of creep increases as the creep load rises; and where the creep load is the same, the increases in confining pressure decrease the amount of creep, which may even be reduced to nil. The creep is minimized by the pre-stretching operation of the 'Tensar' process. Note that the creep is more pronounced in PE and PP than it is in PET.

It has been found that the tensile and creep properties of some nonwoven geotextiles can be improved by confinement in soil (McGown *et al.*, 1982). This has a greater effect on the tensile properties of the mechanically bonded geotextiles and those of the heat-bonded geotextiles. The creep of a geosynthetic is likely to be reduced in

*Table 2.2* Factors of safety (after den Hoedt, 1986).

| Polymers | Factor of safety |
| --- | --- |
| Polypropylene | 4.0 |
| Polyester | 2.0 |
| Polyamide (Nylon) | 2.5 |
| Polyethylene | 4.0 |

soil because of the load transfer to the soil through a significant increase in frictional resistance between the soil and the geosynthetic. However, there is little effect from the confining pressure on the performance of the woven geotextiles. Note that the confined or in-soil testing may model the field behaviour of the geosynthetic more accurately. The results from the creep tests under unconfined environment are conservative with regard to the behaviour of the material in service.

Soil pressure can cause the compressive creep in the geocomposite drains that have an open internal structure to allow flow in the plane of the product. Compressive creep can lead to a reduction in thickness, restriction of the flow, or ultimately to collapse of the geocomposite drain or similar structures.

In some applications, an increase in soil strength is accompanied by the reduction in geosynthetic stress with time. An example of this type of application is the foundation support for a permanent embankment over soft deposits. The phenomenon of the decrease in stress at a constant strain with time is called the *stress relaxation*, which is closely related to the creep.

Dynamic creep and repeated loading behaviour of geosynthetics are of paramount importance in a number of applications. These include reinforcement in paved and unpaved roads, reinforced retaining structures and slopes under large repetitive live loads, such as traffic and wave action. Behaviour of the geogrid under repeated loading is generally different from that of geotextiles (Kabir and Ahmed, 1994). Due care must be paid to such applications.

In the past, constitutive models have been developed to describe direction and time-dependent, non-linear, inelastic stress-strain behaviour. More details on this type of model can be found in the works of Perkins (2000).

## 2.5.2  Abrasion

*Abrasion* of a geosynthetic is defined as the wearing away of any part of it by rubbing against a stationary platform by an abradant with specified surface characteristics. The ability of a geosynthetic to resist wear due to friction or rubbing is called the *abrasion resistance*. The abrasion tester used for determining the abrasion resistance consists of two parallel smooth plates, one of which makes a reciprocating motion along a horizontal axis. Both the plates are equipped with clamps at each end to hold the test specimen and the abrasive medium, generally emery cloth, without any slippage. Under controlled conditions of pressure and abrasive action, the abradant generally attached to the lower plate is moved against the geosynthetic test specimen attached to the upper stationary plate. Resistance to abrasion is expressed as the percentage loss of tensile strength or weight of the test specimen as a result of abrasion. In testing the abrasion resistance of geosynthetics, it is important to simulate the actual type of abrasion, which a geosynthetic would meet in the field. Van Dine *et al.* (1982) and Gray (1982) suggested the test procedures for evaluation of resistance to abrasion caused by different processes such as wear and impact.

Geosynthetics used under pavements, railway tracks or in coastal erosion protection may be subject to dynamic loading, which will lead to the mechanical damage of the product in a manner similar to mechanical damage on installation. While the geosynthetics are susceptible to mechanical fatigue, the principal cause of degradation is abrasion and frictional rubbing.

### 2.5.3  Long-term flow characteristics

*Long-term flow capability* of geosynthetics, generally of geotextiles, with respect to the hydraulic load coming from the upstream soil is of significant practical interest. The compatibility between the pore size openings of a geotextile and retained soil particles in filtration and/or drainage applications can be assessed by the *gradient ratio test*. This test is basically used to evaluate the clogging resistance of the geotextiles with cohesionless soils having a hydraulic conductivity/permeability greater than $5 \times 10^{-4}$ m/s under unidirectional flow conditions. It is best suited for evaluating the movement of finer solid particles in coarse-grained or gap-graded soils where the internal stability from differential hydraulic gradients may be a problem. Figure 2.33 shows the constant-head-type permeameter developed by the US Army Corps of Engineers. This permeameter allows the measurement of the head loss along a soil-geotextile system while passing water through the system at different time intervals. After the test is run for some hours (or days), the piezometer readings stabilize and the so-called gradient ratio (GR) is determined. The GR is defined as the ratio of the hydraulic gradient through the lower 25 mm of the soil plus geotextile thickness to the hydraulic gradient through the adjacent 50 mm of soil alone.

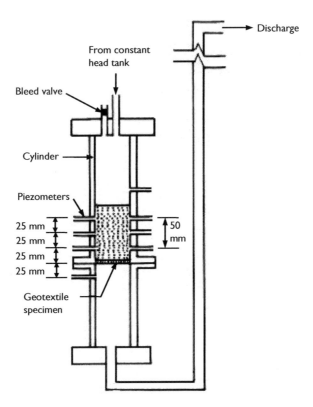

*Figure 2.33* Gradient ratio permeameter developed by US Army Corps of Engineers (after Haliburton and Wood, 1982).

A gradient ratio of one or slightly less is preferred. A value less than one is an indication that some soil particles have moved through the system and a more open filter bridge has developed in the soil adjacent to the geotextile. A continued decrease in the GR indicates piping and may require quantitative evaluation to determine the filter effectiveness. Although the GR values of higher than one mean that some system clogging and flow restriction have occurred, if the system equilibrium is present, the resulting flow may well satisfy the design requirements. Note that the allowable GR values and related flow rates for various soil-geotextile systems will be dependent on the specific site application. One should establish these allowable values on a case-by-case basis. For cohesionless soils, ASTM D5101-12 (ASTM, 2012a) provides the standard test method for evaluation of permeability and clogging behaviour of the soil-geotextile system by the GR. The long-term flow rate behaviour of geotextile filters can also be assessed by an accelerated filtration test method.

The filtration behaviour of soil-geotextile systems with cohesive soils having a hydraulic conductivity/permeability less than or equal to $5 \times 10^{-4}$ m/s can be studied by the hydraulic conductivity ratio test as per ASTM D5567-94 (ASTM, 2011d). This test can be used to determine the hydraulic conductivity ratio (HCR), which is defined as

$$HCR = \frac{k_{sg}}{k_{sgo}}$$

(2.18)

where $k_{sg}$ is the hydraulic conductivity of the soil-geotextile system at any time during the test, and $k_{sgo}$ is the initial hydraulic conductivity of the soil-geotextile system measured at the beginning of the test. The HCR test is used only when water is the permeant liquid. Since the hydraulic conductivity varies with void ratio, which in turn varies with effective stress, the test is carried out as a means of determining the hydraulic conductivity at a controlled level of effective stress to simulate the field conditions. The HCR value indicates the performance of a geotextile as a filter when used with a particular cohesive soil.

A reduction in the HCR with time is representative of significant retention of soil particles. This condition may be desirable in certain drainage applications, or it may be undesirable in other applications. The undesirable development of low-permeability conditions within the geotextile filter, resulting in the filter's inability to perform the intended drainage function, is called the *clogging*. The drainage designer must provide the quantitative definition of clogging on a case-by-case basis. A stable value of HCR indicates that the excessive transport of soil particles up against, or through the geotextile filter does not occur. Note that in the case of continued transport of soil particles through the geotextile filter (a phenomenon known as the *piping*), a stable HCR value can also be obtained. Thus, it becomes necessary to provide a quantitative definition of stabilized filter conditions and the level of acceptable piping in a specific field application.

For soils containing more than 5% non-plastic fines, Richardson and Christopher (1997) suggested a simple field jar test to empirically assess the clogging potential of a geotextile filter. To perform this test, a small amount of soil is placed in a jar (approximately 1/4 full). The jar should preferably have a removable center lid (e.g., a mason jar). The jar is filled with water, and the lid is replaced and secured. It is then

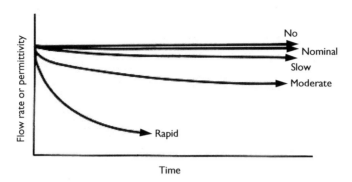

*Figure 2.34* Various degrees of biological clogging.

shaken to form the soil-water slurry. The jar opening is covered with a specimen of the candidate geotextile and secured, and then the jar is allowed to stand for about one minute to allow coarser particles to settle. The liquid is then poured from the jar through the geotextile, tilting the jar such that the trapped air does not impede water flow. If the fines pass through the geotextile, they should not clog. If very little fine soil passes and a significant buildup of fines is observed on the surface of the geotextile, a clogging potential may exist. While certainly not a standardized test, this has been found to be very useful. It is essentially a fine fraction filtration test and permits a qualitative evaluation of the ability of fines to pass through the geotextile.

Some liquids, such as landfill leachates, may create biological activity on the geosynthetic filters, thereby reducing their flow capability. In such applications, the general practice is to determine the potential for, and the relative degree of the biological growth which can accumulate on geosynthetic filters by measuring the flow rates over an extended period of time (e.g., up to 1000 hours) under aerobic or anaerobic conditions on the basis of either a constant head test procedure or a falling head test procedure (ASTM, 2012b). It has been observed that once the biological clogging initiates, the constant head test often passes inadequate quantities of liquid for being measured accurately. It thus becomes necessary to use a falling head test, which works on the basis of the measurement of time of movement of a relatively small quantity of liquid between two selected points on a transparent standpipe. The various degrees of biological clogging are shown in Fig. 2.34. The understanding of the biological clogging helps in making an effective design of filtration and drainage systems and remediation schemes in civil engineering applications such as the landfill projects.

The long-term water flow capacity of geotextiles is also assessed in conjunction with the long-term compressive creep behaviour. In fact, the compressibility of the geotextile over time substantially influences the permittivity and transmissivity of geotextiles in service conditions.

## 2.5.4  Durability

The *durability* of a geosynthetic may be regarded as its ability to maintain the requisite properties against environmental and/or other influences over the selected design life.

It can be thought of as relating to changes over time of both the polymer microstructure and the geosynthetic macrostructure. The former involves the molecular polymer changes and the later assesses the geosynthetic bulk property changes. The durability of a geosynthetic is dependent to a great extent upon the composition of the polymers from which it is made. To quantify the properties of polymers, knowledge of their structures at the chemical, molecular and supermolecular level is necessary (Cassidy *et al.*, 1992). The durability of geosynthetics can be assessed by visual examination, or microscopic examination with a specified magnification factor to give a qualitative prediction of differences between the exposed and unexposed specimens, for example, discolourations, damage to the individual fibres (due to chemical or microbiological attack, surface degradation, or environmental stress cracking), etc. It is traditionally assessed on the basis of mechanical property test results, and not on the microstructural changes that cause the changes in the mechanical properties. It may be assessed in terms of the percentage retained tensile strength $R_T$ and/or the percentage retained strain $R_\varepsilon$, defined as

$$R_T = \frac{T_e}{T_u} \times 100\% \qquad (2.19)$$

where $T_e$ is the mean tensile strength of the exposed geosynthetic specimen, and $T_u$ is the mean tensile strength of the unexposed geosynthetic specimen, and

$$R_\varepsilon = \frac{\varepsilon_e}{\varepsilon_u} \times 100\% \qquad (2.20)$$

where $\varepsilon_e$ is the mean strain at maximum load of the exposed geosynthetic specimen, and $\varepsilon_u$ is the mean strain at maximum load of the unexposed geosynthetic specimen. The durability can also be assessed by determining the changes in the mass per unit area of the geosynthetic.

The object of the durability assessment is to provide the design engineer with the necessary information generally in terms of the property changes or the partial safety factors so that the expected design life can be achieved with confidence. The durability study consists of the following (Standards Australia, 2002):

1   listing significant environmental factors
2   defining the possible degradation phenomena with regard to the selected geosynthetics and the environment
3   estimating the available property as a function of time
4   supplying the designer with suitable reduction factors or available properties at the end of design life of the soil-geosynthetic system.

The effects of a given application environment on the durability of a geosynthetic must be determined through appropriate testing. A selection of appropriate tests for durability assessment requires a consideration of the design parameters, the determination of the primary function(s) and/or performance characteristics of the geosynthetic in the specific field application and the associated degradation processes caused by the application environment. Note that the physical structure of the geosynthetic, the type of the polymer used, the manufacturing process, the application environment, the conditions of storage and installation, and the different loads supported by the geosynthetic are all parameters that govern the durability of the geosynthetic.

From the engineering point of view, the durability of geosynthetics is studied as the *construction survivability* and the *longevity*. *Construction survivability* addresses the geosynthetics survival during the installation. Geosynthetics may suffer a mechanical damage (e.g. abrasion, cuts or holes) during installation due to the placement and the compaction of the overlying fill. In some cases, the installation stresses might be more severe than the actual design stresses for which the geosynthetic is intended. The susceptibility of some geosynthetics to mechanical damage during installation can increase under frost conditions. The severity of the damage increases with the coarseness and angularity of the fill in contact with the geosynthetic and with the applied compactive effort, and it generally decreases with the increasing thickness of the geosynthetic. This damage may reduce the mechanical strength of the geosynthetic, and when the holes are present, it will affect the hydraulic properties as well.

The occurrence and consequences of mechanical damage caused during installation can be assessed by carrying out a field test or by simulating the effects of damage through a trial. The effect of mechanical damage is expressed as the ratio of the mechanical property of the damaged material to that of the undamaged material, as explained earlier. The ratio may be used as a partial safety factor in the design of reinforcement applications. The partial safety factor is used to reduce the characteristic strength of the geosynthetic selected for the application. Note that the installation day is the most difficult in the life of a geosynthetic. In general, the stronger the geosynthetic, the greater its resistance to installation damage, that is, the greater its potential for survivability. The ability of a geosynthetic to survive installation damage is difficult to quantify using the design equations, but one can do it based on the past experience. While selecting the geotextiles, one can follow the M288-06 geotextile specifications presented by the American Association of State Highway and Transportation Officials (AASHTO) (AASHTO, 2013), as described in Section 7.18 (See Chapter 7).

*Longevity* addresses how the geosynthetic properties change over the life of the structure. All geosynthetics are likely to be exposed to weathering during storage and on the construction site before installation. The resistance to weathering is important for the performance of the selected geosynthetic. The weathering of geosynthetics is mainly initiated by climatic influences through the action of solar radiation, heat, moisture and wetting. In most field applications, the geosynthetics are generally covered by soil, while those that remain exposed during their entire life need a far greater degree of resistance. Unless the geosynthetics are to be covered on the day of installation, all geosynthetics should be subjected to an accelerated weathering test. The principle of the accelerated weathering test is to expose the specimens to simulated solar ultraviolet (UV) radiation for different radiant exposures with cycles of temperature and moisture. The strength retained by the geosynthetic at the end of testing, together with the specific application of the geosynthetic, will define the length of time that the geosynthetic may be exposed on the site. Extended artificial weathering tests are required for geosynthetics, which are to be exposed for longer durations. If the geosynthetics are to be used for reinforcement applications, an appropriate partial safety factor should be applied to allow for the reduction in strength.

Generally, as the ambient temperature is increased, the strength, the creep, and the durability characteristics of geosynthetics deteriorate. In fact, the heat exposure causes the change in chemical structure, resulting in changes in physical properties and sometimes in the appearance of a polymer. Geosynthetics are likely to encounter high temperature only in paving applications, where they come in contact with hot asphaltic materials. This application favours the use of PP grids in preference to PE grids, because of their temperature resistance being greater. High temperature extremes should always be avoided. A geosynthetic tested for resistance to oxidation (temperature stability) in accordance with BS EN ISO 13438 (BSI, 2004b) should have the minimum percentage retained strength of 50%.

Geosynthetics may degrade when exposed to the UV component (wavelength shorter than 400 nm) of sunlight. The UV light stimulates oxidation by which the molecular chains are cut off. If this process starts, the molecular chains degrade continuously and the original molecular structure changes, resulting in a substantial reduction of the mechanical resistance and also in the geosynthetic becoming brittle. In most applications, as geosynthetics are exposed to UV light for only a limited time during storage, transport and installation and are subsequently protected by a layer of soil, the UV degradation is not a major cause of concern, provided sensible placement procedures are followed.

Generally, those geosynthetics that are white or grey in colour are likely to be the most vulnerable to UV degradation. Carbon black and other stabilizers are added to many polymers during the production process to provide long-term resistance to UV-induced degradation. To achieve this, carbon black and other stabilizers should be dispersed and distributed uniformly throughout the as-manufactured geosynthetic material. The uniformity of carbon black dispersion can be checked by the microscopic evaluation.

The study of the long-term performance of geosynthetics in sunlight can be carried out by exposing the geosynthetics to natural radiation from outdoor exposures, or by artificial radiation such as carbon (or xenon) arc lighting in laboratory. An outdoor exposure test evaluates the geosynthetics under site-specific atmospheric conditions generally over an 18-month period. The exposure shall begin so as to ensure that the geosynthetic is exposed during the maximum intensity of UV light of the year. A degradation curve in the form of a graph of percent tensile strength retained, percent strain at failure, or modulus versus exposure time, or all of these may be developed for the geosynthetic being evaluated. The durability is generally assessed by comparing the ultimate tensile strength and the elongation at the ultimate tensile strength of exposed specimens with those of unexposed specimens. The artificial exposure test that may be completed even within a week has the advantage of not only accelerating the testing by increasing the mean irradiance level and temperature, and eliminating the cycles of night and day, winter and summer, but also controlling the exposure parameters. Outdoor exposure tests or artificial exposure tests at one location may not be applicable to a project site at another location. The UV degradation test results therefore must be analyzed for practical applications keeping in view geographic location, radiation angle, temperature, humidity, rainfall, wind, air pollution, etc. associated with a particular construction site. The PE and PP products perform worst, with the majority having a 50% strength loss in about 4 to 24 weeks' exposure of UV radiations.

Geosynthetics may come into contact with chemicals/leachates that are not normally a part of the soil environment. Site-specific tests must be performed to know the chemical degradation of the geosynthetic, resulting in a reduction in molecular weight of polymer and in the deterioration of their engineering properties. Index tests are generally used in chemical-resistance studies. Geosynthetic specimens are exposed by immersion to liquids under specific conditions for a specified time, generally 15 days, unless otherwise specified. The durability is assessed by comparing the ultimate tensile strength and the elongation at ultimate tensile strength of exposed specimens with those of unexposed specimens. The chemical-resistance testing of geosynthetics should employ the worst-case scenario conditions. This is necessary to ensure that when in actual use, the geosynthetics will not be subjected to conditions worse than those experienced in the testing laboratory. Accelerated tests should have generally an accepted relationship to the real conditions. The geosynthetic composition should be considered in cases of complex chemical exposure (e.g. leachate) and burial in metal-rich soils. In the absence of actual test data, the chemical resistance can be evaluated, at least initially, by comparing the chemicals anticipated in the application with manufactures' published resistivity charts.

All polymeric materials have a tendency to absorb water over time. The absorbed water causes a chain scission and a reduction in molecular weight of the polymer along with some swelling; this degradative chemical reaction is called the *hydrolysis*. However, the effect of hydrolysis is probably not enough to cause significant changes in mechanical or hydraulic properties of geosynthetics. A geosynthetic consisting solely of PET can be tested for resistance to hydrolysis in accordance with DD ISO/TS 13434 (BSI, 2008). The minimum percentage retained strength should be 50%. A geosynthetic that fulfills this requirement is estimated to have the following minimum retained strengths in saturated soils after 25 years:

- At 25°C: 95%
- At 30°C: 90%
- At 35°C: 80%

For geosynthetics, oxidation and hydrolysis are the most common forms of chemical degradation as they are the processes that involve solvents. Generally, the chemical degradation is accelerated by elevated temperatures because the activation energy for these processes is commonly high. The moderate temperatures associated with most installation environments is, therefore, not expected to promote excessive degradation within the usual service lifetimes of most civil engineering systems. Most of the geosynthetics may be considered as having sufficient durability for a minimum service life of 25 yeas provided that it is used in natural soils with a pH of between 4 and 9, and at a soil temperature less than 25°C. A prudent attention should always be given to unique environments to assess their potential for causing polymer degradation. Resistance to specific chemical attacks (e.g. highly alkaline, pH > 9 or acidic, pH < 4, environments) should be investigated on a site-specific basis. For example, the tests on geotextiles in contact with uncured concrete indicate that the PP products are largely unaffected, whereas the PET products can lose about 50% of their strength in two months of prolonged exposure (Wewerkan, 1982). Since many geosynthetic users are not familiar with polymer chemistry, it would be better to assess the geosynthetic performance on a functional basis and reserve the polymer chemistry for interpreting the unsatisfactory test results or by performing forensic studies, if necessary.

*Macrobiological degradation* is the attack and physical destruction of a geosynthetic by macroorganisms (e.g. insects, rodents, and other higher life forms) leading to a reduction in physical properties. *Microbiological degradation* is the chemical attack of a polymer by enzymes or other chemicals excreted by microorganisms (e.g. bacteria, fungi, algae, yeast, etc.) resulting in a reduction of the molecular weight and changes in physical properties. All geosynthetic resins are very high in molecular weight with relatively few chain endings for the initiation of biological degradation. Therefore, the geosynthetics commonly manufactured from high molecular weight polymers are in general not affected by biological elements. So far there has been no evidence of any biological degradation of geosynthetics. Only those based on natural fibres degrade, as is the intention.

Microbiological degradation cannot be accelerated beyond the optimum soil conditions and temperature; if it is accelerated further, the microorganisms will be destroyed. It can be studied by conducting *soil burial tests* in which the geosynthetic specimens are buried in a prepared microbially active soil bed that is placed in an incubator maintained at a temperature of 28°C and not less than 85% relative humidity for a specified time, generally 14 days, unless otherwise specified. The durability is assessed by comparing the ultimate tensile strength and the elongation of exposed specimens with those of unexposed specimens. Note that there is no need to inoculate the soil with specific bacteria or fungi; all relevant species are assumed to be already present and those that benefit from the nutrients in the geosynthetic, if any, will multiply and accelerate the attack. The soil must be allowed to stabilize before the specimens are placed in it. A sample of untreated cotton is used to test the soil: if the tensile strength of the cotton strips is less than 25% of the original tensile strength after a seven-day exposure, the soil is regarded as biologically active. A good-quality horticultural compost should be sufficient for the soil burial tests (Greenwood *et al.*, 1996). The biological test is generally not required for geosynthetics manufactured from virgin (not recycled) PE, PP, PET and PA. This test may be applied to other materials, including natural fibre-based products, new materials, geocomposites, coated materials and others, which are of doubtful quality.

*Aging* is the alteration of the physical, the chemical, and the mechanical properties of geosynthetics caused by the combined effects of environmental conditions over time. It therefore includes both the polymer degradation and the reduced geosynthetic performance and is dependent on the specific application environment. The resistance of a geosynthetic to aging is referred to as the *durability* as defined earlier. Aging and burial test procedures and results are becoming more critical as the long-term demands on increase in geosynthetic applications. Note that the daily and seasonal variations occur with decreasing intensity as the distance from the ground surface increases. For example, the daily variation in atmosphere temperature and solar radiation is felt to a depth of half a metre. Since the higher temperatures increase the rates of aging and creep of polymers disproportionally, their effect on the geosynthetic behaviour may need to be considered for material installed close to the surface. Aging is an area that requires much research work. The knowledge and understanding of the long-term, in-service behaviour of geosynthetics are vital to the continued growth of the geosynthetic industry and to the science of geosynthetic design.

Most geosynthetics do not suffer from problems with brittle behaviour. However, certain geosynthetic materials may be subject to brittle behaviour as a result of *envi-*

*ronmental stress cracking* (ESC), which is the embrittlement of polymers caused by the combination of mechanical stress and environmental conditions. Semi-crystalline polymers such as PE can be more susceptible to environmental stress cracking. Drawn PET or PP fibres, or the drawn ribs of extruded geogrids, are comparatively resistant to ESC. Susceptibility to ESC can be measured by immersing the notched specimens under load in a bath of liquid and can be accelerated by raising the load or changing the environmental conditions. It is then necessary to carry out tests over a longer term to establish the degree of acceleration.

## 2.6   TEST AND ALLOWABLE PROPERTIES

There are presently a large number of geosynthetics available commercially, each having different properties, but their inclusion in this chapter is beyond the scope of the book. For obtaining the specific values of the various properties of geosynthetics, the users should consult the respective manufacturers or suppliers. Some representative properties of typical commercially available geosynthetics are listed in Table 2.3. A comparison of the properties of woven and nonwoven geotextiles having the same mass per unit area is also given in Table 2.4.

In Chapter 1, you have learnt that a geosynthetic performs one or more functions in a specific field application. A particular function of the geosynthetic is required to be evaluated using some of its properties. Table 2.5 provides a list of important properties related to the basic functions of geosynthetics. These properties are sometimes referred to as the *functional properties*. Note that the data on the soil-geosynthetic interface characteristics are necessary for the reinforcement and separation when the geosynthetic is used in a situation where a differential movement can take place between the geosynthetic and the adjacent material (soil/geosynthetic), which may endanger the stability of the structure. The data on the tensile creep may be required to give an indication of the resistance to the sustained loading, when the geosynthetic fulfills a reinforcement function. The data on the static puncture strength are necessary for the filtration and separation functions when the site loading conditions are such that there is a potential risk of static puncture of the geosynthetic.

Geosynthetics almost always encounter soil and environmental conditions that would be expected to cause reductions in their performance. Their properties can be changed unfavourably by several means such as ageing, mechanical damage, creep, hydrolysis (reaction with water), chemical and biological attack, etc., as described in the previous section. These factors have to be taken into account when a geosynthetic is selected for an application. For instance, a reduction factor has to be taken into account in calculation of the decline of strength caused by these factors.

If the test method for determining the geosynthetic property is not site-specific and completely field simulated, before using this test functional property in calculation of the design factor of safety according to Equation (1.2), it must be modified to an allowable property taking into account of all unfavorable conditions up to the end of the design life as follows:

$$\text{Allowable functional property} = \frac{\text{Test functional property}}{f_1 \times f_2 \times f_3 \times \dots} \qquad (2.21)$$

*Table 2.3* Typical range of some specific properties of commercially available geosynthetics (based on the information compiled by Lawson and Kempton (1995) and Shukla and Yin (2006)).

| Types of geosynthetics | | Tensile strength (kN/m) | Extension at max. load (%) | Apparent opening size (mm) | Water flow rate (volume permeability) (litres/m²/s) | Mass per unit area (g/m²) |
|---|---|---|---|---|---|---|
| Geotextiles | Nonwovens | | | | | |
| | Heat-bonded | 3–25 | 20–60 | 0.02–0.35 | 10–200 | 60–350 |
| | Needle-punched | 7–90 | 30–80 | 0.03–0.20 | 30–300 | 100–3000 |
| | Resin-bonded | 5–30 | 25–50 | 0.01–0.25 | 20–100 | 130–800 |
| | Wovens | | | | | |
| | Monofilament | 20–80 | 20–35 | 0.07–4.0 | 80–2000 | 150–300 |
| | Multifilament | 40–1200 | 10–30 | 0.05–0.90 | 20–80 | 250–1500 |
| | Flat tape | 8–90 | 15–25 | 0.10–0.30 | 5–25 | 90–250 |
| | Knitteds | | | | | |
| | Weft | 2–5 | 300–600 | 0.20–2.0 | 60–2000 | 150–300 |
| | Warp | 20–800 | 12–30 | 0.40–1.5 | 80–300 | 250–1000 |
| | Stitch-bondeds | 30–1000 | 10–30 | 0.07–0.50 | 50–100 | 250–1000 |
| Geogrids | Extruded | 10–200 | 20–30 | 15–150 | NA | 200–1100 |
| | Textile-based | | | | | |
| | Knitted | 20–400 | 3–20 | 20–50 | NA | 150–1300 |
| | Woven | 20–250 | 3–20 | 20–50 | NA | 150–1100 |
| | Bonded cross-laid strips | 30–200 | 3–15 | 50–150 | NA | 400–800 |
| Geomembranes | Natural | | | | | |
| | Reinforced (made from bitumen and nonwoven geotextile) | 20–60 | 30–60 | 0 | 0 | 1000–3000 |
| | Plastomeric (made from plastomers such as HDPE, LDPE, PP or PVC) | | | | | |
| | Unreinforced | 10–50 | 50–200 | 0 | 0 | 400–3500 |
| | Reinforced | 30–60 | 15–30 | 0 | 0 | 600–1200 |
| | Elastomeric (made from elastomers, i.e. rubbers of various types) | | | | | |
| | Reinforced | 30–60 | 15–30 | 0 | 0 | 500–1500 |
| Geocomposites | Geosynthetic clay liners | 10–20 | 10–30 | 0 | 0 | 5000–8000 |
| | Linked structures (Geostrip-based)[1] | 100–1500 | 3–15 | NA | NA | 400–4500 |

Notes:

*NA is not applicable*

[1]Geostrips are geocomposites having tensile strength in the range 20–200 kN and extension at max. load in the range 3–15%. Geobars are geocomposites having tensile strength in the range 20–1000 kN, if reinforced internally and in the range 20–300 kN if reinforced externally, and extension at max. load in the range 3–15% for both cases.

where $f_1$, $f_2$ and $f_3$, etc. are the various reduction factors (i.e. the partial factors of safety) needed to account for differences between the test and the site-specific conditions. These reduction factors reflect appropriate degradation processes, and are equal to or greater than one.

Table 2.4 Comparison of some properties of woven and nonwoven geotextiles having the same mass per unit area.

| Property | Woven | Nonwoven |
|---|---|---|
| Fibre arrangement | Orthogonal | Random |
| Breaking strength | Higher | Lower |
| Breaking elongation | Lower | Higher |
| Initial modulus | Higher | Lower |
| Tear resistance | Lower | Higher |
| Openings | Can be regular | Irregular |
| Filtration | Single layer | Multi-layer |
| Porosity | 35–45% | 55–93% |
| In-plane flow | Lower | Higher |
| Edge | May ravel | Does not ravel |

Table 2.5 Important properties of geosynthetics related to their basic functions.

| Geosynthetic functions | Geosynthetic properties |
|---|---|
| Reinforcement | Strength, stiffness, soil-geosynthetic interface characteristics (frictional and interlocking characteristics), creep, stress relaxation, durability |
| Separation | Characteristic opening size, strength, soil-geosynthetic interface characteristics (frictional and interlocking characteristics), durability |
| Filtration | Characteristic opening size, permittivity, clogging, puncture strength, durability |
| Drainage (fluid transmission) | Characteristic opening size, transmissivity, clogging, durability |
| Fluid barrier | Permittivity, strength, durability, abrasion resistance |
| Protection | Puncture strength, burst strength, stiffness, abrasion resistance, durability |

For example, the laboratory-generated tensile strength $\sigma_f$ is usually an ultimate value $\sigma_{ult}$, which must be reduced for its use in design. This can be carried out using the following:

$$\sigma_{all} = \sigma_{ult} \left( \frac{1}{f_{ID} \times f_{CR} \times f_{CD} \times f_{BD}} \right) \tag{2.22}$$

where $\sigma_{all}$ is the allowable tensile strength to be used in Equation (1.2) for the design purposes; $\sigma_{ult}$ is the ultimate tensile strength obtained from the test; $f_{ID}$ is the reduction factor for installation damage (1.1–3.0 for geotextiles, 1.1–1.6 for geogrids); $f_{CR}$ is the reduction factor for creep (1.0–4.0 for geotextiles, 1.5–3.0 for geogrids); $f_{CD}$ is the reduction factor for chemical degradation (1.0–2.0 for geotextiles, 1.0–1.6 for geogrids); and $f_{BD}$ is the reduction factor for biological degradation (1.0–1.3 for geotextiles, 1.0–1.2 for geogrids).

While dealing with the flow-related problems through or within a geosynthetic, several reduction factors are required to be considered suitably, as mentioned in the following expression for the allowable permittivity $\psi_{all}$:

$$\psi_{all} = \psi_{ult}\left(\frac{1}{f_{CB} \times f_{CR} \times f_{IN} \times f_{CC} \times f_{BC}}\right) \tag{2.23}$$

where $\psi_{ult}$ is the ultimate permittivity obtained from the test; $f_{CB}$ is the reduction factor for soil clogging, blinding and blocking (2.0–10.0 for geotextiles); $f_{CR}$ is the reduction factor for creep reduction of void volume (1.0–3.0 for geotextiles; 1.0–2.0 for geonets); $f_{IN}$ is the reduction factor for intrusion of adjacent materials into the void volume of geotextile (1.0–1.2 for geotextiles; 1.0–2.0 for geonets); $f_{CC}$ is the reduction factor for chemical clogging (1.0–1.5 for geotextiles; 1.0–2.0 for geonets); and $f_{BC}$ is the reduction factor for biological clogging (1.0–10.0 for geotextiles; 1.0–2.0 for geonets).

It is important to note that the values of reduction factors are highly dependent on the area of field application and the prevailing site conditions. For example, in Equation (2.23), the reduction factor for biological clogging can be higher for turbidity and/or microorganism contents greater than 5000 mg/l. The low end of the range for creep reduction factors refer to applications which have relatively short service lifetimes and/or situations where the creep deformations are not critical to the overall system performance. Thus, the designer must use engineering judgment appropriately based on the available information while selecting the reduction factors.

### EXAMPLE 2.4

The ultimate tensile strength of a geogrid, obtained from the wide-width strip tensile test, is 90 kN/m. What will be the allowable tensile strength of the geogrid for the reinforcement application? If the design value of tensile strength is 30 kN/m, determine the factor of safety against the tensile failure.

### SOLUTION

Given: $\sigma_{ult} = 90$ kN/m and $\sigma_{design} = 30$ kN/m.

The values of reduction factors are selected based on the site-specific conditions. The guidelines given in the local codes of practice, if available, should be considered while selecting these factors. For the present problem, consider the following typical values of reduction factors:

$$f_{ID} = 1.1, f_{CR} = 1.6, f_{CD} = 1.3 \text{ and } f_{BD} = 1.1$$

From Equation (2.22),

$$\sigma_{all} = \sigma_{ult}\left(\frac{1}{f_{ID} \times f_{CR} \times f_{CD} \times f_{BD}}\right) = (90)\left(\frac{1}{1.1 \times 1.6 \times 1.3 \times 1.1}\right) = 35.76 \, kN/m$$

From Equation (1.2), the factor of safety against the tensile failure is

$$F = \frac{\sigma_{all}}{\sigma_{design}} = \frac{35.76}{25} = 1.43$$

## 2.7  DESCRIPTION OF GEOSYNTHETICS

The manufacturers' literature generally provides the product information and the relevant properties of geosynthetics. If these property values are being used for design, a modification must be made, as described in the previous section. Geosynthetics, in general, are commercially described as follows:

1  polymer type
2  type of element (e.g. fibre, yarn, strand, rib), if applicable
3  manufacturing process, if essential
4  type of geosynthetic
5  mass per unit area and/or thickness, if applicable
6  additional information/property in relation to specific field applications.

For example, PP staple filament needle-punched nonwoven, 400 g/m$^2$; PET extruded uniaxial geogrid, with 20 mm by 10 mm openings; HDPE roughened sheet geomembrane, 2.0 mm thick, etc. are a few descriptions of geosynthetics. Before unrolling the roll of a geosynthetic at the job site, its identification must be verified properly.

If the geosynthetics are used as a paving fabric, some or all of the following commonly specify them:

1  mass per unit area
2  grab tensile strength in the weakest principal direction
3  elongation
4  bitumen retention
5  fabric storage
6  heat resistance.

Note that the properties used in the specification of geosynthetics are established from the index tests or from the performance tests. As was already discussed, the index tests are used by manufacturers for quality control and by installers for product comparison, material specifications and construction quality assurance. Index tests describe the general strength, and the hydraulic and durability properties of the geosynthetic. The general properties are used to distinguish between polymer type and mass per unit area. Performance tests are used by the designers to establish, where necessary, the design parameters under the site-specific conditions using the soil samples taken from the site.

## 2.8 APPLICATIONS OF GEOSYNTHETICS

Geosynthetics are being used in several areas of civil engineering, especially in geotechnical, transportation, environmental and hydraulic engineering, and in some areas of mining, agricultural and aquacultural engineering. The exact number of applications is difficult to count, and moreover, the applications are growing steadily. Table 2.6 provides a list of some major applications along with the purpose(s) of using geosynthetics, and their basic function(s) and performance characteristics required by the designers. Note that the aim of using geosynthetics in all the applications is to do a better job in a cost-effective, environmentally friendly and sustainable manner. The detailed descriptions of major applications of geosynthetics are presented in Chapters 3 to 6 by categorizing the applications as:

1 Geotechnical applications of geosynthetics (Chapter 3)
2 Hydraulic and geoenvironmental applications of geosynthetics (Chapter 4)
3 Transportation applications of geosynthetics (Chapter 5)
4 Mining, agricultural and aquacultural applications of geosynthetics (Chapter 6)

The general application guidelines and installation requirements as required in most applications are presented in Chapter 7. The details of quality evaluation, performance monitoring and economic evaluation are described in Chapter 8. Synthetic fibres and also many natural and waste fibres are randomly used in soils, resulting in fibre-reinforced soils. This book also includes the details of fibre-reinforced soils and their potential applications in Chapter 9.

## 2.9 DESIGN CONCEPTS

When designing the soil-geosynthetic systems/structures, the most common question is, 'What is the expected lifespan of geosynthetics?' There is no straight answer to this question. However, on the basis of accelerated performance tests in the laboratory and the experiences gained during the past 35–40 years, it is expected that the geosynthetics can have a lifespan of about 120 years, provided they are used appropriately in the field applications, particularly in buried or underwater applications. In fact, it is still a matter of "to believe or not to believe".

When a geosynthetic is used in an application, it is intended to perform particular function(s) for a minimum expected time period, called the *design life*. Geosynthetics are designed to perform a function, or a combination of functions, within the soil-geosynthetic system. Such functions are expected to be performed over the design life

*Table 2.6* Major application areas for geosynthetics.

| Application areas | | Purpose(s) of using geosynthetics | Basic function(s) and performance characteristic (s) |
|---|---|---|---|
| Earth retaining structures | | To reinforce, retain, and protect backfill/soil for improving stability | Reinforcement |
| | | To make wall waterproofing system | Fluid barrier, protection |
| Embankments | | To keep embankment materials separated from soft foundation soil from not being changed in behaviour over the service period | Separation |
| | | To improve stability of embankment edges, to bridge soft foundation soils, to make steep-sided slopes | Reinforcement |
| | | To drain the water from the base of the embankment | Drainage, filtration |
| Foundations | | To improve load-bearing capacity, to reduce settlement. | Reinforcement |
| | | To prevent erosion and scouring around underwater foundations using bags, tubes and mattresses filled with soil, to form underwater foundations | Containment, Screening |
| Unpaved roads | | To improve load-bearing capacity, to reduce degree of rutting, to bridge soft foundation soils/sinkholes | Reinforcement |
| Paved roads and airfields | Overlay base level | To prevent/control water infiltration | Fluid Barrier |
| | | To prevent/control reflective cracking | Cushioning |
| | Subgrade level | To prevent contamination of subbase/base course | Separation |
| | | To provide quick disposal of water to side drains | Filtration, drainage |
| | | To prevent the enlargement of karst sinkholes, to control swelling and shrinkage of expansive soils | Fluid barrier, protection |
| | | To bridge soft foundation soils/sinkholes, to improve performance of the base/subbase materials | Reinforcement |
| | | To prevent frost heave in frost-sensitive soils | Fluid barrier, drainage, insulation, protection |
| Railway tracks | | To prevent ballast contamination | Separation |
| | | To dispose of water to side drains | Filtration, drainage |
| | | To prevent contamination in railroad refueling areas, to prevent upward groundwater movement in a railroad cut | Fluid barrier, protection |
| | | To reinforce track systems and distribute loads | Reinforcement |

(Continued)

Table 2.6 (Continued)

| Application areas | Purpose(s) of using geosynthetics | Basic function(s) and performance characteristic(s) |
|---|---|---|
| Slopes | To protect soil slope against erosion along with slope armor | Filtration |
| | To protect earthen slopes against erosion while vegetation is being established | Vegetative reinforcement, surface stabilization |
| | To prevent erosion and scouring using bags, tubes and mattresses filled with soil | Containment, screening |
| | To prevent soil slope against movement/sliding | Reinforcement |
| Landfills | To prevent leachate from infiltrating into soil | Fluid barrier, protection |
| | To drain leachate | Filtration, drainage |
| Earth dams | To reduce seepage through the dam embankment, to provide upstream face infiltration cut-off | Fluid Barrier, protection |
| | To prevent internal erosion/piping | Filtration, protection |
| | To drain seepage water | Drainage, filtration |
| Containment ponds, reservoirs, and canals | To reduce leakage/seepage of water/liquid into ground | Fluid barrier, protection |
| | To minimize the migration of sediments, to prevent the transportation of solid particles suspended in water | Screening |
| | To prevent erosion of the earthen surfaces while vegetation is being established | Surface stabilization, vegetative reinforcement |
| | To prevent erosion and scouring of earthen surfaces using bags, tubes and mattresses filled with soil | Containment, screening |
| Filters and drains | To protect the drainage medium, to provide drainage medium | Filtration, drainage, separation |
| Tunnels and underground structures | To prevent seepage | Fluid barrier, protection |
| | To provide drainage of seepage water | Drainage |

Note: In most mining, agricultural and aquacultural applications, the geosynthetics are generally used as liners, filters and/or drainage mediums.

of the soil-geosynthetic system, which is typically less than 5 years for short-term use, around 25 years for temporary use, and 50 to 100 years, or more for permanent use. The nature of the application system and the consequences of its failure may influence the design life (e.g. 75 years for retaining walls, 100 years for bridge abutments, and 10 to 30 years for mining structures). Geosynthetics may have a short-term function although the system is permanent; for example an embankment over a weak foundation may require a geosynthetic reinforcement only while the consolidation is occurring and until the weak foundation has gained sufficient strength to support the embankment load. The design life for a soil-geosynthetic system is set by the client or the designer and is decided at the design stage.

The primary responsibility of a designer is to design a facility that fulfils the operational requirements of the owner/operator throughout its design life, complies with accepted design practices as per the relevant standards and codes of practice and meets or exceeds the minimum requirements of the permitting agency. The designer should be aware of the possible construction and maintenance constraints. Also, social conditions, safety requirements, and environmental impact may affect the eventual outcome of the design process. Based on these facts and the main functional objectives of the given structure, a set of technical requirements should be assessed.

This section presents the basic concepts of the analytical approach and methods of design process for major applications of geosynthetics. For more detailed design process, one should follow the relevant standards, the codes of practice and the design manuals, as applicable at the work place.

In general, a designer of the geosynthetic-reinforced soil structures must consider the potential failure mode, and then the appropriate test procedure should be used to evaluate the soil-geosynthetic interaction properties. There are many ways to perform a given test in the laboratory, depending on the case to be designed. The recommended way is the one that best simulates the actual performance of the geosynthetic at the site. Usually, a laboratory test simulates the field situation at only one point of the geosynthetic. When the whole field situation can be simulated in a laboratory test, the test results can be applied to the field situation either directly or using minor mathematical adjustments to deal with the difference in scale between the laboratory and the field situations. In this case, the test is a model test and an analogical method of design is used. This method of design is the simplest one but it can rarely be used, so other methods, such as analytical methods, based on the mathematical theories and the basic parameters of geosynthetics, and empirical methods, based on experience and sometimes systematic testing including the full-scale tests, are needed (Giroud, 1980).

The design of a structure incorporating the geosynthetics aims to ensure its strength, stability and serviceability over its intended life span. There are mainly four design methods/approaches for the geosynthetic-related structures or systems as briefly described below:

1   *Design-by-experience*: This method is based on one's past experience or that of others. This is recommended if the application is not driven by a basic function or it has a nonrealistic test method.

2   *Design-by-cost-and-availability*: In this method, the maximum unit price of the geosynthetic is calculated by dividing the funds available by the area to be covered

by the geosynthetic. The geosynthetic with the best quality is then selected within this unit price limit according to its availability. Being technically weak, this method is nowadays rarely recommended by the current standards of practice.

3   *Design-by-specification*: This method often consists of a property matrix where common application areas are listed along with minimum (or sometimes maximum) property values. Such a property matrix is usually prepared on the basis of local experiences and field conditions for routine applications by most of the governmental agencies and other larger users of geosynthetics. For example, the AASHTO M288-06 specifications (AASHTO, 2013), as described in Section 7.18 (See Chapter 7), provide the designer and the field quality inspector with a very quick method of evaluating and designing geotextiles for common applications such as filters, separators, stabilizers, and erosion control layers.

4   *Design-by-function*: This method is the preferred design approach for geosynthetics. The general approach of this method consists of the following steps:

   a   Assessing the particular application, define the primary function of the geosynthetic, which can be reinforcement, separation, filtration, drainage, fluid barrier or protection.
   b   Make the inventory of loads and constraints imposed by the application.
   c   Define the design life of the geosynthetic.
   d   Calculate, estimate or otherwise determine the required functional property of the geosynthetic (e.g. strength, permittivity, transmissivity, etc.) for the primary function.
   e   Test for or otherwise obtain the allowable property (available property at the end of the design life) of the geosynthetic, as discussed in Section 2.6.
   f   Calculate the factor of safety *FS* using Equation (1.2), reproduced as

$$FS = \frac{\text{Allowable (or test) functional property}}{\text{Required (or design) functional property}} \qquad (2.24)$$

   g   If *FS* from Equation (2.24) is not acceptable, check into geosynthetics with more appropriate properties.
   h   If *FS* from Equation (2.24) is acceptable, check if any other function of the geosynthetic is also critical, and repeat the above steps.
   i   If several geosynthetics are found to meet the required factor of safety, select the geosynthetic on the basis of cost-benefit ratio, including the value of available experience and product documentation.

Note that the design-by-function method bears heavily on identifying the primary function to be performed by the geosynthetic. For any given application, there will be one or more basic functions that the geosynthetic will be expected to perform during its design life. The accurate identification of the geosynthetic function as the primary function(s) is essential. Hence, a special care is required while identifying the primary function(s).

All geosynthetic designs should begin with a criticality and severity assessment of the project conditions for the particular application. The designer should always keep in mind the geosynthetic failure mechanisms that result in unsatisfactory performance (Table 2.7). The properties of geosynthetics should be selected to protect against the excessive reductions in performance under the specific soil and environmental conditions during the whole design life, as shown in Fig. 2.35, and appropriate factors of

*Table 2.7* Some geosynthetic failure mechanisms.

| Function | Failure mode(s) | Possible cause(s) |
|---|---|---|
| Reinforcement | Large deformation of the soil-geosynthetic structure | Excessive tensile creep of the geosynthetic |
| | Reduced tensile resisting force | Excessive stress relaxation of the geosynthetic |
| Separation/ Filtration | Piping of soils through the geosynthetic | Openings in the geosynthetic may be incompatible with retained soil. Openings might have been enlarged as a result of in situ stress or mechanical damage. |
| Filtration | Clogging of the geosynthetic | Permittivity of the geosynthetic might have been reduced as a result of particle buildup on the surface of or within the geosynthetic. Openings might have been compressed as a result of long-term loading. |
| Drainage | Reduced in-plane flow capacity | Excessive compression creep of the geosynthetic |
| Fluid barrier | Leakage through the geosynthetic | Openings may be available in the geosynthetic as a result of puncture or seam failure. |
| Protection | Reduced resistance to puncture | Excessive compression creep of the geosynthetic |

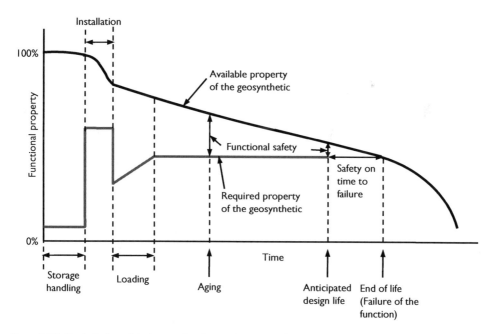

*Figure 2.35* Typical allowable (or test) value and required (design) value of a functional property as a function of time (adapted from Standards Australia, 2002).

safety must be utilized in designs incorporating geosynthetics. Note that the factor of safety is likely to decrease with time if the geosynthetic properties are subject to degradation with time. Especially for the most critical projects, conservative designs are recommended. Because of misconceptions with regard to the functioning of geosynthetics in various constructional and service stages of the project, it is possible that the designer formulates unnecessarily high requirements of geosynthetics. In fact, in most applications, simple design rules are sufficient for a proper choice of geosynthetics. However, the designers should be aware of situations where a more sophisticated approach is necessary, and be able to explain to the client that the difference in approach depends on the specific situations such as type of application, loading conditions, and design life.

The design-by-function approach described here is basically the traditional *working stress design approach* that aims to select the allowable geosynthetic properties so that a nominated minimum total (or global) factor of safety is achieved. In geosynthetic applications, particularly reinforcement applications (e.g. geosynthetic-reinforced earth retaining walls), it is now common to use the *limit state design approach*, rather than working stress design that involves the global safety factors. For the purpose of geosynthetic-reinforced soil design, a limit state is deemed to be reached when one of the following occurs:

1   collapse, major damage or other similar forms of structural failure;
2   deformations in excess of acceptable limits; and
3   other forms of distress or minor damage, which would render the structure unsightly, require unforeseen maintenance or shorten the expected life of the structure.

The condition defined in (1) is the *ultimate limit state*, and the conditions in (2) and (3) are *serviceability limit states*. The practice in reinforced soil is to design against the ultimate limit state and check for the serviceability limit state. In the reinforced soil design, some of the limit states may be evaluated by conventional soil mechanics approaches (e.g. settlement). Margins of safety, against attaining the ultimate limit state, are provided by the use of partial material factors and partial load factors. The limit state design for reinforced soil employs four principal partial factors all of which assume prescribed numerical values of unity or greater. Two of these are load factors applied to dead loads (external dead load – $f_f$ and soil unit weight – $f_{fs}$), and to live loads ($f_q$). The principal material factor is $f_m$ applied to the geosynthetic reinforcement parameters, and $f_{ms}$ applied to the soil parameters. The fourth factor $f_n$ is used to take into account the economic ramifications of failure. This factor is employed, in addition to the material factors, to produce a reduced design strength. Note that it is not feasible to uniquely define values for all these partial factors. The prescribed ranges of values are decided to take account of the type of geosynthetic application, the mode of loading and the selected design life. The partial factors are applied in a consistent manner to minimize the risk of attaining a limit state.

In the limit state design of geosynthetic reinforcement applications, the disturbing forces are increased by multiplying by the prescribed load factors to produce the design loads, whereas the restoring forces (strength test values) are decreased by dividing by the prescribed material factors to produce the design strengths. There is deemed to be an adequate margin of safety against attaining the ultimate limit state if

Design strength (factored down strength)
  $\geq$ Design load (stress due to factored loading)                        (2.25)

In the case of drainage application of geosynthetics, this requirement can be expressed as

Design drainage capacity (factored down drainage capacity)
$\geq$ Design flow (factored up expected flow) (2.26)

For assessing the deformations or strains to determine compliance with the appropriate serviceability limit sate, the prescribed numerical values of load factors are different to those used in assessing the ultimate limit state and usually assume a value of unity. In assessing the magnitudes of total and differential settlements, all partial factors are set to a value of unity, except for those pertaining to the reinforcements (BSI, 2010). With respect to the serviceability limit state, the design requirement for a geosynthetic can be expressed as

Allowable elongation $\geq$ Elongation at serviceability loading (2.27)

In the generalized form, it can be said that the limit state design, considering all possible failure modes and all appropriate partial factors being applied, aims to produce a soil-geosynthetic system that satisfies the following principal equation (Standards Australia, 2002):

Design resistance effect $\geq$ Design action effect (2.28)

for its all design elements.

Note that that relationship (2.28) defines the fundamental principle of limit state design. In the case of internal stability, the design resistance effect may be generated in the soil and the reinforcement, whereas it is generated in the soil only in the case of external stability.

When the safety of people and environment is at great risk because of the failure of the geosynthetic used, or when a reliable method is not available to determine the requirements of the geosynthetic to be used, it becomes desirable to perform suitable practical tests. If the tests are being conducted in the laboratory, special attention must be required to get reliable data to be used for field applications.

The adoption of suitable design and construction method is essential not only to reduce the design and construction costs, but also to minimize the long-term operation, maintenance and monitoring expenses.

Note that many software programs are available for the analysis, design and specification of geosynthetic-related structures/systems. There are internet sources for information about these commercially available software programs. The programs are mostly included under the categories of geosynthetics, reinforced slopes and walls, slope stability or ground improvement. Before using any software program, it is important to have a check for the demonstrated validation to a standard procedure of analysis and design.

## Chapter summary

1 The physical properties of geosynthetics that are of prime interest are specific gravity, mass per unit area, thickness and stiffness. For commonly used geo-

synthetics, the mass per unit area varies typically in the range of 100 g/m² to 1000 g/m².

2   The compressibility of a geosynthetic is measured by the decrease in its thickness at the increasing applied normal pressure, and this is very important property for nonwoven geotextiles and drainage geocomposites.

3   The tensile strength of a geosynthetic is defined as the maximum load that can be applied per unit length along the edge of the geosynthetic in its plane. A woven geotextile may have different tensile strengths in machine (or warp) and cross-machine (or weft) directions. The tensile modulus of a geosynthetic is the slope of its stress-strain or load-strain curve, as determined from the wide-width strip tensile test.

4   The mechanical properties such as tearing strength, static puncture strength, impact strength, bursting strength and fatigue strength are related to survivability/constructability and separation function of geosynthetics.

5   The soil-geosynthetic interaction through interface friction and/or interlocking characteristics is the key element in the performance of the geosynthetic-reinforced soil structures. The direct shear test and the pullout/anchorage test are commonly used to evaluate the soil-geosynthetic interaction.

6   The hydraulic properties (porosity, permittivity and transmissivity) of geosynthetics influence their ability to function as filters and drains. If a geotextile has an $O_{95}$ value of 300 µm, then 95% of geotextile pores are 300 µm or smaller.

7   Creep behaviour, abrasion resistance, long-term flow capability and durability (construction survivability and longevity) are the endurance and degradation properties of geosynthetics, which are related to their behaviour during service conditions, including the time. Creep is an important factor in the design and performance of some geosynthetic-reinforced structures (retaining walls, steep-sided slopes, embankments over weak foundations, etc.).

8   The durability of a geosynthetic may be regarded as its ability to maintain the requisite properties against environmental and/or other influences over the selected design life. All polymeric materials have a tendency to absorb water over time; this phenomenon is called the hydrolysis. The macrobiological degradation is the attack and physical destruction of a geosynthetic by macroorganisms, while the microbiological degradation is the chemical attack of a polymer by enzymes or other chemicals exerted by microorganisms. Aging is the alteration of physical, chemical, and mechanical properties of geosynthetics caused by the combined effects of environmental conditions over time.

9   The allowable functional property is the test functional property divided by the product of partial safety factors for installation damage, creep, chemical degradation, biological degradation, etc. as applicable.

10  Geosynthetics, in general, are commercially described in terms of polymer type, type of element, manufacturing process, type of geosynthetic, mass per unit area and/or thickness, etc., as applicable.

11  Geosynthetics are being used in several areas of civil engineering, especially in geotechnical, transportation, environmental and hydraulic engineering, and also in some areas of mining, agricultural and aquacultural engineering.

12  The design of a structure incorporating the geosynthetics aims to ensure its strength, stability and serviceability over its intended life span. The design-by-function approach, which is the preferred design approach based on the

primary function, is basically the traditional working stress design approach that aims to select the allowable geosynthetic properties so that a nominated minimum total (or global) factor of safety is achieved. The limit state design, considering all possible failure modes and all appropriate partial factors being applied, aims to produce a soil-geosynthetic system such that the design resistance effect is equal to or greater than the design action effect for all design elements.

## Questions for practice

(Select the most appropriate answer to the multiple-choice questions from Q 2.1 to Q 2.16.)

2.1 The most useful geosynthetic physical property, which is closely related to the engineering performance, is
  (a) thickness
  (b) mass per unit area
  (c) strength
  (d) stiffness.

2.2 The base polymer of a geosynthetic can be identified by determining
  (a) mass per unit area
  (b) thickness
  (c) strength
  (d) specific gravity.

2.3 The compressibility is relatively high for
  (a) geogrids
  (b) geonets
  (c) needle-punched nonwoven geotextiles
  (d) woven geotextiles.

2.4 If the tensile strength of a geotextile in a technical report is written as 100/40 kN/m, then its strength in the cross-machine direction will be
  (a) 40 kN/m
  (b) 60 kN/m
  (c) 100 kN/m
  (d) 140 kN/m.

2.5 Which one of the following depicts the deformation required to develop a given stress in the geosynthetic?
  (a) strength
  (b) stiffness
  (c) compressibility
  (d) none of the above.

2.6 The preferred measure of in-plane water flow capacity of a geotextile is
  (a) permeability
  (b) permittivity

(c) transmissivity

(d) volume rate of flow.

2.7 The nonwoven geotextiles generally have
   (a) high permittivity
   (b) high tensile strength
   (c) high modulus of elasticity
   (d) none of the above.

2.8 If a geotextile has an $O_{95}$ value of 300 μm, then
   (a) 5% of geotextile pores are 300 μm or smaller
   (b) 95% of geotextile pores are 300 μm or larger
   (c) 95% of geotextile pores are 300 μm or smaller
   (d) none of the above.

2.9 Which one of the following statements is incorrect?
   (a) Geogrids have relatively high dimensional stability, high tensile strength and high tensile modulus at low strain levels.
   (b) It is generally observed that the modulus of a geosynthetic confined in soil is likely to be lower than when tested in isolation.
   (c) For filter applications of a woven geotextile, the percentage open area (POA) should be higher to avoid any clogging phenomenon to occur throughout the design life of the particular application.
   (d) None of the above.

2.10 Which one of the following tests can be used to evaluate the clogging resistance of geotextiles with cohesionless soils (having a hydraulic conductivity/permeability greater than $5 \times 10^{-4}$ m/s) under unidirectional flow conditions?
   (a) Gradient ratio test
   (b) Hydraulic conductivity ratio test
   (c) Field jar test
   (d) None of the above.

2.11 A geosynthetic tested for resistance to oxidation (temperature stability) should have the minimum percentage retained strength of
   (a) 10%
   (b) 20%
   (c) 35%
   (d) 50%.

2.12 If the combined reduction factor is 3.3, then the allowable tensile strength of a geogrid having an ultimate tensile strength of 210 kN/m will be
   (a) 36.6 kN/m
   (b) 63.6 kN/m
   (c) 210 kN/m
   (d) 693 kN/m.

2.13 In earth retaining structures, geosynthetics are used mainly to function as
   (a) separation
   (b) filtration

(c) reinforcement

(d) protection

2.14 The design life of a geosynthetic-reinforced soil wall is generally

(a) 5 years

(b) 25 years

(c) 50 to 100 years

(d) more than 100 years.

2.15 Which one of the following is the most preferred design approach for soil-geosynthetic systems?

(a) design-by-function

(b) design-by-experience

(c) design-by-cost-and-availability

(d) design-by-specification.

2.16 The principal partial factor of safety generally employed in the limit state design for reinforced soil structures can be

(a) less than 1

(b) equal to 1

(c) greater than 1

(d) both (b) and (c).

2.17 What do you mean by the conditioning of geosynthetics? Explain its importance.

2.18 Why is the unit mass (or weight) of a geosynthetic is measured in terms of mass (or weight) per unit area as opposed to mass (or weight) per unit volume?

2.19 If the thickness of an HDPE geomembrane is 2 mm, and the specific gravity of the polymeric compound is 0.96, then determine the mass per unit area of the geomembrane.

2.20 How will you measure the stiffness of a geosynthetic?

2.21 What are the tensile properties of a geosynthetic? Explain the test results of the wide-width tensile test and discuss their limitations.

2.22 What is the purpose of conducting the multi-axial tensile strength test on geosynthetics?

2.23 What is the minimum average roll value of a geosynthetic property? How is it related to minimum, average and maximum values?

2.24 Draw a typical load-strain curve for geotextiles. Differentiate between the off-set modulus and the secant modulus using this typical curve.

2.25 In geosynthetic testing,

(a) What is an index test?

(b) What is a performance test?

2.26 List the survivability properties of geosynthetics. Give some examples of geo-synthetic applications where one or all of these properties will essentially be required to be checked.

2.27 Suggest some field situations showing puncturing and bursting of geosynthetics.

2.28 What are the currently used test methods to evaluate the soil geosynthetic interface characteristics? Explain the basic principles of these methods by means of neat sketches.

2.29 Discuss the typical results from the direct shear test on geotextiles.

2.30   The direct shear test was conducted in the laboratory to study compacted fly ash – nonwoven geotextile interface characteristics. The following data were obtained:

| Normal stress (kPa) | Shear strength (kPa) |
| --- | --- |
| 100 | 42 |
| 200 | 68 |
| 300 | 95 |

Plot the Mohr failure envelope and obtain the fly ash-geotextile interface characteristics, that is, the angle of interface shear resistance and adhesion.

2.31   What is the pullout capacity formula? What are your observations about the accuracy of this formula?

2.32   Calculate the porosity of the geotextile with the following details:
Thickness, $\Delta x = 1.8$ mm
Mass per unit area, $m = 150$ g/m$^2$
Density of polymer solid, $\rho_s = 910$ kg/m$^3$

2.33   For geotextiles, the pore size is measured directly, rather than using particle size as an estimate of pore size, as is done with soils. Is there any specific reason for it? Justify your answer.

2.34   List the various test methods developed for measuring the size of openings in the geotextiles. Compare these methods by giving the relevant details.

2.35   Draw the pore size distribution curves for typical woven and nonwoven geotextiles. Do you observe any specific difference in the two curves? If yes, list the differences.

2.36   Differentiate between permittivity and transmissivity.

2.37   It is preferable to determine and report the permittivity and transmissivity values of geotextiles rather than their coefficients of in-plane and cross-plane permeability, respectively. Explain the reasons for this.

2.38   Why is permittivity used in filtration and transmissivity used in drainage, rather than just the respective coefficients of permeability?

2.39   What are the long-term normal stress and environmental implications for flow rate capability of geosynthetics?

2.40   How does the confining pressure affect the thickness, permittivity and transmissivity of needle-punched nonwoven geotextiles?

2.41   Calculate the transmissivity of a geonet using the following laboratory-based data:
Flow rate per unit width, $q = 0.72 \times 10^{-4}$ m$^2$/s
Hydraulic gradient, $i = 0.05$

2.42   In a laboratory constant head cross-plane permeability test on a 50-mm diameter geotextile specimen, the following parameters are measured:
Nominal thickness, $\Delta x = 2.1$ mm
Flow rate of water normal to the plane of the geotextile, $Q_n = 0.317$ l/s
Head loss across the geotextile, $\Delta h = 300$ mm
Calculate the permittivity and the cross-plane coefficient of permeability of the geotextile.

2.43   In a laboratory constant head in-plane permeability test on a 300-mm length (flow direction) by 200-mm width geotextile specimen, the following parameters are measured:

Nominal thickness, $\Delta x = 2.6$ mm

Flow rate of water in the plane of the geotextile, $Q_p = 68$ cm$^2$/min

Head loss in the plane of the geotextile, $\Delta h = 150$ mm

Calculate the transmissivity and the in-plane coefficient of permeability of the geotextile.

2.44   In a laboratory determination of transmissivity by radial method on a geotextile specimen (outer radius = 150 mm, inner radius = 25 mm), the following parameters are measured:

Nominal thickness, $\Delta x = 2.6$ mm

Flow rate of water in the plane of the geotextile, $Q_p = 1620$ cm$^3$/min

Head loss in the plane of the geotextile, $\Delta h = 300$ mm

Calculate the transmissivity and the in-plane coefficient of permeability of the geotextile.

2.45   State the Fick's first law of diffusion. What is the use of this law for geomembranes?

2.46   Do you think that creep is an important factor in the design and performance of some geosynthetic-reinforced structures? If yes, provide the list of such geosynthetic-reinforced structures along with a proper justification in support of your answer.

2.47   What is the role of geosynthetic creep in the selection of allowable load to be used in designs? Explain.

2.48   What do mean by geosynthetic stress relaxation? Is it related to geosynthetic creep? If yes, then how?

2.49   Describe the field jar test to empirically assess the clogging potential of a geotextile filter.

2.50   What are 'HCR' and 'GR'? Explain their significance in applications of geosynthetics.

2.51   What are the various degrees of biological clogging? Explain the importance of their study.

2.52   In the absence of ultraviolet-light degradation, what causes a polymer structure to age?

2.53   What are the major causes of degradation of the geosynthetics?

2.54   What factors would affect the durability of geosynthetics embedded in soils?

2.55   What is the objective of the durability assessment of geosynthetics? How can you assess durability for a specific application of the geosynthetic?

2.56   Differentiate between macrobiologial degradation and microbiological degradation?

2.57   Using the reduction factors, how can you estimate the allowable functional property of geosynthetic from the typical laboratory test values for a specific application?

2.58   Based on the market survey in your locality, make an attempt to compare strength, modulus, durability and costs of geotextiles with those of geogrids of similar mass per unit area.

2.59   If the ultimate tensile strength of a woven geotextile from an index-type test is 50 kN/m, then determine the allowable tensile strength to be used in the design of a geotextile-reinforced retaining wall.

2.60   If the ultimate permittivity of a nonwoven geotextile from an index-type test is 1.6 s$^{-1}$, then determine the allowable permittivity value to be used in the design of a paved road.

2.61 Name the properties of geosynthetics related to the following basic functions:
(a) reinforcement
(b) filtration
(c) fluid Barrier

2.62 What is the reason due to which the test property values of a geosynthetic are not directly considered as the design property values?

2.63 In your opinion, is there a necessity for the certification of laboratories that do testing of geosynthetics? If yes, why?

2.64 What is the main aim of using geosynthetics in civil engineering projects?

2.65 What is the fastest growing application area in the geosynthetic engineering?

2.66 What is the expected lifespan of geosynthetics?

2.67 What are the different design approaches for the geosynthetic-related structures or systems? Explain briefly.

2.68 Describe the general approach of design-by-function method for geosynthetics. What are the limitations of this method?

2.69 What should be the role of a designer for geosynthetic applications?

2.70 What is the fundamental principle of limit state design for geosynthetics?

2.71 What are the advantages of limit state design approach over the traditional working stress design approach for geosynthetics?

2.72 If several geosynthetics are found to meet the required factor of safety for an application, then how will you select one of them for use in that particular application?

## References

AASHTO (American Association of State Highway and Transportation Officials) (2013). *Geotextile Specification for Highway Applications*, Designation: M 288-06 (2011), *Standard Specifications for Transportation Materials and Methods of Sampling and Testing*, Thirty-third Edition, Part 1B: Specifications M 280-R 63, American Association of State Highway and Transportation Officials, Washington, DC, USA.

Andrawes, K.Z., McGown, A. and Murray, R.T. (1986). The load-strain-time-temperature behaviour of geotextiles and geogrids. *Proceedings of the 3rd International Conference on Geotextiles*. Vienna, Austria, pp. 707–712.

ASTM (American Society for Testing and Materials) (2010). ASTM D5617-04, *Standard Test Method for Multi-Axial Tension Test for Geosynthetics*. ASTM International, West Conshohocken, PA, USA.

ASTM (2011a). ASTM D6364-06, *Standard Test Method for Determining Short-Term Compression Behaviour of Geosynthetics*. ASTM International, West Conshohocken, PA, USA.

ASTM (2011b). ASTM D4595-11, *Standard Test Method for Tensile Properties of Geotextiles by the Wide-Width Strip Tensile Test*. ASTM International, West Conshohocken, PA, USA.

ASTM (2011c). ASTM D5886-95, *Standard Guide for Selection of Test Methods to Determine Rate of Fluid Permeation through Geomembranes for Specific Applications*. ASTM International, West Conshohocken, PA, USA.

ASTM (2011d). ASTM D5567-94, *Standard Test Method for Hydraulic Conductivity Ratio (HCR) Testing of Soil/Geotextile Systems*. ASTM International, West Conshohocken, PA, USA.

ASTM (2012a). ASTM D5101-12, *Standard Test Method for Measuring the Filtration Compatibility of Soil-Geotextile Systems*. ASTM International, West Conshohocken, PA, USA.

ASTM (2012b). ASTM D1987-07, *Standard Test Method for Biological Clogging of Geotextile or Soil/Geotextile Filters*. ASTM International, West Conshohocken, PA, USA.

ASTM (2013). ASTM D4632/4632M-08, *Standard Test Method for Grab Breaking Load and Elongation of Geotextiles*. ASTM International, West Conshohocken, PA, USA.

Bhatia, S.K., Smith, J.L. and Christopher, B.R. (1994). Interrelationship between pore openings of geotextiles and methods of evaluation. *Proceedings of the 5th International Conference on Geotextiles, Geomembranes and Related Products*. Singapore, pp. 705–710.

Bonaparte, R. and Berg, R. (1987). Long term allowable tension for geosynthetic reinforcement. *Proceedings of the Geosynthetics '87*. New Orleans, Louisiana.

BSI (British Standards Institution) (2004a). BS EN ISO 13738, *Geotextiles and Geotextile-Related Products: Determination of Pullout Resistance in Soil*. British Standards Institution, London, UK.

BSI (2004b). BS EN ISO 13438, *Geotextiles and Geotextile-Related Products: Screening Test Method for Determining the Resistance to Oxidation*. British Standards Institution, London, UK.

BSI (2008). DD ISO/TS 13434, *Geosynthetics: Guidelines for the Assessment of Durability*. British Standards Institution, London, UK.

BSI (2010). BS 8006-1, *Code of Practice for Strengthened/Reinforced Soils and Fills*. British Standards Institution, London, UK.

Cassidy, P.E., Mores, M., Kerwick, D.J., Koeck, D.J., Verschoor, K.L. and White, D.F. (1992). Chemical resistance of geosynthetic materials. *Geotextiles and Geomembranes*, 11, 1, pp. 61–98.

Chang, D.T.T., Chen, C.A. and Fu, Y.C. (1996). The creep behaviour of geotextiles under confined and unconfined conditions. *Proceedings of the International Symposium on Earth Reinforcement*. Fukuoka, Japan, pp. 19–24.

den Hoedt, G. (1986). Creep and relaxation of geotextiles fabrics. *Geotextiles and geomembranes*, 4, 2, pp. 83–92.

El-Fermaoui, A. and Nowatzki, E. (1982). Effect of confining pressure on performance of geotextiles in soils. *Proceedings of the 2nd International Conference on Geotextiles*. Las Veggas, USA, pp. 799–804.

Gray, C.G. (1982). Abrasion resistance of geotextiles fabrics. *Proceedings of the 2nd International Conference on Geotextiles*. Las Vegas, USA, pp. 817–821.

Giroud, J.P. (1980). Introduction to geotextiles and their applications. *Proceedings of the 1st Canadian Symposium on Geotextiles*, pp. 3–31.

Giroud, J.P. (1984). Geotextiles and geomembranes definitions, properties and design, IFAI, St Paul, Minnesota.

Giroud, J.P. (2004). Poisson's ratio of unreinforced geomembranes and nonwoven geotextiles subjected to large strains. *Geotextiles and Geomembranes*, 22, 4, pp. 297–305.

Greenwood, J.H. and Myles, B. (1986). Creep and stress relaxation of geotextiles. *Proceedings of the 3rd International Conference on Geotextiles*. Vienna, Austria, pp. 821–826.

Greenwood, J.H., Trubiroha, P., Schroder, H.F., Frank, P. and Hufenus, R. (1996). Durability standards for geosynthetics: the tests for weathering and biological resistance. *Proceedings of the 1st European Geosynthetics Conference*. Eurogeo 1, Maastricht, Netherlands, pp. 637–641.

Haliburton, T.A. and Wood, P.D. (1982). Evaluation of the U.S. Army Corps of Engineer gradient ratio test for geotextile performance. *Proceedings of the 2nd International Conference on Geotextiles*. Las Vegas, USA, pp. 97–101.

Ingold, T.S. and Miller, K.S. (1988). *Geotextiles Handbook*. Thomas Telford Ltd., London, U.K.

Jewell, R.A. (1986). Material properties for the design of geotextile reinforced slopes. *Geotextiles and Geomembranes*, **2**, 2, pp. 83–109.

John, N.W.M. (1987). *Geotextiles*. Blackie, London.

Kabir, M.H. and Ahmed, K. (1994). Dynamic creep behaviour of geosynthetics. *Proceedings of the 5th International Conference on Geotextiles, Geomembranes and Related Products*. Singapore, pp. 1139–1144.

Kutay, M.E., Guler, M. and Aydilek, A.H. (2006). Analysis of factors affecting strain distribution in geosynthetics. *Journal of Geotechnical and Geoenvironmental Engineering, ASCE*, **132**, 1, pp. 1–11.

Lawson, C.R. and Kempton, G.T. (1995). Geosynthetics and their use in reinforced soil. Terram Ltd., U.K.

Lopes, M.L. and Ladeira, M. (1996). Role of the specimen geometry, soil height, and sleeve length on the pullout behaviour of geogrids. *Geosynthetics International*, **3**, 6, pp. 701–719.

McGown, A., Andrawes, K.Z. and Kabir, M.H. (1982). Load-extension testing of geotextiles confined in soil. *Proceedings of the 2nd International Conference on Geotextiles*. Las Vegas, USA, pp. 793–796.

Mikki, H., Hayashi, Y., Yamada, K., Takasago and Shido, H. (1990). Plane strain tensile strength and creep of spun-bonded nonwovens. *Proceedings of the 4th International Conference on Geotextiles, Geomembranes and Related Products*. The Hague, pp. 667–672.

Myles, B. and Carswell, I.G. (1986). Tensile testing of geotextiles. *Proceedings of the 3rd International Conference on Geotextiles*. Vienna, Austria, pp. 713–718.

Narejo, D., Hardin, K. and Ramsey, B. (2001). Geotextile specifications: those vexing qualifiers. *Geotechnical Fabrics Report*, pp. 24–27.

Perkins, S.W. (2000). Constitutive modeling of geosynthetics. *Geotextiles and Geomembranes*, **18**, 5, pp. 273–292.

Richardson, G.N. and Christopher, B. (1997). Geotextiles in drainage systems. *Geotechnical Fabrics Report*, April, pp. 17–28.

Shamsher, F.H. (1992). *Ground Improvement with Oriented Geotextiles and Randomly Distributed Geogrid Micro-Mesh*. Ph.D. Thesis submitted to Indian Institute of Technology Delhi, New Delhi.

Shukla, S.K. and Yin, J.H. (2006). *Fundamentals of Geosynthetic Engineering*. Taylor and Francis, London.

Shukla, S. K., Sivakugan, N. and Mahto, S. (2009). A simple method for estimating Poisson's ratio of geosynthetics at zero strain. *Geotechnical Testing Journal, ASTM*, **32**, 2, pp. 181–185.

Standards Australia (2002). HB 154, *Technical Handbook: Geosynthetics – Guidelines on Durability*. Standards Australia International Limited, Sydney, NSW, Australia.

Standards Australia (2012a). AS 3706.3, *Determination of Tearing Strength of Geotextiles – Trapezoidal Method*. Standards Australia International Limited, Sydney, NSW, Australia.

Standards Australia (2012b). AS 3706.4, *Determination of Burst Strength of Geotextiles – California Bearing Ratio (CBR): plunger method*. Standards Australia International Limited, Sydney, NSW, Australia.

Standards Australia (2012c). AS 3706.9, *Determination of Permittivity, Permeability and Flow Rate of Geotextiles*. Standards Australia International Limited, Strathfield, NSW, Australia.

Standards Australia (2014). AS 3706.5, *Determination of Puncture Resistance of Geotextiles – Drop Cone Method*. Standards Australia International Limited, Strathfield, NSW, Australia.

Van Dine, D., Raymond, G. and Williams, S.E. (1982). An evaluation of abrasion tests for geotextiles. *Proceedings of the 2nd International Conference on Geotextiles*. Las Vegas, USA, pp. 811–816.

Van Santvoort, G. (ed.) (1995). *Geosynthetics in Civil Engineering*. A.A. Balkema, Rotterdam, The Netherlands.

Wesseloo, J., Visser, A.T. and Rust, E. (2004). A mathematical model for the strain rate dependent stress-strain response of HDPE geomembranes. *Geotextiles and Geomembranes*, 22, 4, pp. 273–295.

Wewerka, M. (1982). Practical experience in the use of geotextiles. *Proceedings of the Symposium on Recent Developments in Ground Engineering Techniques*. Bangkok, pp. 167–175.

## Answers to selected questions

2.1   (b)
2.3   (c)
2.5   (b)
2.7   (a)
2.9   (b)
2.11  (d)
2.13  (c)
2.15  (a)
2.19  1920 g/m$^2$
2.32  90.8%
2.41  $1.44 \times 10^{-3}$ m$^2$/s
2.44  $2.57 \times 10^{-5}$ m$^2$/s, 0.01 m/s
2.59  15.15 kN/m (Note: Assumed reduction factors are $f_{ID} = 1.1$, $f_{CR} = 2.0$, $f_{CD} = 1.5$ and $f_{BD} = 1.0$)

# Chapter 3

# Geotechnical applications of geosynthetics

## 3.1 INTRODUCTION

In Chapter 2, you learnt about the properties of geosynthetics, areas of geosynthetic applications and basic design concepts of geosynthetic-related structures/systems. This chapter presents the details of major geotechnical applications of geosynthetics, especially use of geosynthetics in retaining walls, embankments, shallow foundations and slopes. In view of the objectives of using geosynthetics in these geotechnical structures, the description presented here mainly focuses on their strengthening and stabilization with an emphasis on the basic details, the analysis and design concepts, the application guidelines, and some case studies.

## 3.2 RETAINING WALLS

### 3.2.1 Basic description

Retaining walls are required where construction of slopes is uneconomical or not technically feasible. A retaining wall prevents the backfill material from assuming its natural slope. Geosynthetic-reinforced soil retaining walls consist of geosynthetic layers as the reinforcing elements within the backfill to help resist the lateral earth pressures. A geosynthetic-reinforced soil retaining wall has thus three basic components (Fig. 3.1):

1  soil backfill, which is usually specified to be a clean, free-draining, non-plastic/granular soil;
2  reinforcement layers, which are generally woven geotextile or geogrid layers;
3  facing, which is not necessary, but often used to maintain the appearance and to avoid soil erosion between the reinforcement layers.

If the porous geotextile layers are used as the reinforcement layers, the cohesive soils can also be used as the backfill material. However, arrangements must be made for the vertical drainage using the granular material or the geotextile. The fines (particles smaller than 0.075 mm sieve size) in the granular backfill soil should generally have a plasticity index of less than 6 and angle of internal friction greater than 30°, and their percentage should not exceed 15%. Particles in the granular backfill material should generally be smaller than 19 mm sieve size. If the particles larger than

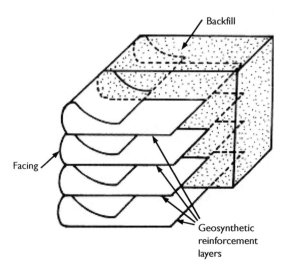

*Figure 3.1* Schematic diagram of a geosynthetic-reinforced soil retaining wall.

19 mm sieve size are present in the backfill, then the geosynthetic strength reduction due to the installation damage must be considered properly in the design.

A geosynthetic is mainly used to function as a reinforcement. It resists the lateral earth pressure and thus maintains the stability of the backfill. Its presence also causes a reduction in the load-carrying requirements of the wall-facing elements, resulting in material and time savings. Filtration and drainage are secondary functions to be served by the geosynthetic in retaining walls.

Woven geotextiles and geogrids with a high modulus of elasticity are generally used as the soil reinforcing elements in the geosynthetic-reinforced retaining walls. The facing can dictate the type of geosynthetic reinforcement. Because of the permanent reinforcement function, high demands are made upon the durability of the geosynthetic reinforcement. The force is mainly transmitted to the geotextile layers through the friction between their surfaces and the backfill soil, and to the geogrid layers through the passive soil resistance on the grid transverse members as well as through the friction between the soil and their horizontal surfaces. Note that the long-term load transfer is greatly governed by the durability and the creep characteristics of geosynthetics, and hence they must be assessed suitably as a part of the design.

Note that the galvanized steel or other metal strips (both deformed/ribbed and flat)/meshes/grids are also used in the construction of soil retaining walls as the inextensible reinforcement in place of geosynthetic layers, which generally behave as the extensible reinforcement. High modulus stiff reinforcement allows very little strain of the soil to occur, which results in the internal failure wedge becoming modified from the classical Coulomb/Rankine soil failure wedge. In many parts of the world such as Canada and the United States, the reinforced soil walls are frequently referred to as *mechanically stabilized earth (MSE) walls/structures*. They are commonly used in Canada as flexible retaining structures, ranging in height from approximately one metre to tens of metres (CGS, 2006).

The performance of a geosynthetic-reinforced soil wall is highly dependent on the type of facing elements used and the care with which it is designed and constructed. Facing elements can be installed as the wall is being constructed or after the wall is built. Geosynthetic wraps, precast reinforced concrete segmental panels, segmental/modular concrete blocks (MCBs), full-height precast concrete panels, wire mesh panels, gabion baskets, and treated timber panels are some different facing types. Geosynthetic layers are attached directly to these facing elements. Figure 3.2 shows the schematic diagrams of geosynthetic-reinforced retaining walls with different facing elements, which are commonly used in practice.

The wraparound wall face tends to exhibit a relatively large deformation at the wall face and a significant settlement at the crest adjacent to the wall face. It is also not aesthetically appealing, since it gives an impression of a relatively low-quality structure. However, it is the most economical facing and it was used on many early retaining walls. The wraparound facings are usually sprayed with bitumen emulsion, concrete mortar, or gunite (material similar to mortar)/shotcrete in lifts to produce a thickness in the order of 150–200 mm (Fig. 3.3). A wire mesh anchored to the geotextile wraparound facing may be necessary to keep the coating on the face of the wall. This coating provides protection against ultraviolet (UV) light exposure, potential vandalism and possible fire. If the facing elements are required to be installed at the end of wall construction, then shotcrete, cast-in-place concrete panels, precast concrete panels and timber panels can be attached to the steel bars placed or driven between the layers of geosynthetic-wrapped wall face. The selection of facing type depends on several factors, including the intended use of the wall, permissible (total and differential) settlements, required durability, drainage, connection type, loads

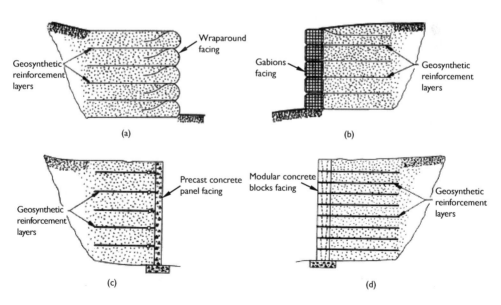

*Figure 3.2* Side views of geosynthetic-reinforced retaining walls: (a) with wraparound geosynthetic facing; (b) with gabions facing; (c) with full-height precast concrete panel facing; (d) with segmental/modular concrete blocks (MCBs) facing.

generated during handling, temperature and shrinkage effects, seismic movements, life-cycle cost, future maintenance, level of security and aesthetics (CGS, 2006).

Geogrids along with filter layer (nonwoven geotextile or conventional granular blanket) can also be used for the wraparound facings (Fig. 3.4). With proper ultraviolet light stabilizer, the geogrids can be left uncovered for a number of years, even for a design life of 50 years or more, provided they are heavy and stiff (Wrigley, 1987).

MCBs may have some kind of keys or inserts, which provide a mechanical interlock with the layer above. They provide flexibility with respect to the layout of curves and corners. They can tolerate larger differential settlements than the conventional structures. MCBs are manufactured from cement concrete and produced in different sizes, textures and colours; therefore they provide a varied choice to the engineer

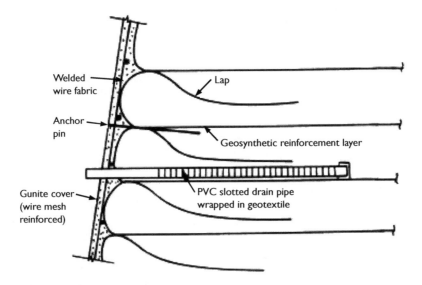

*Figure 3.3* Protection of the geotextile wraparound facing.

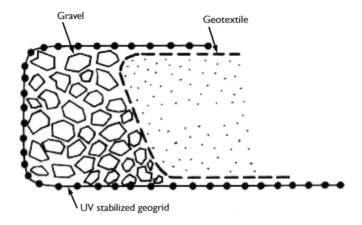

*Figure 3.4* Geogrid wraparound facing.

(Fig. 3.5). Typically all the blocks shown in Fig. 3.5 are 250–450 mm in length, 250–500 mm in width, and 150–200 mm in height. The mass of each block varies typically from 25 to 48 kg.

In a permanent geosynthetic-reinforced retaining wall (or steep-sided embankment), the geosynthetic load remains constant throughout the life of the structure, and therefore it is an example of a time-independent reinforcement application (Fig. 3.6).

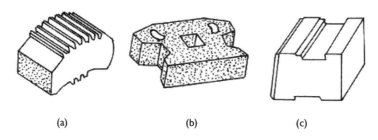

(a)                          (b)                          (c)

*Figure 3.5* Examples of MCB units used in the UK: (a) porcupine; (b) keystone; (c) geoblock (after Dikran and Rimoldi, 1996).

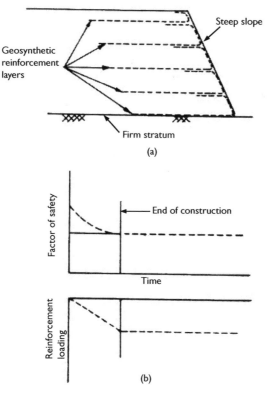

*Figure 3.6* An example of a time-independent reinforcement application: (a) geosynthetic-reinforced wall; (b) typical variation of slope factor of safety and reinforcement loading with time (after Paulson, 1987).

In this case, the creep strain may be very high and, therefore, the factor of safety against the creep should not be compromised.

Geosynthetic-reinforced retaining walls are generally an economical alternative to conventional gravity or cantilever retaining walls, especially for higher retaining walls in fill sections, as found in a large number of retaining wall projects completed successfully worldwide in the past. They can usually be sited on or near the ground surface, which avoids excavation and replacement, costly deep foundation construction and use of ground improvement techniques. In the case of a very weak foundation soil, a geosynthetic-reinforced base can be economically provided to the reinforced wall (Fig. 3.7). An even greater economy can be achieved through the use of low-quality backfills that may be available near the construction sites. Being relatively more flexible, the geosynthetic-reinforced retaining walls are very suitable for sites with poor foundation soils and for seismically active areas.

## 3.2.2   Analysis and design concepts

The approach for analysis and design of geosynthetic-reinforced soil retaining walls is quite well established. A number of approaches have been proposed; however, the most commonly used simple approach is based on concepts of classical earth pressure theory with determination of factor of safety against failure of the soil and the reinforcement at limit equilibrium. The analysis and design consist of the following three major parts:

a   Internal stability analysis (i.e. 'local stability analysis' or 'tieback analysis'): An assumed Rankine failure surface is used, with consideration of possible failure modes of geosynthetic-reinforced soil mass, such as geosynthetic rupture, geosynthetic pullout, connection (and/or facing elements) failure (Fig. 3.8) and exces-

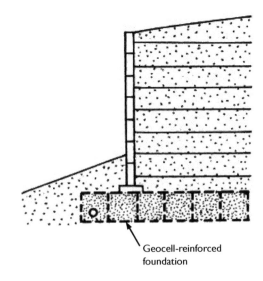

Geocell-reinforced foundation

*Figure 3.7* Reinforced soil wall with a reinforced base.

sive geosynthetic creep. The analysis is mainly aimed at determining tension and pullout resistance in the geosynthetic reinforcement, length of reinforcement, and integrity of the facing elements.

b   External stability analysis (i.e. 'global stability analysis'): The overall stability of the geosynthetic-reinforced soil mass is checked, including sliding, overturning, load-bearing capacity failure, and deep-seated slope failure (Fig. 3.9). For this analysis, the volume of retained soil that is reinforced by the horizontal layers of geosynthetic reinforcement and the facing column can be imagined to act as monolithic block of material. This homogenization of the reinforced zone for modular facing systems is assured by keeping the spacing between reinforcement layers to be not more than twice the width of the facing units (CGS, 2006).

c   Analysis for the facing system, including its attachment to the reinforcement: The facing elements are designed to resist the horizontal force in the soil reinforcements at the reinforcement to facing connection and the potential compaction stresses that may occur near the wall face during erection of the wall. The connection capacity must be at least 1.5 times the design tensile load. Geosynthetic-reinforced soil walls are often constructed with a small facing batter for aesthetic reasons and to accom-

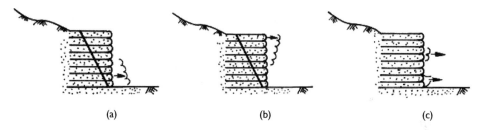

(a)                               (b)                               (c)

*Figure 3.8* Internal failure modes of geosynthetic-reinforced soil retaining walls: (a) geosynthetic rupture; (b) geosynthetic pullout; (c) connection (and/or facing elements) failure.

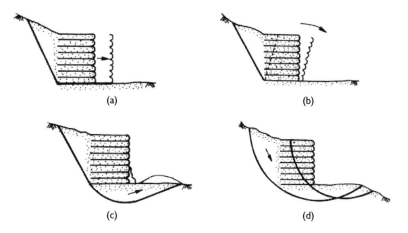

(a)                               (b)

(c)                               (d)

*Figure 3.9* External failure modes of geosynthetic-reinforced soil retaining walls: (a) sliding; (b) overturning; (c) load-bearing capacity failure; (d) deep-seated slope failure.

modate the outward wall deformations, which develop due largely to the extensible properties of the geosynthetic reinforcement, soil creep and facing construction.

Figure 3.10(a) shows a geotextile-reinforced retaining wall of height $H$ with a geotextile wraparound facing without any surcharge and live load. The backfill is a homogeneous granular soil. According to Rankine active earth pressure theory, the active earth pressure $\sigma_a$ at any depth $z$ is given by

$$\sigma'_a = K_a \sigma'_v = K_a \gamma_b z \tag{3.1}$$

where $K_a$ is the Rankine active earth pressure coefficient, and $\gamma_b$ is the total unit weight of the granular backfill. Note that, with a proper drainage arrangement, the backfill remains drained, and hence $\sigma'_v = \gamma_b z$. The value of $K_a$ can be estimated from

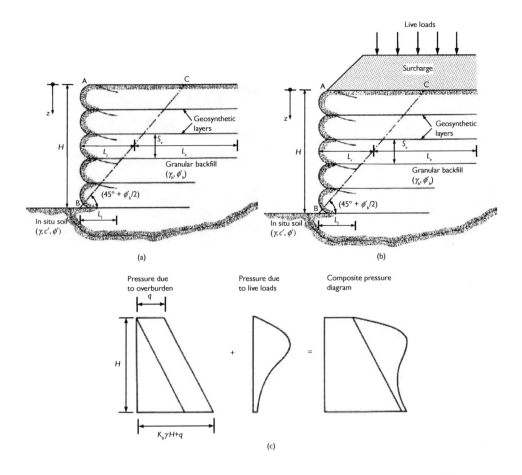

*Figure 3.10* (a) Geosynthetic-reinforced retaining wall without surcharge and live load; (b) geosynthetic-reinforced retaining wall with surcharge and live load; (c) lateral earth pressure distribution.

$$K_a = \tan^2\left(45° - \frac{\phi_b'}{2}\right) \tag{3.2}$$

where $\phi_b'$ is the effective angle of shearing resistance of the granular backfill.

The factor of safety $F_R$ against the tensile failure of the reinforcement (i.e. the reinforcement rupture) at any depth $z$ may be expressed as

$$F_R = \frac{\sigma_{all}}{\sigma_a' S_V} \tag{3.3}$$

where $\sigma_{all}$ is the allowable tensile strength of the reinforcement (kN/m), and $S_v$ is the vertical spacing of the reinforcement layers at any depth $z$ (m). Since for the retaining walls the geosynthetic reinforcement needs to provide stability throughout the whole life of the structure, the long-term sustained load test data, that is, the creep test data, should be used for design purposes.

The magnitude of the $F_R$ is generally taken as 1.3–1.5. Equation (3.3) with Equation (3.1) can be rearranged as

$$S_V = \frac{\sigma_{all}}{\sigma_a' F_R} = \frac{\sigma_{all}}{(K_a \gamma_b z) F_R} \tag{3.4}$$

The geotextile layer at any depth $z$ will fail by the pullout (i.e. the bond failure) if the frictional resistance developed along its surfaces is less than the force to which it is being subjected. This type of failure occurs when the length of geotextile reinforcement is not sufficient to prevent its slippage with respect to the soil. The effective length $L_e$ of a geotextile layer along which the frictional resistance is developed, may be conservatively taken as the length that extends beyond the limits of the Rankine active failure zone (ABC in Fig. 3.10(a))

The factor of safety $F_p$ against the reinforcement pullout at any depth $z$ may be expressed as

$$F_p = \frac{2L_e \sigma_v' \tan\phi_i'}{\sigma_a' S_V} \tag{3.5}$$

where $\phi_i'$ is the effective angle of shearing resistance of soil-reinforcement interface, and is approximately taken as $2\phi_b'/3$.

The magnitude of the $F_p$ is generally taken as 1.3–1.5. Using Equation (3.1), Equation (3.5) can be rearranged as

$$L_e = \frac{S_V K_a F_p}{2\tan\phi_i'} \tag{3.6}$$

The length $L_r$ of the reinforcement layer within the Rankine failure zone ABC can be calculated as

$$L_r = \frac{H-z}{\tan(45° + \phi_b'/2)} \tag{3.7}$$

From Equations (3.6) and (3.7), the total length of reinforcement layer at any depth $z$ is

$$L = L_e + L_r = \frac{S_V K_a F_p}{2\tan\phi_i'} + \frac{H-z}{\tan(45° + \phi_b'/2)} \tag{3.8}$$

Note that the mixed types of failure, that is, combinations of reinforcement rupture and pullout failure can also occur depending on the geometry of the structure, external loads, etc. Usually, in the lower parts of the retaining wall, the reinforcement is destroyed through the rupture due to a lack of strength, and in the upper parts, the reinforcement is destroyed through the pullout due to an insufficient resisting length.

When designing the facing system, it can be assumed that the stress at the face is equal to the maximum horizontal stress in the reinforced backfill. This assumption makes the design conservative because some stress reduction generally occurs near the face. In fact, the maximum stresses are usually located near the potential failure surface and then they decrease in both directions: towards the free end of the reinforcement and towards the facing. Values of the stress near the facing depend on its flexibility. In the case of a rigid facing, the stresses near the facing and those at the potential failure surface do not differ significantly. In the case of a flexible facing, the stress near the facing is lower than that at the potential failure surface (Sawicki, 2000). If a wraparound facing is to be provided, the lap length $L_l$ can be determined using the following expression:

$$L_l = \frac{S_V K_a F_p}{4\tan\phi_i'} \tag{3.9}$$

The design procedure for the geosynthetic-reinforced retaining wall with a wraparound vertical face and without any surcharge is given in the following steps:

*Step 1:* Establish the wall height ($H$).
*Step 2:* Determine the properties of granular backfill soil, such as total unit weight ($\gamma_b$) and angle of shearing resistance ($\phi_b'$).
*Step 3:* Determine the properties of foundation soil, such as total unit weight ($\gamma$) and shear strength parameters ($c$ and $\phi$, or $c'$ and $\phi'$ as applicable).
*Step 4:* Determine the angle of shearing resistance of the soil-geosynthetic interface ($\phi_i'$).
*Step 5:* Estimate the Rankine active earth pressure coefficient from Equation (3.2).
*Step 6:* Select a reinforcement that has the allowable tensile strength equal to $\sigma_{all}$.
*Step 7:* Determine the vertical spacing of the reinforcement layers at various levels from Equation (3.4).
*Step 8:* Determine the length $L$ of reinforcement layer at various levels from Equation (3.8).
*Step 9:* Determine the lap length $L_l$ at any depth $z$ from Equation (3.9).

*Step 10:* Check the factors of safety against external stability, including sliding, overturning, load-bearing capacity failure, and deep-seated slope failure as carried out for the conventional retaining wall design, assuming that the geosynthetic-reinforced soil mass acts as a rigid body in spite of the fact that it is really quite flexible. The minimum values of factors of safety against sliding, overturning, load-bearing failure, and deep-seated failure are generally taken as 1.5, 2.0, 2.0 and 1.5, respectively.

*Step 11:* Check the requirements for backfill drainage and surface runoff control.

*Step 12:* Check both total and differential settlements of the retaining wall along the wall length. This can be carried out as per the conventional methods of settlement analysis. The post-construction settlement of the foundation soil should not exceed 100 mm for discrete panels/blocks, which result in flexible structures. Typical differential settlements of 1 in 100 are considered as safe for discrete concrete panel facings (1 in 500 for full height panels) (IRC, 2014).

For the critical structures, especially the permanent ones, efforts must be made to consider the dead and the live load surcharges (Figs. 3.10(b) and (c)) as well as the seismic loading in the design analysis as per the site location and field conditions. The readers can refer to the chapter contributed by Bathurst *et al.* (2012) in the book by Shukla (2012a) for more details on the seismic aspects of geosynthetic-reinforced soil walls and slopes. The pseudo-static seismic stability analysis of reinforced soil wall using the horizontal slice method has shown that the stability of a reinforced soil wall is largely affected by the horizontal seismic forces. The tensile resistance which has to be mobilized by the reinforcement to maintain the stability of wall, increases with an increase in surcharge and horizontal seismic forces, whereas the same decreases with an increase in the cohesion of soil (Chandaluri *et al.*, 2015).

Design factors of safety should be decided on the basis of local codes, if available, and on the basis of experience gained from safe and economical designs. Due to flexibility of the geosynthetic-reinforced retaining walls, the design factors of safety are generally kept lower than normally used for the rigid retaining structures. Unless the foundation is very strong, a minimum embedment depth must be provided as is done with most foundations.

The level of compaction of backfill influences the facing design. The connection stresses are caused by the settlement of the wall resulting from poor compaction of the backfill near the wall face. They can also be created by heavy compaction near the wall face. Therefore, an optimum level of compaction of granular fill is recommended near the wall face. The facing connection must be designed to resist the lateral pressures, the gravity forces and the seismic forces along with connection stresses, if there is any possibility of their occurrence.

When the geosynthetic-reinforced retaining walls are chosen as a design alternative, readily available on-site soils are often figured into the design from the start. The suitability of soils as a backfill material can be decided on the basis of three key parameters, namely effective angle of shearing resistance, shear strength when compacted and saturated, and frost-heave potential. Table 3.1 provides some guidelines on the suitability of backfill materials using these parameters. It should be noted that as the quality of fill decreases, lower angles of shearing resistance are present, resulting in higher lateral earth pressures and flatter failure surfaces. Consequently, the

*Table 3.1* Retaining wall backfill (after NCMA, 1997).

| Unified soil classification | Effective angle of shearing resistance (degrees) | Shear strength when compacted and saturated | Frost-heave potential | Comments |
|---|---|---|---|---|
| GW, GP | 37–42 | Excellent to good | Low | Recommended for backfill |
| GM, SW, SP | 33–40 | Excellent to good | Moderate | Recommended for backfill |
| GC, SM, SC, ML, CL | 25–32 | Good to fair | Moderate to high | Recommended for backfill with additional criteria |
| MH, CH, OH, OL | N/A | Poor | High | Generally not recommended for backfill |
| PT | N/A | Poor | High | Not recommended for backfill |

reinforcement strength and the length increase. Fine-grained soils are recommended as a backfill material only when the following four additional design criteria are implemented (Wayne and Han, 1998):

1   Internal drainage must be designed and installed properly.
2   Only soils with low to moderate frost-heave potential should be considered.
3   The internal cohesive shear strength parameter $c$ is conservatively ignored for long-term stability analysis.
4   The final design is checked by a qualified geotechnical engineer to ensure that the use of cohesive soils does not result in unacceptable, time-dependent movement of the retaining wall.

## EXAMPLE 3.1

Consider a geotextile-reinforced soil retaining wall with the following details:
   Height of the retaining wall, $H = 8$ m
   For the granular backfill
      Total unit weight, $\gamma_b = 17$ kN/m$^3$
      Angle of internal friction, $\phi'_b = 35°$
   Allowable tensile strength of geotextile reinforcement, $\sigma_{all} = 20$ kN/m
   Factor of safety against geotextile rupture = 1.5
   Factor of safety against geotextile pullout = 1.5

Calculate the spacing and length of geotextile layers, and the lap length at depths $z = 2$ m, 4 m and 8 m from the top of the wall.

## SOLUTION

From Equation (3.2), the coefficient of active earth pressure is

$$K_a = \tan^2\left(45° - \frac{\phi'_b}{2}\right) = \tan^2\left(45° - \frac{35°}{2}\right) = 0.271$$

At z = 2 m:
From Equation (3.4), the spacing of geotextile layers is

$$S_V = \frac{\sigma_{all}}{(K_a \gamma z)F_R} = \frac{20}{(0.271)(17)(2)(1.5)} = 1.447 \text{ m}$$

From Equation (3.8), the length of the geotextile layers is

$$L = L_e + L_r = \frac{S_V K_a F_P}{2\tan\phi'_i} + \frac{H-z}{\tan(45° + \phi'_b/2)}$$

$$= \frac{(1.447)(0.271)(1.5)}{2\tan[(2/3)\times35°]} + \frac{8-2}{\tan[45° + (35°/2)]} = 0.682 \text{ m} + 3.123 \text{ m} = 3.805 \text{ m}$$

From Equation (3.9), the lap length is

$$L_l = \frac{S_V K_a F_P}{4\tan\phi'_i} = \frac{(1.447)(0.271)(1.5)}{(4)\tan[(2/3)\times35°]} = 0.341 \text{ m}$$

At z = 4 m:
From Equation (3.4), the spacing of geotextile layers is

$$S_V = \frac{\sigma_{all}}{(K_a \gamma z)F_R} = \frac{20}{(0.271)(17)(4)(1.5)} = 0.724 \text{ m}$$

From Equation (3.8), the length of the geotextile layers is

$$L = L_e + L_r = \frac{S_V K_a F_P}{2\tan\phi'_i} + \frac{H-z}{\tan(45° + \phi'_b/2)}$$

$$= \frac{(0.724)(0.271)(1.5)}{2\tan[(2/3)\times35°]} + \frac{8-4}{\tan[45° + (35°/2)]} = 0.341 \text{ m} + 2.082 \text{ m} = 2.423 \text{ m}$$

From Equation (3.9), the lap length is

$$L_l = \frac{S_V K_a F_P}{4\tan\phi'_i} = \frac{(0.724)(0.271)(1.5)}{(4)\tan[(2/3)\times35°]} = 0.170 \text{ m}$$

At z = 8 m:
From Equation (3.4), the spacing of geotextile layers is

$$S_V = \frac{\sigma_{all}}{(K_a \gamma z) F_R} = \frac{20}{(0.271)(17)(8)(1.5)} = 0.362 \text{ m}$$

From Equation (3.8), the length of the geotextile layers is

$$L = L_e + L_r = \frac{S_V K_a F_P}{2\tan\phi_i'} + \frac{H - z}{\tan(45° + \phi_b/2)}$$

$$= \frac{(0.362)(0.271)(1.5)}{2\tan[(2/3) \times 35°]} + \frac{8 - 8}{\tan[45° + (35°/2)]} = 0.170 \text{ m} + 0 \text{ m} = 0.170 \text{ m}$$

From Equation (3.9), the lap length is

$$L_l = \frac{S_V K_a F_P}{4\tan\phi_i'} = \frac{(0.362)(0.271)(1.5)}{(4)\tan[(2/3) \times 35°]} = 0.085 \text{ m}$$

Note: Keeping the field aspects and construction simplicity in view, one can use $S_v$ = 0.5 m, $L$ = 5 m, $L_l$ = 1 m for z ≤ 4 m, and $S_v$ = 0.3 m, $L$ = 2.5 m, $L_l$ = 1 m, for z > 4 m.

Note that in most reinforced soil walls constructed in the past, the minimum length of backfill soil reinforcement has been taken typically at 70% of the height of the wall as measured from the top of the leveling pad. The typical ratios of length of reinforcement zone to height of wall are 0.5 to 0.7 (CGS, 2006). The reinforcement length may be increased as required for surcharges and other external loads, or for soft foundation soils. In general, a minimum reinforcement length of 2.4 m, regardless of the wall height, may be recommended, primarily due to size limitations of conventional spreading and compaction equipment. Shorter reinforcement lengths, on the order of 1.8 m, but not less than 70% of the wall height, can be considered if smaller compaction equipment is used, facing panel alignment can be maintained, and minimum requirements for wall external stability are met. For walls on rock or very competent foundation soil (e.g. SPT value > 50), the bottom reinforcements may be shortened to a minimum of 0.4H with the upper reinforcements lengthened to compensate for external stability issues in lieu of removing rock or competent soil for construction. For conditions of marginal stability, consideration must be given to ground improvement techniques to improve the foundation stability, or to lengthening of reinforcement (AASHTO, 2007).

Note that the typical vertical reinforcement spacing for geotextile-wrapped walls varies between 0.2 to 0.5 m to protect against the bulging. For spacings greater 0.6 m, unless the wall has a rigid face, the intermediate/secondary geotextile layers may be required to prevent excessive bulging of the wall face between the primary geotextile layers. To provide a coherent reinforced soil mass, the vertical spacing of primary reinforcement should not exceed 0.8 m, in all types of reinforcement (IRC, 2014). According to CGS (2006) guidelines, regardless of the facing type, the reinforcement spacing should not exceed 1 m.

### 3.2.3  Application guidelines

In the actual construction, the geosynthetic-reinforced soil retaining walls have continued to demonstrate excellent performance characteristics and exhibit many advantages over conventional retaining walls. To achieve a better performance, the following points must be considered on the site-specific basis:

1  Any unsuitable foundation soils should be replaced with compacted granular backfill materials. The foundation treatment, if required, should be first completed to ensure that the foundation design parameters are attained.

2  The foundation soil should be excavated and compacted to the required embedment depth and width, to a dry unit weight of 95% of the modified Proctor maximum dry unit weight (IRC, 2014).

3  An initial levelling pad of about 150 mm-thick cement concrete having suitable width should be placed.

4  The minimum embedment depth of the bottom of the reinforced soil mass, measured from the top of the leveling pad should be decided based on the load-bearing capacity, settlement and stability requirements as generally considered for the embedment depths of most shallow foundations (Shukla, 2015).

5  The geosynthetic layer should be installed with its principal strength (warp strength) direction perpendicular to the wall face.

6  With the geosynthetic reinforcement layer, it is necessary to take care not to tear it in the direction parallel to the wall face because a partial tear of this type will reduce the amount of tensile force carried out by the geotextile reinforcement layer.

7  The overlap along the edge of the geosynthetic layer should generally exceed 200 mm. If there is a possibility of large foundation settlements, then sewn or other suitable joints (see Chapter 7) may be recommended between the adjacent geosynthetic layers.

8  There should not be any wrinkles or slack in the geosynthetic layer as they can result in the differential movement.

9  Granular soil backfill should generally be compacted to at least 95% of the modified Proctor maximum dry unit weight or 80% relative density. The compacted lift thickness should vary from 200 to 300 mm. Efforts should be made to compact uniformly to avoid any differential settlement.

10  The backfill should be compacted, taking care not to get the compactor very close to the facing element, so that it is not highly stressed, resulting in pullout or excessive lateral displacement of the wall face. It is therefore recommended to use lightweight hand-vibratory compactors within 1–1.5 m of the wall face.

11  The wraparound geosynthetic facing can be constructed using temporary formwork as shown in Fig. 3.11. The lap length should be generally not less than 1 m.

12  A construction system for the permanent geosynthetic-reinforced soil retaining wall (GRSRW), widely used in Japan, can be adopted. This system uses a full-height rigid (FHR) facing that is cast in place using staged construction procedures (Fig. 3.12). This system has several special features such as the use of relatively short reinforcement layers and use of the low-quality on-site soil as the backfill.

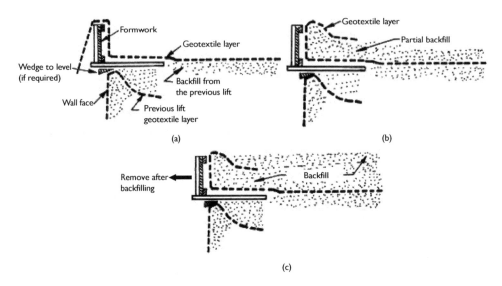

*Figure 3.11* Lift construction sequence for a geotextile-reinforced soil wall: (a) step 1; (b) step 2; (c) step 3 (after Steward *et al.*, 1977).

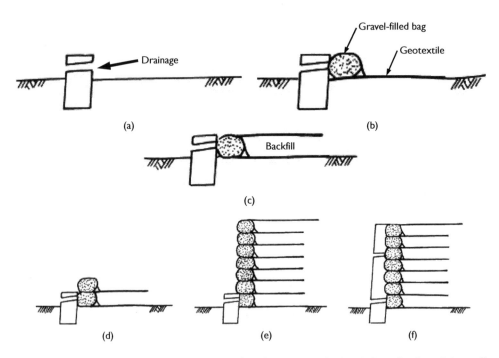

*Figure 3.12* Standard staged construction procedure for a geosynthetic-reinforced soil retaining wall (GRSRW): (a) concrete base; (b) geotextile and gravel-filled bag placement; (c) backfill and compaction; (d) placement of the second layer of geotextile and gravel-filled bag; (e) all layers constructed; (f) concrete facing constructed (after Tatsuoka *et al.*, 1997).

*Figure 3.13* Construction procedure for a geogrid-reinforced retaining wall: (a) earth work for the geogrid-reinforced retaining wall; (b) placement of the geogrid layers; (c) placement of the geotextile filter layer near the face of the retaining wall; (d) connection between the folded geogrid sheet and the next geogrid sheet; (e) a view of the completed retaining wall.

13  In the case of facing made with segmental or modular concrete blocks (MCBs), full-height precast concrete panels, welded wire panels, gabion baskets, or treated timber panels, the facing connections should be made prior to placing the backfill and must be carefully checked as per the design guidelines because the success of the geosynthetic-reinforced retaining wall is highly dependent on the facing connections.

14  The required thickness of drainage material should be placed at the back of facing block/panel and within the space in the facing blocks. The backfill material and the drainage material should be separated using the nonwoven geotextile layer as the filter.

15  It is necessary to have tight construction specifications and quality inspection to insure that the wall face is constructed properly; otherwise an unattractive wall face, or a wall face failure, can result.

16  Geogrid-reinforced retaining walls can be constructed with the geotextile filters near the face. Major constriction steps are shown in Fig. 3.13.

### 3.2.4  Case studies

Geosynthetic-reinforced soil walls are gaining considerable attention as retaining structures and are providing a valuable alternative to the traditional concrete walls. No footing of any special kind is required in the case of geosynthetic-reinforced retaining walls, and the lowest geosynthetic reinforcement layer is placed directly on the foundation soil. They demonstrate the possibility of using soils of poor mechanical characteristics with an ample safety margin. Compared to the concrete walls, they present a low cost-benefit ratio and a low environmental impact.

### Case study 1

Gourc and Risseeuw (1993) reported a case study on a geotextile-reinforced wraparound faced wall built in late 1982 in Prapoutel, France. Figure 3.14 shows the overall cross-section of the wall. Its length is 170 m and its height ranges between 2 and 9.6 m. The backfill soil consisted of the following properties: $\gamma_b = 18\,\text{kN}/\text{m}^3$; $\phi = 30°$ and $c' = 33\,\text{kPa}$. The selected geotextile was Stabilenka 200, a woven polyester (PET) product, having a mass per unit area of 450 g/m² and a tensile strength of 200 kN/m with a strain at failure of 8% in the roll direction. Although the soil-geotextile interface friction angle $\phi_i'$ measured in the laboratory was equivalent to the angle of internal friction $\phi'$ of soil backfill, it was considered preferable to adopt the value $\phi_i' = 0.8\phi'$. The design was based on an extremely steep embankment using the anchor-tied wall design method proposed by Broms (1980) and considering a factor of safety of 1.3 in calculating the anchor force. The design principle assumed a uniform active earth pressure over the entire height of the embankment. It leads generally to a factor of safety less than those obtained with other design methods, which indicates that the method used was more conservative than others. The design calculation gave a vertical geotextile spacing of 1.20 m for an embankment height of 9.60 m. It is important to note that this wall was preferred to a reinforced concrete retaining wall or reinforced earth wall essentially for cost reasons; the reinforced earth wall was estimated to be 40% more expensive.

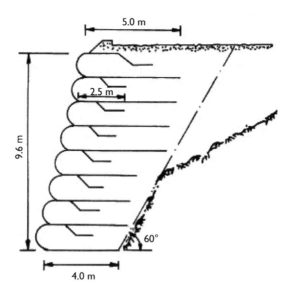

*Figure 3.14* A cross-section of the geotextile-reinforced wraparound faced wall, Prapoutel, France (after Gourc and Risseeuw, 1993).

The backfill material available at the site was graded in layers 1.20 m thick, with successive layers being separated by a geotextile sheet. The geotextile reinforcement lengths adopted as per the design were 5 m in the upper layers and 4 m in the lower layers. Specially made angle formworks were used for placing the successive 1.20 m thick layers. The work rate was approximately 50 linear metres each layer per day. It was reported that 10 years after completion, the structure gave no cause for criticism regarding its overall stability. The only reservation to be made concerned the facing of the wall. Because of the steep slope, the plant growth after hydraulic seeding proved to be difficult, and a bituminous protection with mulch was subsequently provided. However, it was reported to have some geotextile deterioration. At a few locations on the face, large stones punctured the geotextile, with no significant soil loss. A slight loss of soil from the facing occurred at certain joints between adjacent sheets. Joints were made simply by overlapping, probably over an insufficient width.

## Case study 2

A geosynthetic-reinforced soil retaining wall with segmental facing panels was constructed on the Mumbai-Pune Expressway (Panvel bypass – package I) by the Maharashtra State Road Development Corporation Limited, Mumbai, India. The height of the retaining wall varied from 2.5 to 13 m. The design was carried out using the tie-back wedge method considering both the internal and the external stability as per the site conditions. Tensar 40 RE, 80RE, 120 RE and 16 RE geogrids were used as reinforcements. Modular blocks (400 mm × 220 mm × 150 mm) and segmental panels (1400 mm × 650 mm × 180 mm) were used as the facing elements. The extensive use of the Tensar connectors gave the perfect connection between the wall-facing panels

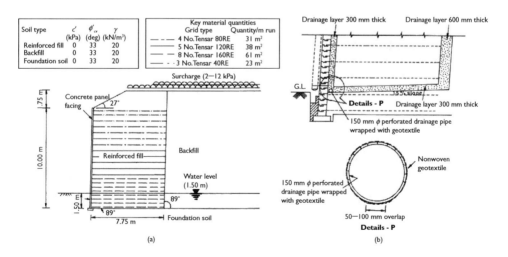

*Figure 3.15* Geosynthetic-reinforced retaining wall built on the Mumbai-Pune Expressway (Panvel Bypass – package I), India: (a) cross-section with the details of soil and reinforcement; (b) details of the drainage system (courtesy of Netlon India, 2001; Shukla and Yin, 2006).

*Figure 3.16* A portion of retaining wall on the Mumbai-Pune Expressway (Panvel Bypass-Package I), India during its construction stage (courtesy of Netlon India, 2001; Shukla and Yin, 2006).

and the Tensar geogrids. A nonwoven geotextile was used to wrap over the perforated pipe to allow free drainage. The construction sequences adopted were based on large model experiments, the experiences and the technical justifications. The construction work was completed in the year 2001. Figure 3.15 shows the details of the wall at one of its cross-sections, along with soil and reinforcement characteristics. A portion of the wall during construction stage is shown in Fig. 3.16. The retaining wall has been performing well without any noticeable problem since its completion.

## 3.3 EMBANKMENTS

### 3.3.1 Basic description

The construction of embankments over weak/soft foundation soils is a challenging task for geotechnical engineers. In the conventional method of construction, the soft soil is replaced by a suitable soil or it is improved (by preloading, dynamic consolidation, lime/cement mixing, or grouting) prior to the placement of the embankment. Other options such as staged construction with sand drains, use of stabilizing berms, and pile foundations are also available for application. These options can be either time consuming, expensive, or both. The alternate option is to place a geosynthetic (geotextile, geogrid or geocomposite) layer over the soft foundation soil and construct the embankment directly over it (Fig. 3.17(a)). More than one geosynthetic layer may be required, if the foundation soil has voids or weak zones caused by sinkholes, thawing ice, old streams, etc., or weak pockets of silt, clay or peat (Fig. 3.17(b)). In such situations, the geosynthetic layer is often called the *basal geosynthetic layer*. In some cases, the most effective and economic solution may be some combination of a conventional ground improvement and/or construction alternative together with a geosynthetic layer. For example, taking into account the strength gain that occurs with staged embankment construction, lower strength and therefore lower-cost geosynthetic can be utilized.

The geosynthetic as a basal layer in an embankment over the soft foundation soil can serve one of the following basic functions or a combination thereof:

1 reinforcement
2 drainage
3 separation/filtration.

The reinforcement function usually aims for a temporary increase in the safety factor of embankment, which is associated with a faster rate of construction or the use of steeper slopes that would not be possible in the absence of reinforcement. The drainage function is associated with the increase in the rate of consolidation to give a more stable embankment or staged construction. In fact, the geosynthetic allows for free drainage of the foundation soils to reduce pore water pressure buildup below the embankment (Fig. 3.18(a)). The consolidation of soft foundation soil can be further accelerated by installing vertical drains along with the basal drainage blanket (Fig. 3.18(b)). The separation function helps in preventing the mixing of the embankment material and the soft foundation soil; thus reducing the consumption of the embankment material.

The use of a geosynthetic basal layer is generally attractive for low ratios between the foundation soil thickness and the embankment base width (say, less than 0.7). For thick foundation soils, the contribution of the reinforcement can be less significant (Palmeira, 2012).

The following factors may be of major concern when choosing the basal geosynthetic to function as a reinforcement:

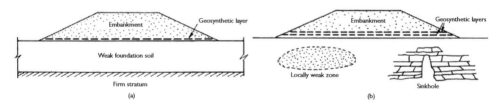

*Figure 3.17* Embankment over weak foundation soils: (a) embankment on uniform weak foundation soil; (b) embankment on locally weak foundation soil with lenses of clay or peat, or with sinkholes (after Bonaparte and Christopher, 1987).

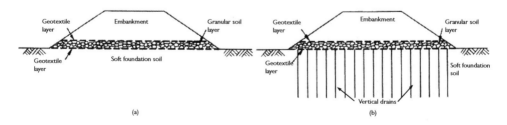

*Figure 3.18* Embankment over weak foundation soil: (a) with basal drainage layer; (b) with vertical drains and basal drainage layer.

- tensile strength and stiffness
- soil-reinforcement bond characteristics
- creep characteristics
- geosynthetic resistance to mechanical damage
- durability.

The geosynthetic used as the basal layer in the embankment should have high tensile strength, low elongation, low creep and high durability. Geogrids or high strength woven geotextiles made from polyester are the most suitable geosynthetics as the basal geosynthetic layer. Reinforcement-drainage-separation geocomposites can also be used for the basal layer where the drainage function is required. A nonwoven geotextile bonded to a geogrid provides in-plane drainage while the geogrid functions as the tensile reinforcement. Currently the geosynthetic reinforcements with tensile strength even higher than 1000 kN/m are available in the market. Use of geocells having adequate tensile strength can be a good option for the basal layer; but the benefits must be assessed properly.

In most cases, the geosynthetic reinforcement is required beneath an embankment only during the embankment construction and for a short period afterwards, because the consolidation of the soft foundation soil results in an increase in the load-bearing capacity of the foundation soil in due course of time. When a basal geosynthetic is used beneath a permanent embankment, the strain becomes fairly constant, once most of the settlement has taken place. In such a situation, there may be a loss of tensile stress experienced by the geosynthetic with time (Fig. 3.19). The phenomenon of the decrease in stress, at a constant strain, with time is called the *stress relaxation*, which is closely related to creep. Fortunately, during this period the underlying soil is consolidating and increasing in strength. The subsoil is therefore able to offer a greater resistance to failure as the time passes. The factor of safety should not be compromised if the rate at which the geosynthetic loses its stress is greater than the rate of strength gain occurring in the foundation soil.

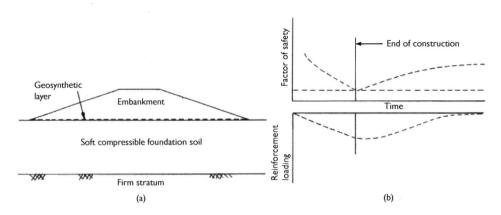

*Figure 3.19* An example of time-dependent reinforcement application: (a) geosynthetic-reinforced embankment; (b) typical variation of slope factor of safety and reinforcement loading with time (after Paulson, 1987).

If the consolidation of the foundation soil is required to be accelerated for a consequent gain in strength, nonwoven geotextiles may be recommended. Where the settlement criteria require high-strength and high modulus geosynthetic, the geocomposites may be used to provide the drainage function. Note that at some very soft soil sites, especially where there is no vegetative layer, a geogrid layer, if laid, may require a lightweight nonwoven geotextile layer as a separator/filter to prevent the contamination of the first lift, especially if it is an open-graded soil. The geotextile layer is not required if a sand layer is placed as the first lift, which meets the soil filtration criteria.

## 3.3.2   Analysis and design concepts

The basic design approach for an embankment over the soft foundation soil with a basal geosynthetic layer is to design against the mode (or mechanism) of failure. The potential failure modes are the following:

1   Overall slope stability failure (Fig. 3.20(a))
2   Lateral spreading (Fig. 3.20(b))
3   Embankment settlement (Fig. 3.20(c))
4   Overall bearing failure (Fig. 3.20(d))
5   Pullout failure (Fig. 3.20(e)).

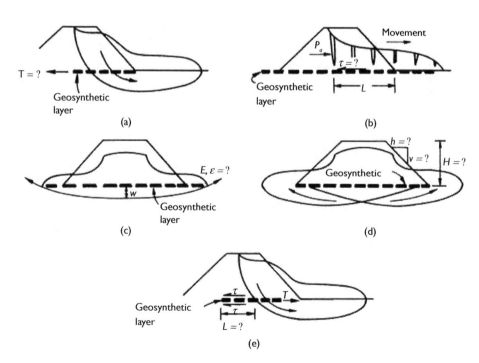

*Figure 3.20* Design models for embankments with a basal geosynthetic layer over the soft foundation soils (after Fowler and Koerner, 1987).

These failure modes indicate the types of analysis that are required. Each failure mode generates the required or design value for the embankment geometry or the tensile strength of the geosynthetic. The conventional geotechnical design procedures, based on the total stress approach, are generally utilized with a modification for the presence of the geosynthetic layer.

### 1 Overall slope stability failure

This is the most commonly considered failure mechanism, where the failure mechanism is characterized by a well-defined failure surface cutting the embankment fill, the geosynthetic layer and the soft foundation soil (Fig. 3.20(a)). This mechanism can involve the tensile failure of the geosynthetic layer or the bond failure due to insufficient anchorage of the geosynthetic extremity beyond the failure surface. The analysis proceeds along the usual steps of conventional slope stability analysis with the geosynthetic providing an additional stabilizing force $T$ at the point of intersection with the failure surface being considered. The geosynthetic thus provides an additional resisting moment required to obtain the minimum required factor of safety. Figure 3.21 shows such a conventional circular slope stability model, usually preferred for the preliminary routine analyses. Opinions are divided on the calculation of the resisting (or stabilizing) moment $\Delta M_g$ due the tensile force $T$ in the geosynthetic layer: $\Delta M_g = TR$ or $\Delta M_g = Ty$ or any other value, where $R$ is the radius of critical slip arc, and $y$ is the moment arm of the geosynthetic layer. However, for the circular failure arcs and the horizontal geosynthetic layers, it is conservative to assume $\Delta M_g = Ty$ and to neglect any other possible effects on soil stresses. This approach is conservative because it neglects any possible geosynthetic reinforcement along the alignment of the failure surface, as well

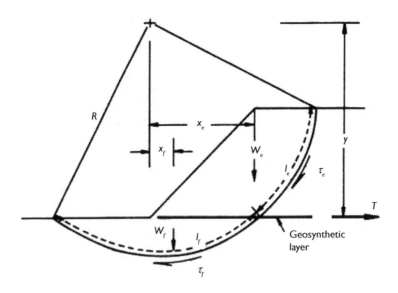

*Figure 3.21* Overall slope stability analysis.

as any confining effect of the geosynthetic. The factor of safety against the overall stability failure is then given as:

$$FS = \frac{\text{Resisting Moment}}{\text{Driving Moment}} = \frac{\tau_e l_e R + \tau_f l_f R + Ty}{W_e x_e + W_f x_f}$$ (3.10)

where $\tau_e$, $\tau_f$ are the shear strengths of embankment and foundation materials, respectively; $l_e$, $l_f$ are the arc lengths within embankment and foundation soil, respectively; $W_e$, $W_f$ are the weights of soil masses within the embankment and foundation soil, respectively; and $x_e$, $x_f$ are the moment arms of $W_e$ and $W_f$, respectively, to their centres of gravity.

Trials are made to find the critical failure arc for which the necessary force in the geosynthetic is maximum. Usually a target value for the safety factor is established and the maximum necessary force is determined by searching the critical failure arc. Note that different methods of analysis or forms of definition of the safety factor may affect the result obtained. The design must consider the fact that a geosynthetic can resist creep if the working loads are kept well below the ultimate tensile strength of the geosynthetic. The recommended working load should not exceed 25% of the ultimate load for polyethylene (PE) geosynthetics, 40% of the ultimate load for polypropylene (PP) geosynthetics and 50% of the ultimate load for polyester (PET) geosynthetics.

### 2 Lateral spreading

The presence of a tension crack through the embankment isolates a block of soil, which can slide outward on the geosynthetic layer (Fig. 3.20(b)). The horizontal earth pressure acting within the embankment mainly causes the lateral spreading. In fact, the horizontal earth pressure causes the horizontal shear stress at the base of the embankment, which must be resisted by the foundation soil. If the foundation soil does not have an adequate shear resistance, the failure can result. The lateral spreading can therefore be prevented if the restraint provided by the frictional bond between the embankment and the geosynthetic exceeds the driving force resulting from the active soil pressure (and/or hydrostatic pressure in the case of water-filled cracks) within the embankment. For the conditions as sketched in Fig. 3.22, the resultant active earth pressure $P_a$ and the corresponding maximum tensile force $T_{max}$ are calculated as

$$P_a = \frac{1}{2}\gamma H^2 K_a$$ (3.11)

and

$$T_{max} = \frac{\tau_r B}{2} = \frac{(\gamma H \tan\phi_i)B}{2}$$ (3.12)

where $\gamma$ is the total unit weight of the embankment material; $H$ is the embankment height; $B$ is the embankment base width; $K_a$ is the active earth pressure coefficient for

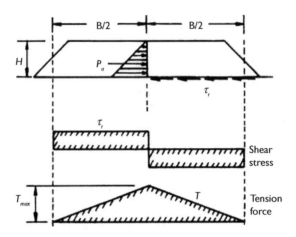

*Figure 3.22* Block sliding analysis.

the soil; $\tau_r$ is the resisting shear stress; and where $\phi_i$ is the angle of shearing resistance of soil-geosynthetic interface.

For no lateral spreading, one can get

$$\frac{T_{max}}{P_a} \geq 1 \tag{3.13}$$

or

$$\tan\phi_i \geq \frac{HK_a}{B} \tag{3.14}$$

It is a general practice to consider a minimum safety factor of 1.5 with respect to strength and a geosynthetic strain limited to 10%. The required geosynthetic strength $T_{req}$ and modulus $E_{req}$ therefore are obtained as

$$T_{req} = 1.5T_{max} \tag{3.15}$$

$$E_{req} = \frac{T_{max}}{\varepsilon_{max}} = 10T_{max} \tag{3.16}$$

The lateral spreading failure mechanism becomes important only for the steep embankment slopes on reasonably strong subgrades and very smooth geosynthetic surfaces. Thus, it is not the most critical failure mechanism for the situations of soft foundation soils.

## EXAMPLE 3.2

A 4-m high and 10-m wide embankment is to be built on the soft ground with a basal geotextile layer. The embankment material has a total unit weight of 18 kN/m³ and the angle of shearing resistance of 35°. Calculate the geotextile strength and modulus required to prevent the block sliding on the geotextile. Assume the soil-geotextile interface angle of shearing resistance is two thirds of the angle of shearing resistance of the embankment material.

## SOLUTION

From Equation (3.12), the maximum tensile force in the geotextile is

$$T_{max} = \frac{(\gamma H \tan \phi_i) B}{2} = \frac{(18)(4)\left[\tan\left(\frac{2}{3} \times 35°\right)\right](10)}{2} = 155.29 \text{ kN/m}$$

From Equation (3.15), the required geosynthetic strength is

$$T_{req} = 1.5 T_{max} = 1.5 \times 155.29 = 232.94 \text{ kN/m}$$

From Equation (3.16), the required geotextile modulus is

$$E_{req} = 10 T_{max} = 10 \times 155.29 = 1552.9 \text{ kN/m}$$

### 3 Embankment settlement

Embankment settlement takes place because of consolidation of the foundation soil (Fig. 3.20(c)). The settlement can also occur due to the expulsion of the foundation soil laterally. This mechanism may occur for heavily reinforced embankments on thin soft foundation soil layers (Fig. 3.23). The factor of safety against the soil expulsion $F_e$ can be estimated as (Palmeira, 2012)

$$F_e = \frac{P_P + R_B + R_T}{P_A} \tag{3.17}$$

where $P_p$ is the passive reaction force against the block movement, $R_T$ is the force at the top of the soil block, $R_B$ is the force at the base of the soil block, and $R_A$ is the active thrust on the soil block.

The active and passive forces can be evaluated by the earth pressure theories, while the forces at the base and top of the soil block can be estimated as a function of the undrained shear strength $S_u$ at the bottom of the foundation soil and adherence between the reinforcement layer and the surface of the foundation soil, respectively.

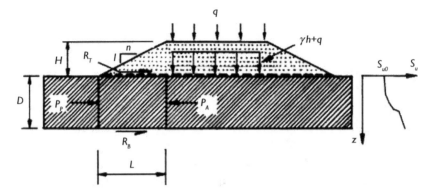

*Figure 3.23* Embankment settlement due to the lateral expulsion of foundation soil.

The geosynthetic layer may reduce the differential settlement of the embankment somewhat, but little reduction of the magnitude of its total final settlement can be expected, because the compressibility of the foundation soils is not altered by the geosynthetic, although the stress distribution may be somewhat different. As the inclusion of geosynthetic reinforcement alone does not reduce the total settlement of the foundation, the total settlement can be calculated using the conventional settlement analysis. The embankment settlement can result in an excessive elongation of the geosynthetic reinforcement, resulting in increase in tensile strain and load in the reinforcement. The acceptable settlement of foundation is a serviceability function in the case of pavements on embankments. However, it is a general practice to limit the total strain in a geosynthetic to 10% in order to minimize the settlements within the embankment. Therefore, the modulus of the geosynthetic to be selected should be $10T_{req}$, where $T_{req}$ is based on the overall stability calculation. For getting this benefit significantly from the geosynthetic layer, its edges must be folded back similar to 'wraparound' in retaining walls or anchored in trenches properly or weighted down by berms. Prestressing the geosynthetic in the field, if possible, along with the edge anchorage can further reduce both the total and the differential settlements within the embankment (Shukla and Chandra, 1996a; Shukla and Raghavendra, 2008).

### 4 Overall bearing failure

The bearing capacity of an embankment foundation soil is essentially unaffected by the presence of a geosynthetic layer within or just below the embankment (Fig. 3.20(d)). Therefore, if the foundation soil cannot support the weight of the embankment, then the embankment cannot be built. The overall bearing capacity can only be improved if a mattress like the reinforced surface layer of larger extent than the base of the embankment will be provided. The overall bearing failure is usually analyzed using the classical soil mechanics bearing capacity analyses. These analyses may not be appropriate if the soft foundation soil is of limited depth, that is, its depth is small compared to the base width of the embankment. In such a situation, a lateral squeeze analysis should be performed. This analysis compares the shear forces developed under the embankment with the shear strength of the corresponding soil. The overall bearing failure check helps in knowing the height of the embankment as well as the side-slope angles that can be adopted on a given foundation soil. Construction of an embankment higher than the estimated value would require the staged construction that allows the underlying soft soils time to consolidate and gain strength.

## 5 Pullout failure

The forces transferred to the geosynthetic layer to resist a deep-seated circular failure, that is, the overall stability failure, must be transferred to the soil behind the slip zone as shown in Fig. 3.20(e). The pullout capacity of a geosynthetic is a function of its embedment length behind the slip zone. The minimum embedment length $L$ can be calculated as

$$L = \frac{T_a}{2(c_a + \sigma_v \tan \phi_i)} \tag{3.18}$$

where $T_a$ is the force mobilized in the geosynthetic per unit length, $c_a$ is the adhesion of soil to geosynthetic, $\sigma_v$ is the average vertical stress, and $\phi_i$ is the shear resistance angle of soil-geosynthetic interface.

If a high strength geosynthetic is used, then the embedment length required is typically very large. However, in confined construction areas, this length can be reduced by folding back the edges of the geosynthetic similar to 'wraparound' in retaining walls or anchored in trenches properly or weighted down by berms.

The design procedure for the embankment with basal geosynthetic layer(s) is given in the following steps:

*Step 1:* Define the geometrical dimensions of the embankment (embankment height *H*; width of crest *b*; side slope, vertical to horizontal as 1:*n*).

*Step 2:* Define the loading conditions (surcharge, traffic load, dynamic load). If there is a possibility of frost action, swelling and shrinkage, and erosion and scour, then the loading caused by these processes must be considered in the design.

*Step 3:* Determine the engineering properties of the foundation soil (shear strength parameters, consolidation parameters). Chemical and biological factors that may deteriorate the geosynthetic must be determined. Though most geosynthetics have a very high resistance to chemical and biological degradations, in unusual situations such as very low (<3) or very high (>9) pH soils, or other unusual chemical environments such as mine or other waste dump sites, the chemical compatibility of the geosynthetic polymer should be checked to assure that it will retain the design strength at least until the underlying subsoil is strong enough to support the embankment without the reinforcement (IRC, 2013).

*Step 4:* Determine the engineering properties of embankment fill materials (compaction characteristics, shear strength parameters, biological and chemical factors that may deteriorate the geosynthetic). The first few lifts of fill material just above the geosynthetic layer should be free draining granular materials.

This requirement provides the best frictional interaction between the geosynthetic and fill, as well as providing a drainage layer for the excess pore water to dissipate from the underlying soils.

*Step 5:* Establish the geosynthetic properties (strength and modulus, soil-geosynthetic friction). Also establish the tolerable geosynthetic deformation requirements. The geosynthetic strain can be allowed up to 2–10%. According to BS: 8006, the maximum strain in the reinforcement should not exceed 5% for short-term applications and 5–10% for long-term applications. When choosing the maximum allowable reinforcement strain, the strain compatibility of the reinforcement with the soft soil must be ensured (IRC, 2013). The selection of geosynthetic should also consider drainage, constructability (survivability) and environmental requirements.

*Step 6:* Check against all the modes of failure, as described earlier. If the factors of safety are sufficient, then the design is satisfactory, otherwise the steps should be repeated by making appropriate changes, wherever possible.

### 3.3.3 Application guidelines

For the construction of an embankment with a basal layer over a very soft foundation soil, a specific construction sequence must be followed to avoid any possibility of failures (geosynthetic damage, non-uniform settlements, embankment failure, etc.) during construction. The following guidelines may help achieve this objective in practice.

1 A geosynthetic layer is placed over the foundation soil, generally with minimal disturbance of the existing materials. Small vegetative cover, such as grass and reeds, should not be removed during the subgrade preparation. There can be several alternatives with regard to the installation of the geosynthetic layer inside the embankment. Some of them are as follows:

   a   a geosynthetic layer inside the embankment (Fig. 3.24(a))
   b   several geosynthetic layers along the embankment height (Fig. 3.24(b))
   c   a geocell at the base of the embankment (Fig. 3.24(c))
   d   a geosynthetic layer at the base of the embankment with folded ends (Fig. 3.24(d))
   e   combination of a geosynthetic layer with berms (Fig. 3.24(e))
   f   a geosynthetic layer (or layers) with vertical piles. (Fig. 3.24(f))

Each alternative has its own advantages. A geosynthetic layer inside the embankment (Fig. 3.24(a)), rather than along the interface between the fill material and the foundation soil, favours a better reinforcement anchorage length, particularly for geogrids due to interlocking effect. If different functions are to be achieved, then several geosynthetic layers of different types can be installed along the embankment height (Fig. 3.24(b)). A combination of geosynthetic layers creates a stiffer mass, which tends to reduce the differential settlements. This effect can also be achieved with the use of geocells filled with embankment material (Fig. 3.24(c)). The increase of geosynthetic anchorage can be achieved by the use of folded edges (Fig. 3.24(d)) and/or berms (Fig. 3.24(e)). If the settlements of the embankment are to be limited, then the geosynthetic layer (or layers) can be installed along with the vertical piles (Fig. 3.24(f)).

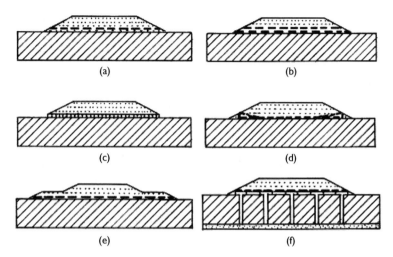

*Figure 3.24* Geosynthetic installation: (a) a geosynthetic layer inside the embankment; (b) several geo-synthetic layers along the embankment height; (c) a geocell at the base of the embankment; (d) a geosynthetic layer at the base of the embankment with folded ends; (e) combination of a geosynthetic layer with berms; (f) a geosynthetic layer (or layers) with vertical piles.

2 The geosynthetic layer is usually placed with its strong direction (warp/machine direction) perpendicular to the centreline of the embankment (Fig. 3.25). It should be unrolled as smoothly as possible, without dragging, transverse to the centreline of the embankment. Additional reinforcement with its strong direction oriented parallel to the centreline may also be required at the ends of the embankment. The geosynthetic should be pulled taut to remove wrinkles, if any. Lifting by wind can be prevented by putting weights (sand bags, stone pieces, etc.).

3 Seams should be avoided perpendicular to the major principal stress direction, which is generally along the width of the embankment (Fig. 3.25). Since for the surcharge/areal fills, a major principal stress direction cannot be defined, in such situations, the seams should be made by sewing.

4 Narrow horizontal geosynthetic strips may be placed along the side slopes with wraparound to enhance compaction at the edges (Fig. 3.26). The edge geosynthetic strips also help reduce erosion and may assist in the establishment of vegetation.

5 The embankment should be built using the low ground pressure construction equipment. Slack/wrinkles in the reinforcement layer should be removed manually. Direct movement of vehicles on the reinforcement should be prevented.

6 When possible, the first few lifts of fill material (0.5 to 1 m) just above the geosynthetic should be free draining granular materials; then the rest of the embankment can be constructed to grade with any locally available materials. This is required to have the best frictional interaction between the geosynthetic and the fill, as well as to have a drainage layer for the dissipation of excess pore water dissipation from the underlying foundation soils. The granular material should be compacted to specified design value of the modified Proctor maximum dry unit weight.

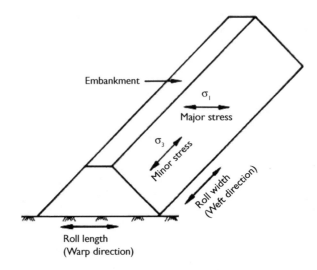

*Figure 3.25* Geosynthetic orientation for the linear embankments.

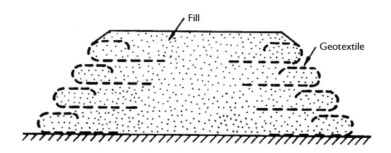

*Figure 3.26* Embankment with wraparound side slopes.

7　In the case of extremely soft foundations (when a mud-wave forms), the geosynthetic-reinforced embankments should be constructed as per the sequence of construction shown in Fig. 3.27. A perimeter berm system can be constructed to contain the mud-wave.

8　The first lift should be compacted only by tracking in place with dozers or endloaders. Once the embankment is at least 60 cm above the original ground, subsequent lifts can be compacted with a smooth drum vibratory roller or other suitable compactor. Traffic on the first lift should be parallel to the centreline of the embankment.

9　A minimum number of instruments, such as piezometers, settlement plates, and inclinometers, can be installed in order to verify the design assumptions and control the construction. If the piezometer indicates an excessive pore water pressure, the construction should be halted until the pressures drop to a predetermined safe value. The settlement plates installed at the geosynthetic level can help monitor the settlement during the construction and thus adjusting the fill requirements

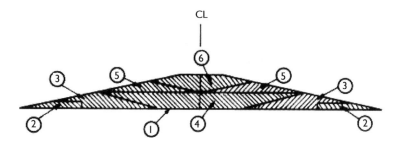

1 Lay the geosynthetic in continuous transverse strips; sew the strips together.

2 End the dump access roads.

3 Construct the outside sections to anchor the geosynthetic.

4 Construct the interior section to set the geosynthetic.

5 Construct the interior sections to tension the geosynthetic.

6 Construct the final central section.

*Figure 3.27* Construction sequence for the geosynthetic-reinforced embankments over the soft foundation soils (after Haliburton *et al.*, 1977).

appropriately. Inclinometers should be considered at the embankment toes to monitor the lateral displacement.

### 3.3.4   Case studies

#### Case study I

The construction of a 3.5-km long embankment through the tidal area of Deep Bay in the New Territories district of Hong Kong was constructed between April and end of November 1982 (see Fig. 3.28). The stability of slopes of the 3.5-m high embankment was achieved by reinforcement of the base using a heavy-duty polyester woven geotextile between the fill and the foundation mud. The undrained shear strength of the foundation soil was estimated to be as low as 5–10 kPa in the top 6 m of unconsolidated marine deposits, that is, a very soft clay. It was decided not to increase the base area for reducing the pressure on the ground because of the scarcity of land in Hong Kong. Also, the embankment went along the existing fish ponds. The dikes of these ponds were not to be damaged in any way during the construction of the embankment. Additionally, the whole area was flooded twice a day due to tides.

Construction work started by installing a 5-m wide geotextile (Stabilenka 200, manufactured and supplied by Akzo Industrial Systems). The reinforcing geotextiles were sewn together on site utilizing the local labour. The muddy conditions presented the labourers with some difficulties as they struggled to carry and pull the 30-m lengths of the geotextile into position. Once the geotextiles were positioned and sewn together, access was not so difficult. It was possible for the contractor to move in

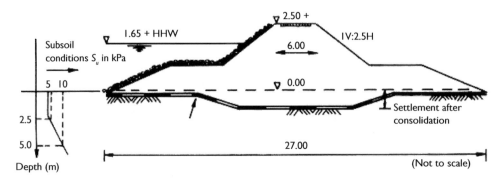

*Figure 3.28* Cross-section of the geotextile-reinforced embankment over tidal mud area of the Deep Bay, New Territories, Hong Kong (after Risseeuw and Voskamp, 1993).

immediately. After a 1-m layer of fill was laid with the aid of light Komatsu dozers, the contractor was able to use heavily loaded 35-ton wheelbase trucks to dump the fill close to the front. Then, a fleet of large and small Komatsu swamp dozers was used to push the fill material into position. During the continued filling operation, the freshly deposited mud top layer (300–500 mm thick) squeezed out. The underlying stratum, however, was confined by the reinforcing geotextile. The average work output to bring the 3.5-km long embankment up to its 3.5-m height was 150 m per week. This was considered a good rate of construction, taking into account the fact that the installation of the geotextile and filling was restricted by the state of the tide.

The primary benefit of the geotextile reinforcement was to contribute to the short-term stability, whereas it was calculated that the embankment would be stable without the need for reinforcement after a prolonged period of subsoil consolidation. The field monitoring up to 1993 indicated that no decrease of stability or excessive subsidence had been encountered.

### Case study 2

A 100-year old railway track from Magdeburg to Berlin had to be partially reconstructed for train velocities of 160 km/h. The railway line traversed deep deposits of soft organic soils. The organic soil consisted, to a great extent, of peat with a water content of 300–600% well above the liquid limit, undrained shear strength $c_u < 10$ kPa and a constrained modulus $E_s = 0.2$–$0.8$ MPa besides the old railway embankment. Below the old railway embankment the organic soil was preconsolidated and the constrained modulus was on the order of $E_s = 2$–$6$ MPa with the undrained shear strength $c_u > 15$ kPa. The soft soil layers were not uniform. Apart from peat, sandy to clayey organic silts were encountered with varying amounts of plants residues and shells. The old railway tracks had suffered considerable settlement in the past, so it was necessary to improve the bearing capacity and deformational behaviour of the ground in this area.

Over a total length of 2.1 km, the geogrid-reinforced structure was erected in several sections from 1994 to 1995. The structural system is sketched in Fig. 3.29. It consists of the geogrid reinforced embankment, the precast concrete pile caps, the

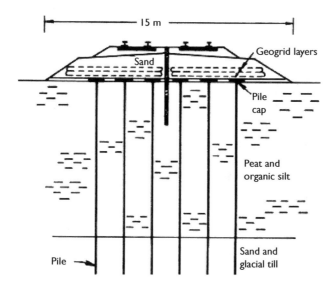

*Figure 3.29* Geogrid-reinforced railway embankment on piles, Germany (after Zanzinger and Gartung, 2002).

ductile cast iron piles, the soft soil that prevents buckling of the piles, and, finally, the sand layer or glacial till at depth with sufficient bearing capacity to carry the total load. Based on the structural analyses, three layers of knitted polyester geogrids were selected. Their short-term tensile strength in machine direction and in cross-machine direction was 150 kN/m each. Under the design loading, the geogrids experienced no more than 20% of their short-term tensile strength.

The section, shown in Fig. 3.29, contains a sheet pile wall at the centre line of the embankment. This structural element served the purpose of securing half of the embankment for railway traffic, while the other half was being reconstructed. The sheet piling was pulled after completion of the new embankment. Trains were passing over the first section of the new structure as from May 1994, and the entire construction was completed in December 1995. The performance of the structure was monitored by two extensive instrumented sections and in addition by conventional surveying. The data obtained over a period of more than seven years from displacement and strain measurements had demonstrated that the structure was safe and performed adequately with respect to serviceability. The railway personnel operating trains on the railway line had repeatedly reported that the trains were running more smoothly on the sections with the reinforced piled embankment than on sections with a conventional replacement of the soft organic soil with compacted sand (Zanzinger and Gartung, 2002b).

The following points regarding the reinforced embankments are worth mentioning:

1   For the embankments over soft soils, the geosynthetic reinforcement is needed only during construction and foundation consolidation; hence a short-term constant rate of strain tensile test can be used for design purposes.

2    For reinforced surcharge/areal fills, used as parking lots, storage yards, and construction pads, the applied loads are close to axi-symmetric, therefore, the design strengths and strain considerations are generally the same in all directions. The analysis for geosynthetic requirements remains the same as those discussed here. However, special seaming techniques must often be considered to meet the required strength requirements.

## 3.4    SHALLOW FOUNDATIONS

### 3.4.1    Basic description

Geosynthetic-reinforced foundation soils are being used to support footings of many structures including warehouses, oil drilling platforms, platforms of heavy industrial equipment, parking areas and bridge abutments. In a usual construction practice, one or more layers of geosynthetic (geotextile, geogrid, geocell or geocomposite) are placed inside a controlled granular fill beneath the footings (Fig. 3.30). Such reinforced foundation soils provide improved load-bearing capacity and reduced settlements by distributing the imposed loads over a wider area of weak subsoil. In the conventional construction techniques without any use of the reinforcement, a thick granular layer is needed which may be costly or may not be possible, especially in the sites of limited availability of good-quality granular materials.

The geosynthetics, in conjunction with foundation soils, may be considered to perform mainly reinforcement and separation functions. The reinforcement function of geosynthetics can be observed in terms of their several roles, as discussed in Section 1.6.1 (see Chapter 1). Geosynthetics (particularly, geotextiles, but perhaps also geogrids) also improve the performance of the reinforced soil system by acting as a separator between the soft foundation soil and the granular fill. In many situations, the separation can be an important function compared to the reinforcement function. In general, the improved performance of a geosynthetic-reinforced foundation soil can

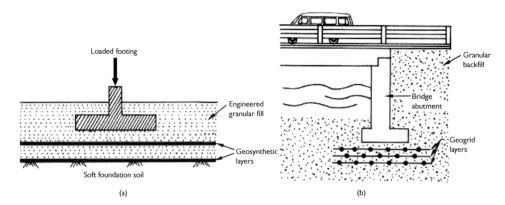

*Figure 3.30* Reinforced foundation soils supporting footings of structures: (a) loaded footing; (b) bridge abutment.

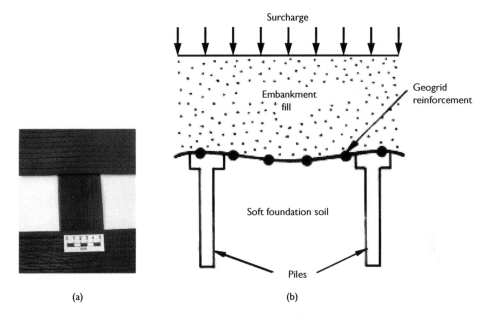

*Figure 3.31* 'Paralink' geogrid: (a) pictorial view; (b) use over piles.

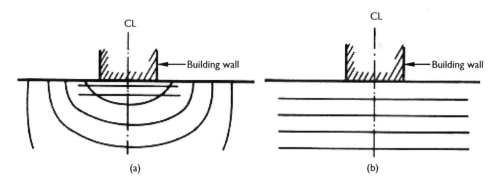

*Figure 3.32* Arrangement of reinforcement layers beneath a footing: (a) ideal arrangement (after Basset and Last, 1978); (b) practical arrangement.

be attributed to an increase in shear strength of the foundation soil from the inclusion of the geosynthetic layer(s). The soil-geosynthetic system forms a composite material that inhibits the development of the soil-failure wedge beneath shallow spread footings. Geosynthetic products like 'Paralink' as shown in Fig. 3.31(a) can be very effective for use over soft foundation soils as well as over voids and piles (Fig. 3.31(b)).

The ideal reinforcing pattern has the geosynthetic layers placed horizontally below the footing, which becomes progressively steeper farther from the footing (Fig. 3.32(a)). It means that the reinforcement should be placed in the direction of major principal strain. However, for practical simplicity, the geosynthetic sheets are often laid horizontally as shown in Fig. 3.32(b).

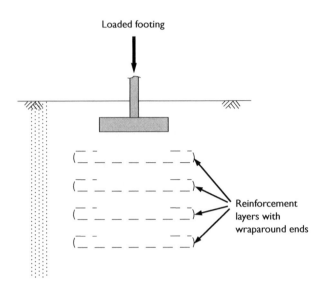

*Figure 3.33* A loaded footing resting on geosynthetic-reinforced granular fill with reinforcement layers having wraparound ends.

The most recent experimental studies have shown that the wraparound ends of the reinforcement (Figure 3.33) bring additional improvement in load-bearing capacity as well as stiffness in terms of the modulus of subgrade reaction. The wraparound ends technique also reduces the width of land required for the construction of the foundation, thus saving the cost of the project significantly, especially in urban areas (Kazi *et al.*, 2015a, b). The wraparound reinforcement technique has been used earlier by the author in some field projects successfully. The geosynthetic reinforcement can be prestressed to cause further benefits (Shukla and Chandra, 1994a, b; Shukla, 1995; Lovisa *et al.*, 2010). The prestressed ends may be anchored in trenches for maximizing the benefits.

### 3.4.2   Analysis and design concepts

The basic design approach for geosynthetic-reinforced foundation soils must consider their modes (or mechanisms) of failure. The possible potential failure modes are as follows:

1   *Bearing capacity failure of the soil above the uppermost geosynthetic layer* (Fig. 3.34(a)): This type of failure appears likely to occur if the depth of the uppermost layer of reinforcement ($u$) is greater than about 2/3 of the width of footing ($B$), that is, $u/B > 0.67$, and if the reinforcement concentration in this layer is sufficiently large to form an effective lower boundary into which the shear zone will not penetrate. This class of bearing capacity problems corresponds to the bearing capacity of a footing on the shallow soil bed overlying the strong rigid boundary.

2   *Pullout of the geosynthetic layer* (Fig. 3.34(b)): This type of failure is likely to occur for shallow and light reinforcement ($u/B < 0.67$) and number of reinforcement layers, $N < 3$.

3   *Breaking of the geosynthetic layer* (Fig. 3.34(c)): This type of failure is likely to occur with long, shallow, and heavy reinforcement ($u/B < 0.67$, $N > 3$ or 4). The reinforcement layers always break approximately under the edge or towards the centre of the footing. The uppermost layer is most likely to break first, followed by the next deep layer and so forth.

4   *Creep failure of the geosynthetic layer* (Fig. 3.34(d)): This failure may occur due to a long-term settlement caused by sustained surface loads and subsequent geosynthetic stress relaxation.

The first three modes of failure were first reported by Binquet and Lee (1975a) in the case of a footing resting on the sand bed reinforced with metallic reinforcement on the basis of observations made during the laboratory model tests (Binquet and Lee, 1975b). The fourth mode of failure, i.e., the creep failure, was discussed by Shukla (2002a, 2012b) and Koerner (2005).

A large number of studies have been carried to evaluate the beneficial effects of reinforcing the soils with geosynthetics as related to the load-carrying capacity and

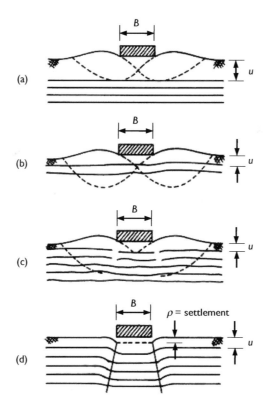

*Figure 3.34* Possible modes of failure of geosynthetic-reinforced shallow foundations: (a) bearing capacity failure of soil above the uppermost geosynthetic layer; (b) pullout of the geosynthetic layer; (c) breaking of the geosynthetic layer; (d) creep failure of the geosynthetic layer (after Binquet and Lee, 1975a; Shukla, 2002a, 2012b; Koerner, 2005).

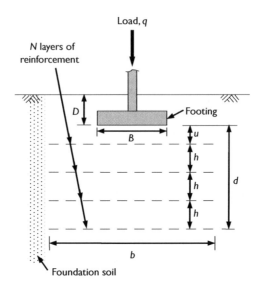

*Figure 3.35* Geometrical parameters of geosynthetic-reinforced foundation soil.

the settlement characteristics of shallow foundations (Shukla, 2002a, 2012b). They all point to the conclusion that the geosynthetic reinforcement increases the load-carrying capacity of the foundation soil and reduces the depth/thickness of the granular fill for the same settlement level. Through laboratory model tests, Guido *et al.* (1985) and many other researchers have studied the parameters affecting the load-bearing capacity of geosynthetic-reinforced foundation soils. All these parameters can be summarized as follows (Fig. 3.35):

- depth of footing, $D$
- width of footing, $B$
- strength of foundation soil, $\tau_f$
- depth below the footing of the first geosynthetic layer, $u$
- depth below the footing of the bottommost geosynthetic layer, $d$
- number of geosynthetic layers, $N$
- vertical spacing of the geosynthetic layers, $h$
- width of geosynthetic reinforcement layers, $b$
- tensile strength of geosynthetic reinforcement, $\sigma_G$

The parameters, $u$, $N$ and $h$ cannot be considered separately, as they are dependent on each other. It has been reported that more than three geosynthetic layers are not beneficial, and the optimum size of the geosynthetic layer is about three times the width of the footing, $B$. For beneficial effects, the geosynthetic layers should be laid within a depth equal to the width of footing. The optimum vertical spacing of the geosynthetic reinforcement layers is between $0.2B$ and $0.4B$. For a single layer reinforced soil, the optimum embedment depth is approximately $0.3B$.

For expressing the improvement conveniently, as well as for comparing the test data from the studies, the *bearing capacity ratio* (BCR), a term introduced by Binquet and Lee (1975a,b), is commonly used. This term is defined as

$$BCR = \frac{q_R}{q_{uU}} \qquad (3.19)$$

where $q_{uU}$ is the ultimate load-bearing capacity of the unreinforced soil, and $q_R$ is the load-bearing capacity of the geosynthetic-reinforced soil at a footing settlement corresponding to the settlement $\rho_u$ of the footing on unreinforced soil at the ultimate load-bearing capacity $q_{uU}$. Figure 3.36 shows the typical load-settlement curves for a soil with and without reinforcement, and illustrates how $q_{uU}$ and $q_R$ are determined.

Several researchers carried out the load-bearing capacity analysis considering the limited roles of geosynthetics in improving the load-bearing capacity and taking different sets of assumptions. For more details, the readers can refer to the books by Shukla (2002b; 2012a). If a strip footing rests on a geosynthetic-reinforced soil slope, the load-bearing capacity may be determined analytically using the method proposed by Jha *et al.* (2013).

Among the reinforcement practices for buildings, roads, and embankments constructed on soft ground; the use of a geocell foundation mattress is a unique method, in which the mattress is placed upon the soft foundation soil of insufficient bearing capacity so as to withstand the weight of the superstructure. The geocell foundation mattress is a honeycombed structure formed from a series of interlocking cells (Fig. 3.37). These cells are fabricated directly on the soft foundation soil using the uniaxial-polymer geogrids in a vertical orientation connected to a biaxial base geogrid and then filled with granular materials resulting in a structure usually 1 m deep. This arrangement forms not only a stiff platform, which provides a working area for the

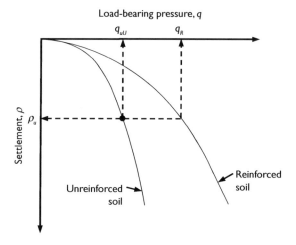

*Figure 3.36* Typical load-settlement curves for unreinforced and geosynthetic-reinforced soils.

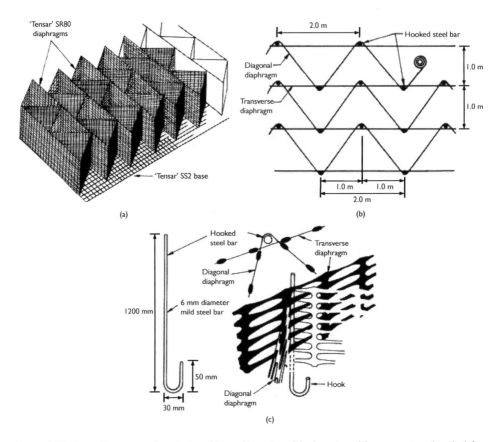

(a)

(b)

(c)

*Figure 3.37* Geocell mattress foundation: (a) configuration; (b) plan view; (c) connection details (after Bush *et al.*, 1990).

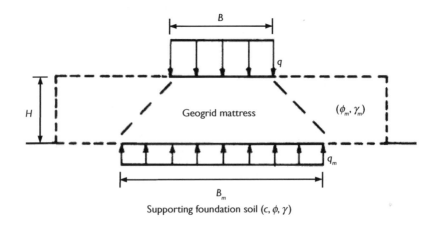

*Figure 3.38* Load-bearing capacity analysis of the geogrid-mattress foundation.

workers to push forward the construction of the geocell itself, and supports the subsequent structural loads, but also a drainage blanket to assist the consolidation of the underlying soft foundation soil. The incorporation of a geocell foundation mattress provides a relatively stiff foundation to the structure and this maximizes the bearing capacity of the underlying weak soil layer. The geocell mattress is self-contained and, unlike constructions with horizontal layers of geosynthetics, it needs no external anchorage beyond the base of main structure. As a consequence of the flexible interaction with the supporting foundation soil underneath, even locally or unevenly applied vertical load propagates within the mattress and is transmitted widely to the supporting foundation soil.

Ochiai *et al.* (1994) described a conventional approach for the assessment of the improvement of bearing capacity due to placement of the geogrid-mattress foundation, as discussed earlier. In this approach, a vertical load of intensity $q$ and width $B$, applied on the mattress, is transmitted widely to the supporting foundation soil with the corresponding intensity $q_m$ and width $B_m$ (Fig. 3.38). The ultimate bearing capacity $q_u$ without the use of the mattress may be given by Terzaghi's equation as (Shukla, 2014; 2015):

$$q_u = cN_c + \frac{1}{2}\gamma B N_\gamma \qquad (3.20)$$

where $c$ is the cohesion and $\gamma$ is the unit weight of the supporting foundation soil, and $N_c$ and $N_\gamma$ are the bearing capacity factors. On the other hand, assuming that the placement of the geogrid mattress has a surcharge effect on the bearing capacity of the supporting foundation, the ultimate bearing capacity $q_{um}$ with the use of mattress may be given as:

$$q_{um} = cN_c + \gamma_m H N_q + \frac{1}{2}\gamma B_m N_\gamma \qquad (3.21)$$

where $\gamma_m$ is the unit weight of the mattress, $H$ is the thickness of the mattress, and $N_q$ is the bearing capacity factor. Therefore, the increase in the bearing capacity $\Delta q_u$ due to the placement of the mattress can be given as:

$$\Delta q_u = q_{um} - q_u = \gamma_m H N_q + \frac{1}{2}\gamma(B_m - B)N_\gamma \qquad (3.22)$$

It is therefore found that the evaluation of the bearing capacity improvement requires the estimation of the width $B_m$. The experimental studies have revealed that the width of the supporting foundation soil over which the vertical stress is distributed becomes larger as the thickness of the geogrid-mattress becomes greater, and as the vertical stiffness of the supporting foundation soil becomes lower. It was suggested, from the design point of view, that the width of the geogrid-mattress should be at least large enough to accommodate the vertical stress distribution, which takes place under the mattress.

In addition to the load-bearing capacity analysis of the geosynthetic-reinforced foundation soil, the designer must carry out the settlement analysis for the structural and functional safety of the structures resting on the geosynthetic-reinforced foundation soils. One can make such an analysis using the mechanical foundation model presented by Shukla and Chandra (1994a) and Shukla (1995) (see Fig. 3.39). This model allows the study of time-dependent settlement behaviour of the geosynthetic-reinforced granular fill–soft soil system. In this model, the geosynthetic reinforcement and the granular fill are represented by the stretched rough elastic membrane and the Pasternak shear layer, respectively. The general assumptions are that the geosynthetic reinforcement is linearly elastic, rough enough to prevent slippage at the soil interface and has no shear resistance. A perfectly-rigid plastic friction model is adopted to represent the behaviour of the soil-geosynthetic interface in shear. The compressibility of the granular fill is represented by a layer of Winkler springs attached to the bottom of the Pasternak shear layer. The saturated soft foundation soil is idealized by the Terzaghi's spring-dashpot system. The spring represents the soil skeleton and the dashpot simulates the dissipation of the excess pore water pressure. The spring constant is assumed to have a constant value with depth of the foundation soil and also with time. Yin (1997a, b) further improved the mechanical foundation model by incorporating a nonlinear constitutive model for the granular fill and a nonlinear spring model for the soft soil. In the process of developing the simple foundation models, Shukla and Yin (2003) suggested a model based on the Timoshenko beam concept for the time-dependent settlement analysis of a geosynthetic-reinforced granular fill-soft soil system when the granular fill is relatively dense.

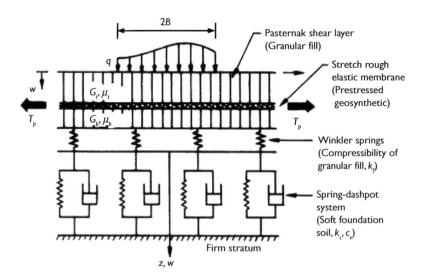

*Figure 3.39* Mechanical foundation model (after Shukla and Chandra, 1994b; Shukla, 1995).

The equations governing the response of the generalized mechanical foundation model are (Shukla, 1994b; Shukla, 1995) the following:

$$q = \overline{X_1}\frac{k_f k_s w}{k_s + k_f U} - \left\{G_t H_t + \overline{X_2}(T_p + T)\cos\theta + \overline{X_1}G_b H_b\right\}\frac{\partial^2 w}{\partial x^2} \tag{3.23}$$

$$\frac{\partial T}{\partial x} = -\overline{X_3}\left(q + G_t H_t \frac{\partial^2 w}{\partial x^2}\right) - \overline{X_4}\left(\frac{k_f k_s w}{k_s + k_f U} - G_b H_b \frac{\partial^2 w}{\partial x^2}\right) \tag{3.24}$$

where

$$\overline{X_1} = \frac{1 + K_{0R}\tan^2\theta - (1 - K_{0R})\mu_b \tan\theta}{1 + K_{0R}\tan^2\theta + (1 - K_{0R})\mu_t \tan\theta} \tag{3.25a}$$

$$\overline{X_2} = \frac{1}{1 + K_{0R}\tan^2\theta + (1 - K_{0R})\mu_t \tan\theta} \tag{3.25b}$$

$$\overline{X_3} = \mu_t \cos\theta(1 + K_{0R}\tan^2\theta) - (1 - K_{0R})\sin\theta \tag{3.25c}$$

$$\overline{X_4} = \mu_b \cos\theta(1 + K_{0R}\tan^2\theta) + (1 - K_{0R})\sin\theta \tag{3.25d}$$

Note that $q$ is the applied load intensity; $w(x, t)$ is the vertical surface displacement; $T(x, t)$ is the tensile force per unit length mobilized in the membrane; $T_p$ is the pretension per unit length applied to the membrane; $G_t$ and $H_t$ are the shear modulus and thickness of the upper shear layer, respectively; $G_b$ and $H_b$ are the shear modulus and thickness of the lower shear layer, respectively; $\mu_t$ and $\mu_b$ are the interface friction coefficients at the top and bottom faces of the membrane, respectively; $k_f$ is the modulus of subgrade reaction of the granular fill; $k_s$ is the modulus of subgrade reaction of the soft foundation soil; $K_{0R}$ is the coefficient of lateral stress at rest at an overconsolidation ratio $R$, which is defined here as the ratio of the maximum stress, to which the granular fill is subjected through compaction, to the existing stress under the working load; $\theta$ is the slope of the membrane; $U$ is the average degree of consolidation of soft foundation soil; $c_v$ is the coefficient of consolidation of the soft foundation soil; $x$ is the distance measured from the centre of the loaded region along the x-axis; $B$ is the half width of loading; and $t$ is any particular instant of time measured from the instant of loading. It should be noted that Equations (3.23), (3.24) and (3.25) governing the model response are applicable for the plane-strain loading conditions. For the axisymmetric loading conditions, the readers can refer to the work of Shukla (1995), and Shukla and Chandra (1998).

The parameters of the mechanical foundation models can be determined as per the guidelines suggested by Selvadurai (1979), and Shukla and Chandra (1996b). The parametric studies carried out by Shukla and Chandra (1994b) show the effects of various parameters on the settlement response of the geosynthetic-reinforced granular fill – soft soil system. Fig. 3.40 shows the settlement profiles for a typical set of parameters at various stages of consolidation of the soft foundation soil. The trend of results obtained using the generalized Shukla model (Shukla, 1995) is in good agreement with other reported works.

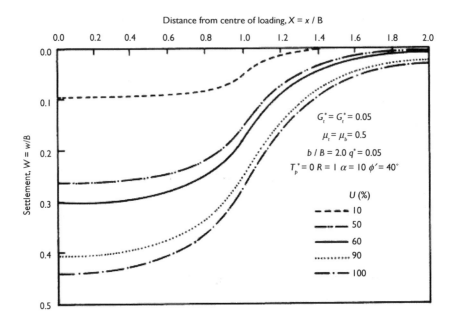

*Figure 3.40* Settlement profiles of the geosynthetic-reinforced granular fill-soft soil system at various stages of consolidation of the soft foundation soil (Shukla and Chandra, 1994b; Shukla, 1995).

Note: $G_t^* = G_t H_t / k_s B^2$; $G_b^* = G_b H_b / k_s B^2$; $q^* = q / k_s B$; $T_p^* = T_p / k_s B^2$; $\alpha = k_f / k_s$; $\phi'$ is the angle of shearing resistance of the granular fill; $b$ is the half width of the geosynthetic reinforcement.

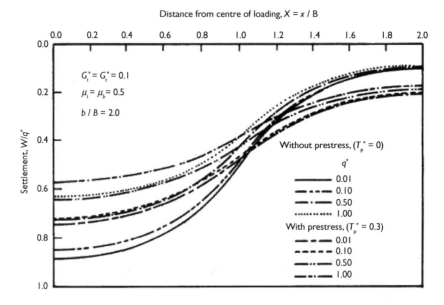

*Figure 3.41* Settlement profiles of the geosynthetic-reinforced granular fill-soft soil system, indicating the effect of prestressing the geosynthetic reinforcement (Shukla and Chandra, 1994a; Shukla, 1995).

It is now well established that the geosynthetics, particularly the geotextiles, show their beneficial effects only after relatively large settlements, which may not be a desirable feature for many structures resting on the geosynthetic-reinforced foundation soils. Hence there has been a need for a technique that can make the geosynthetic reinforcement more beneficial without the occurrence of large settlements. Prestressing the geosynthetic reinforcement can be one of the techniques to achieve this goal. The study, carried out by Shukla and Chandra (1994a), has shown that an improvement in the settlement response increases with an increase in prestress in the geosynthetic reinforcement within the loaded footing and is most significant at the centre of the loaded footing with a reduction in the differential settlement (Fig. 3.41).

Note that the level of compaction of the granular fill affects the settlement behaviour of the geosynthetic-reinforced soil. For reduced settlements, a higher degree of compaction is always desirable; however, the beneficial effects of the geosynthetic layer decrease with an increase in the degree of compaction of the granular fill (Shukla and Chandra, 1994c; Shukla, 1995; Shukla and Chandra, 1997).

### 3.4.3 Application guidelines

The guidelines for geosynthetic installation and compaction of granular fills are highly governed by the type of the footing, the applied load, and the foundation soil characteristics. In most shallow foundation applications, the geosynthetic layer(s) will be installed at the base of the foundation trench followed by placement of a compacted granular fill.

### 3.4.4 Case studies

#### Case study 1

A section of Federal Highway B 180, more than 20 m long, at Neckendorf near Eisleben, Germany was destroyed across its entire width in 1987 by a sink-hole of diameter of about 8 m located almost on the road axis below 30 m depth. Although the hole was filled with fill material, the danger of a new cave-in due to caverns deep underground still existed. To allow the roadway back into operation, the opening had to be bridged-over sufficiently to allow no more subsidence than 10 cm over 30 m of roadway even under heavy truck trafficking. The 20-m long weak section was bridged over with a geogrid-reinforced gravel/sand layer (see Fig. 3.42). The layer was about 60 cm thick by 60 m long and approximately 11 m wide. This layer supported the entire road surface.

The geogrid reinforcement was installed in three layers. The bottom layer consisted of two 5-m geogrid strips laid longitudinally side-by-side. The second layer consisted of a transverse geogrid strip, completely encapsulated and overlapped, resulting in a third layer. The design provided effective reinforcement against longitudinal and transverse deflection as well as torsion. The flexible *Fortrac 1200/50-10* geogrid is composed of very low elongation, low creep Aramid fibres with total tensile strength of 1200 kN/m and only 3% elongation. The mesh size is $10 \times 10$ mm. The reinforced layer was prepared within a few days in October 1993.

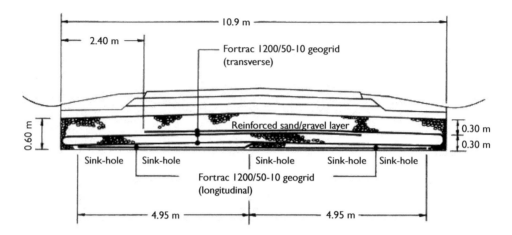

*Figure 3.42* Geogrid-reinforced gravel layer bridges over the sink-hole on Federal Highway B 180 near Eisleben, Germany (courtesy of HUESKER Synthetic GmbH & Co., Germany).

### Case study 2

Alston *et al.* (2015) have reported the construction of geogrid-reinforced granular soil pads as the foundations for two 31-m diameter liquid storage tanks at the Emerald Street slip, which lies on the south side of Hamilton Harbour, Ontario, Canada. The tanks apply a load of 120 kPa to the pads. The soil profile consists of a pavement structure from ground surface to 0.5 m, mostly silty-sandy fill from 0.5 m to 12 m (*N* value widely varying from 5 to 55), stiff silty clay from 12 to 24 m (*N* value in the range of 3–15) and dense to very dense clayey silt and sand (*N* value greater than 50) followed by shale bedrock. The construction started with removal of the existing asphaltic concrete pavement as well as the immediately underlying zones of the upper fill layer to a depth of 1.5 m below the base of the tank foundation. This excavation phase was followed by the dense compaction of the material exposed in the base of the excavation using a 10-t smooth drum roller. The design thickness of the engineered granular base was 1.5 m, and it consisted of sand, gravel, ash and cinders present in varying proportions. The selected reinforcement consisted of two layers of uniaxial PET geogrid (Mirafi 5XT) that has the tensile strength of 68 kN/m in the primary direction in the engineered fill pad. The main direction of the uniaxial geogrid was laid radially towards the centre of the tank, and perpendicular to the tank wall. The lower level of the geogrid was positioned at the base of the fill pad, and the second layer 300 mm above the base. The geogrid was extended outside the footprint of the wall footing for a width of 3 m. The post-construction monitoring of settlement of the tank foundation shows that the settlement of the foundation has been within the tolerable limits. Note that the geogrid-reinforced pads have been found cost-effective compared to other possible options of stone columns and end-bearing pile foundations.

## 3.5 SLOPES – STABILIZATION

### 3.5.1 Basic description

Slopes can be natural or man-made (cut slopes or embankment slopes). Several natural and man-made factors, which have been identified as the causes of instability to slopes, are well known to the civil engineering community (Shukla, 1997). Many of the problems of the stability of natural slopes (i.e. hillsides) are radically different from those of man-made slopes (i.e. artificial slopes) mainly in terms of nature of soil materials involved, the environmental conditions, location of groundwater level, and stress history. In the man-made slopes, there are also essential differences between cuts and embankments. The latter are structures which are (or at least can be) built with relatively well-controlled materials. In the cuts, however, this possibility does not exist. The failures of slopes, called the landslides, may result in loss of property and lives, and create inconvenience in several forms to our normal activities (Fig. 3.43).

Several slope stabilization methods are available to improve the stability of unstable slopes (Broms and Wong, 1990; Abramson *et al.*, 2002). The slope stabilization methods generally reduce the driving forces, increase the resisting forces, or both. The advent of geosynthetic reinforcement materials has brought a new dimension of efficiency to stabilize the unstable and failed slopes by constructing various forms of structures, such as reinforced slopes, retaining walls, etc. mainly due to their corrosive resistance and long-term stability. In recent years, the geosynthetic-reinforced soil slopes have provided innovative and cost-effective solutions to the slope stabilization problems, particularly after a slope failure has occurred or if a steeper than safe unreinforced slope is desirable. The geosynthetic-reinforced slopes provide a wide array of design advantages as mentioned below (Simac, 1992):

*Figure 3.43* A severe landslide causing inconvenience to traffic movement (after Shukla and Baishya, 1998).

- reduce the land requirement to facilitate a change in grade;
- provide additional usable area at toe or crest of slope;
- use available on-site soil to balance earthwork quantities;
- eliminate the import costs of select fill or export costs of unsuitable fill;
- meet the steep changes in grade, without the expense of retaining walls;
- eliminate the concrete face treatments, when not required for surficial stability or erosion control;
- provide a natural vegetated face treatment for environmentally sensitive areas;
- provide a noise abatement for high traffic areas and minimize vandalism;
- offer a design that is easily adjustable for surcharge loadings from buildings and vehicles.

Construction of reinforced slopes may highlight some of the above advantages in the following applications:

- repair of failed slopes;
- construction of new embankments;
- widening of existing embankments; and
- construction of alternatives to retaining walls.

Reinforced slopes are basically compacted fill embankments that incorporate the geosynthetic tensile reinforcement arranged in horizontal planes. The tensile reinforcement holds the soil mass together across any critical failure plane to ensure the stability of the slope. Facing treatments ranging from vegetation to armour systems are applied to prevent the raveling and the sloughing of the face.

The tensile reinforcement should, to be effective, be placed in the direction of tensile normal strains, ideally in the direction and along the line of action of the major principal tensile strain. Figure 3.44(a) shows the ideal reinforcement layout. As can be seen, although the horizontal layers of reinforcement would be correctly aligned under the crest of the slope, they would have inappropriate inclinations under the batter, especially at the toe. Even though an idealized reinforcement layout might be determined it would be impractical if it took the form shown in Fig. 3.44(a). Consequently the geosynthetic layers are usually placed in the horizontal layers within the slope as shown in Fig. 3.44(b).

Figure 3.45 illustrates the two basic applications of slope reinforcement for stability enhancement in relation to the slope angle, which also represents the angle of repose ($\beta$), defined to be the steepest slope angle that may be built without reinforcement, that is, the factor of safety, $FS = 1$. The geosynthetic tensile reinforcement may be used to improve the stability of slopes having a slope angle equal to or less than $\beta$, that are at or slightly greater than $FS$ of 1.0 (Fig. 3.45(a)). This would be typical of a landslide repair, where the grades are established but the soil has failed. Alternatively, the tensile reinforcement may be incorporated into a slope having a slope angle greater than $\beta$ (Fig. 3.45(b)) that otherwise could not be built or stand on its own. This creates a sloped earth retaining structure for steep changes in grade that previously required a retaining wall.

Figure 3.46(a) shows an active zone of the soil slope where the instability may occur and the restraint zone in which the soil may remain stable. The required function of any reinforcing system would be to maintain the integrity of the active zone and effectively anchor this to the restraint zone, to maintain an overall integrity of the

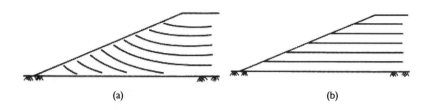

*Figure 3.44* Reinforcement orientations: (a) idealized orientation; (b) practical orientation (after Ingold, 1982).

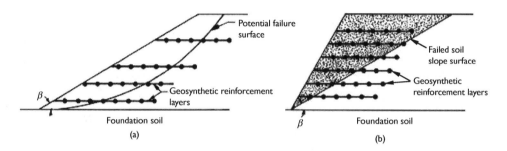

*Figure 3.45* Role of reinforcement in slopes: (a) increase the factor of safety; (b) stabilize the steepened portion of slope (after Simac, 1992).

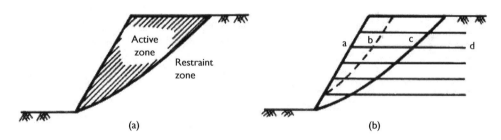

*Figure 3.46* Modes of slope reinforcement failure (after Ingold, 1982).

soil slope. This function may be achieved by the introduction of a series of horizontal reinforcements or restraining members as indicated in Fig. 3.46(b). This arrangement of reinforcement is associated with three prime modes of failure, namely, the following:

* tensile failure of the reinforcement
* pullout from the restraint zone, and
* pullout from the active zone.

Using the horizontal reinforcement, it would be difficult to guard against the latter mode of failure. There may be the problem of obtaining adequate bond lengths. This can be illustrated by reference to Fig. 3.46(b), which shows a bond length a-c for the entire active zone. This bond length may be adequate to generate the required restoring force for the active zone as a rigid mass; however, the active zone contains infinite prospective failure surfaces. Many of these may be close to the face of the batter as typified by the broken line in Fig. 3.46(b) where the bond length would be reduced to length a-b and as such be inadequate to restrain the more superficial slips. This reaffirms the soundness of using encapsulating reinforcement or facing elements where a positive restraining effect can be administered at the very surface of the slope by the application of normal stresses (Fig. 3.47). The shallow or surficial soil slope failure (Fig. 3.48(a)) can also be prevented by installing shorter, more closely spaced, surficial reinforcement layers in addition to the primary reinforcement layers (Fig. 3.48(b)). A second purpose of the surficial reinforcement is to provide a lateral resistance during compaction of the soil.

In the past, limited experimental studies were conducted to understand the behaviour of reinforced soil slopes. Das *et al.* (1996) presented the results of bearing capacity tests for a model strip foundation resting on a biaxial geogrid-reinforced clay slope. The geometric parameters with usual notations of the test model are shown in Fig. 3.49. Based on this study, the following conclusions can be drawn:

1  The first layer of the geogrid reinforcement should typically be located at a depth of $0.4B$ ($B$ = width of footing) below the footing for the maximum increase in the ultimate bearing capacity derived from the reinforcement.
2  The maximum depth of reinforcement, which contributes to the bearing capacity improvement, is about $1.72B$.

Gill *et al.* (2013) presented the results of a series of plane-strain laboratory model load tests carried out on both reinforced and unreinforced fly ash embankment slopes. The tests were conducted by varying the parameters such as embedment ratio, length and number of reinforcement layers, and edge distance from slope crest. A numerical study using the finite-element analysis was also carried out to verify the model test results. The study shows that insertion of a single geogrid reinforcement layer within a depth of 2.5 times the footing width in the fly ash slope considerably improves the load-carrying capacity of footings resting on such slopes. In general, the bearing

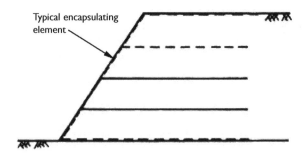

*Figure 3.47* Encapsulating reinforcement (after Ingold, 1982).

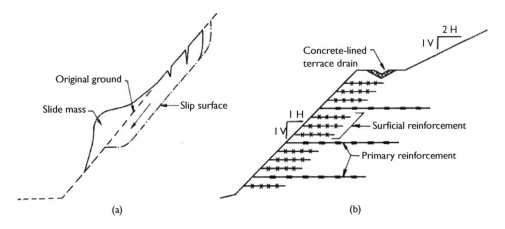

*Figure 3.48* (a) Typical surficial soil slope failure; (b) typical cross-section of reinforced soil slope (after Collin, 1996).

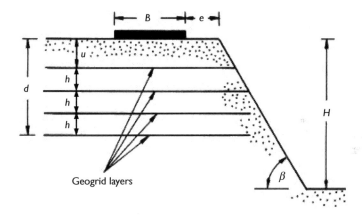

*Figure 3.49* Geometric parameters for a surface strip foundation on the geogrid-reinforced clay slope (after Das *et al.*, 1996).

capacity of the footing increases with an increase in the edge distance up to 3 times the footing width. The optimum number of reinforcement layers has been found to be 4.

Geotextiles, both woven and nonwoven, and geogrids, are being used more and more for reinforcing the steep slopes. Geotextiles, especially the nonwoven ones, exhibit a considerable strain before breaking. Also, a nonwoven geotextile is much less stiff than the ground. Hence the deformation of a geotextile-reinforced soil slope is dominated not by the geotextile but by the soil slope. Due to the large extensibility of nonwoven geotextiles, relatively low stresses are induced in them. Their function, however, is to provide an adequate deformability and to redistribute the forces from the areas of high stresses to the areas of low stresses, thus avoiding the crushing of the soil material. Further, the nonwoven geotextiles facilitate a better drainage and help prevent the build-up of pore pressures that cause a reduction in shear strength.

## 3.5.2 Analysis and design concepts

A higher strength of the reinforced soil structure allows for the construction of steep slopes. Compared with other alternatives, the geosynthetic-reinforced soil slope structures are a cost-effective option for slope stabilization.

From stability considerations, a given or proposed slope should meet the safety requirements, viz. the soil mass under given loads should have an adequate safety factor with respect to the shear failure, and the deformation of the soil mass under the given loads should not exceed the required tolerable limits. The analyses are generally made for the worst conditions, which seldom occur at the time of investigation. The methods, originally developed for analyzing the unreinforced slopes, have been extended to analyze reinforced slopes taking care of the presence of reinforcements. There are basically four methods for analyzing the geosynthetic-reinforced soil slopes (Shukla, 2002b, Shukla 2012a):

1 limit equilibrium method
2 limit analysis method
3 slip line method and
4 finite element method.

The *limit equilibrium method* is most widely used to design the geosynthetic-reinforced soil slopes. Various limit equilibrium methods have been used in different studies (Ingold, 1982; Murray, 1982; Leshchinsky and Volk, 1985, 1986; Schmertmann *et al.*, 1987; Jewel, 1990; Wright and Duncan, 1991). In these methods of analysis, it is considered that the slope failure occurs along an assumed or a known failure surface. At the moment of failure, the shear strength is fully mobilized all the way along the failure surface, and the overall slope and its each part are in static equilibrium. The shear strength required to maintain a condition of limiting equilibrium is compared with the available shear strength, giving the average factor of safety along the failure surface as

$$FS = \frac{\text{Shear strength available}}{\text{Shear strength required for stability}} \qquad (3.26)$$

The shear strength of the soil is normally estimated by using the Mohr-Coulomb strength criterion. The allowable tensile strength of the geotextile layers is taken into account while calculating the available shear strength. Several slip surfaces are considered and the most critical one is identified; the corresponding (smallest) factor of safety is then taken to be the factor of safety of the slope. It should generally be greater than 1.3. The problem is generally considered in two dimensions, that is, the conditions of plane strain are used. A two-dimensional analysis is found to give a conservative result compared to a three-dimensional analysis with a dish-shaped surface.

For an assumed circular arc failure plane within the shallow slope (inclination, $\beta \leq 45°$) reinforced with horizontal geosynthetic layers (Fig. 3.50), the factor of safety, in terms of soil shear strength parameters and allowable tensile strength of the geosynthetic, can be obtained by following the method of slices, commonly used for the slope stability analysis of unreinforced soil slopes, as:

$$FS = \frac{\text{Moment of shear strength of soil and allowable tensile strength of geosynthetic along the failure arc}}{\text{Moment of weight of failure mass}} = \frac{\sum_{i=1}^{n} \left( N_i \tan\phi + c\Delta l_i \right) R + \sum_{j=1}^{m} T_j \, y_j}{\sum_{i=1}^{n} \left( w_i \sin\theta_i \right) R}$$

(3.27)

where $w_i$ is the weight of the $i$th slice; $\theta_i$ is the angle made by the tangent to the failure arc at the centre of the $i$th slice with horizontal; $N_i = w_i \cos\theta_i$; $\Delta l_i$ is the arc length of $i$th slice; $R$ is the radius of circular failure arc; $c$ and $\phi$ are the shear strength parameters of soil, cohesion and angle of shearing resistance (total or effective depending upon the field situations), respectively; $T_j$ is the allowable geosynthetic tensile strength for the $j$th layer; $y_j$ is the moment arm for $j$th geosynthetic layer; $n$ is the number of slices; and $m$ is the number of geosynthetic layers.

The stability of steep reinforced slopes (inclination, $\beta > 45°$) can be analyzed by the tieback wedge analysis approach used for vertical retaining walls, as described in Section 3.3.2.

The limit equilibrium methods do not furnish any information on the soil deformations. Nevertheless, these methods have been very useful in solving the slope stability problems and need less computational efforts. By means of suitable factors of safety, whose choice is largely governed by the experience, the amount of deformation can be limited. It is required to consider separate factors of safety for the soil and the geosynthetic reinforcement because their deformational characteristics are different.

The *limit analysis* is a universal method for correct and accurate solution of the slope stability problems (Sawicki and Lesniewska, 1989; Michalowski and Zhao, 1995; Zhao, 1996; Jiang and Magnan, 1997; Porbaha *et al.*, 2000). It is based on the plasticity theory. This method can be applied to slopes (and also other structures) of arbitrary geometry, complicated loading conditions and homogeneous as well as heterogeneous plastic materials. Using the limit theorems, it is possible to bracket the

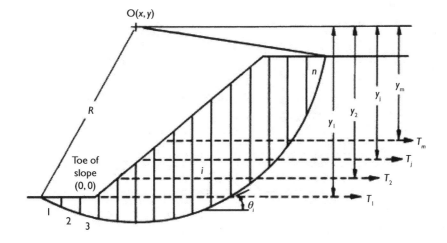

*Figure 3.50* Details of the method of slices for the circular slip analysis.

collapse load even if it cannot be determined exactly. In the lower bound approach, we determine whether there exists an equilibrium stress field, which is in equilibrium with the applied load and with which the plastic yield condition is nowhere violated in the slope. If such a stress field exists, it can be ascertained that the applied load is less than the limit load and no plastic failure will occur in the slope. In the upper bound approach, we search for a kinematically admissible velocity field; we then calculate the corresponding internal and external plastic power dissipations. If the external power dissipation is higher than the internal one, the load can be said to be greater than the limit load. In this way, the limit load can be defined as the load under which there exists a statically admissible stress field; yet a free plastic flow can occur. An efficient and accurate numerical technique like the finite element method is vital to make the limit analysis applicable to the complicated problems of slope stability.

The *slip line method* is based on the derived failure criterion describing the failure of a homogenized geosynthetic-reinforced soil composite and the application of the method of stress characteristics (Anthoine, 1989; de Buhan *et al.*, 1989). The derivation of the failure criterion for a geosynthetic-reinforced soil composite was presented by Michalowski and Zhao (1995). The limit loads on the geosynthetic-reinforced soil slopes can be calculated using the slip line method described by Zhao (1996). This approach is expected to have a wider application in the analysis of slopes with less conventional reinforcements such as continuous filaments or for fibre-reinforced soil slopes.

The *finite element method* of analysis is generally based on a quasi-elastic continuum mechanics approach in which the stresses and the strains are calculated. Since the geosynthetic-reinforced soil slopes exhibit large deformations during the stage construction process, it is appropriate to adopt a nonlinear soil model for the stress-strain analysis with a suitable failure criterion (e.g. Mohr-Coulomb criterion). Such models of varying degrees of complexity have been developed. They require additional parameters, but these can usually be furnished by the standard triaxial test if shear and volumetric strain measurements can be carried out with sufficient accuracy. The geosynthetics are also required to be modeled by an appropriate constitutive model. More details on this method can be found in the works of Rowe and Soderman (1985), Almeida *et al.* (1986) and Ali and Tee (1990).

Among the available methods of analyzing the stability of geosynthetic-reinforced slopes, the limit equilibrium methods are most popular. Essentially, in each method, a failure mechanism is assumed and some of the limit equilibrium requirements are satisfied. Most of the limit equilibrium methods, with their inappropriately oriented slip surfaces, are not correct from the viewpoint of mathematical theory of plasticity, and they do not furnish any information on soil deformations. Ideally, other methods of slope stability analysis, described briefly here, are attractive, but they are really only suited to research studies.

Stabilization of slopes is one of the most challenging tasks for geotechnical engineers. Standardization is not possible due to a variety of cases observed under the field conditions. Use of geosynthetics allows a reduction of earthwork by changing the geometry and also allows the utilisation of soils with average mechanical properties.

Geosynthetic-reinforced slopes are designed to provide the following three basic modes of stability (Simac, 1992):

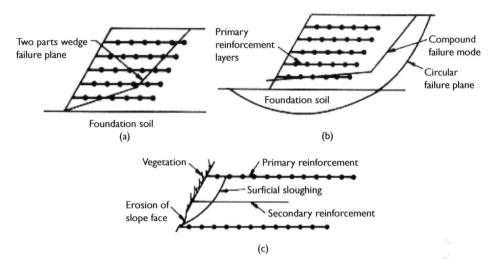

*Figure 3.51* Modes of failure: (a) internal stability; (b) global stability; (c) surficial stability (after Simac, 1992).

1 internal stability (Fig. 3.51(a))
2 global stability (Fig. 3.51(b))
3 surficial stability (Fig. 3.51(c))

The factor of safety must be adequate for both the short-term and the long-term conditions and for all possible modes of failure, similar to those for the unreinforced slopes.

The *internal stability* controls the quantity, strength, length and vertical spacing of the primary reinforcement elements. As the system moves towards failure, the soil deforms, creating a tension in the reinforcement across the failure plane. A limit equilibrium analysis can be carried out to determine the amount of reinforcement tension necessary to maintain the equilibrium with a reasonable factor of safety (minimum $FS = 1.3$).

The conventional slope stability analysis for the potential failures completely around the reinforced soil mass is designated as the *global stability*. The global stability of the entire reinforced soil mass is usually controlled by the foundation upon which it rests. The influence of foundation soil strength, groundwater conditions, soil stratigraphy, imposed surcharge loadings and slope geometry must be analyzed to ensure satisfactory performance. A compound failure mode should also be analyzed, where the potential failure passes partially through the reinforced mass and the soil that is being retained by the reinforced slope.

The *surficial stability* determines the secondary reinforcement requirements of the reinforced slope to preclude surficial sloughing of the slope face during and after construction. Depending on the magnitude of erosive forces imposed on the slope face, the erosion control measures can range from temporary to permanent armoured systems. The available experience suggests a maximum of 60 cm vertical spacing between the secondary reinforcement layers, recommended to be a minimum of 120 cm long. The secondary reinforcement can also be recommended at each compaction lift, about 22–30 cm intervals, in case of some cohesionless soils.

Field and numerical model test results indicate that the limit equilibrium approach to the reinforced slope design provides a suitable though conservative design approach (Christopher *at al.*, 1990). A step-by-step design procedure incorporating the circular arc approach may be found in Christopher and Leshchinsky (1991). Simplified charts for the design of geosynthetic-reinforced slopes have been proposed by some researchers (Jewell and Woods, 1984; Jewel *et al.*, 1984; Christopher and Holtz, 1985). These charts can be used to evaluate the preliminary stability of the geosynthetic-reinforced slopes before more thorough design procedures are performed. The critical reinforced slopes as well as the permanent slopes (having design life greater than 1 to 3 years) should be designed using the comprehensive slope stability analyses. The factor of safety against the slope stability should be taken from the critical surface requiring the maximum reinforcement. The major steps for the design of a reinforced slope can be given as follows:

*Step 1:* Define the geometrical dimensions of the slope (slope height, $H$; slope angle, $\beta$).

*Step 2:* Define the loading conditions (surcharge load, temporary live load and dynamic load).

*Step 3:* Determine the engineering properties (permeability, shear strength and consolidation parameters) of the foundation soils and the slope soils.

*Step 4:* Locate the groundwater table. For the slope and landslide repair projects, identify the cause of instability and locate the previous failure surface.

*Step 5:* Determine the properties (gradation and plasticity index, compaction characteristics, shear strength parameters and chemical composition that may affect the durability of geosynthetic reinforcement) of available reinforced fill.

*Step 6:* Establish the geosynthetic properties (strength and modulus, and soil-geosynthetic interface friction). Also establish the tolerable geosynthetic deformation requirements. The geosynthetic strain can be allowed up to 2–10%. The selection of the geosynthetic should also consider drainage, constructability (survivability) and environmental requirements.

*Step 7:* Determine the factor of safety of the unreinforced slope and determine the geosynthetic reinforcement requirement (vertical spacing and length) based on the internal stability analysis.

*Step 8:* Check the factors of safety against external stability, including sliding, load-bearing failure, foundation settlement, deep-seated slope failure and dynamic stability as carried out for the conventional retaining wall designs assuming that the geosynthetic-reinforced soil mass acts as a rigid body in spite of the fact that it is really quite flexible. The minimum values of factor of safety against sliding, load-bearing failure, deep-seated failure and dynamic loading are generally taken as 1.5, 2, 1.3 and 1.1, respectively.

*Step 9:* Check the requirements for surface and subsurface water control, where the surface water runoff and the drainage are critical for maintaining the slope stability.

The design concepts of some popular stabilization methods are described in the following section along with the application guidelines. Note that the reinforced soil slopes essentially are mechanically stabilized structures with similar behaviour properties and design criteria as those of vertical-faced, gravity mechanically stabilized earth (MSE) walls, such as a reinforced earth or a concrete segmental unit combined with the geosynthetic reinforcement. Therefore, for convenience in designs, one can also consider the reinforced slopes as gravity earth retaining structures with a sloped face.

The reinforced slope design can ideally be carried out using a conventional slope stability computer program modified to account for stabilizing effect of the reinforcement. However, to facilitate the complete design of the geosynthetic-reinforced slopes, many types of software are commercially available, though some are limited to specific soil and reinforcement conditions. Leshchinsky (1997) mentioned 'Reslope' software. For a given problem, including the ultimate strength of the reinforcement layers, the software yields the optimal length and spacing of the geosynthetic layers. This layout satisfies the various specified factors of safety input by the user.

### 3.5.3  Application guidelines

Like the conventional soil slopes, reinforced slopes are generally constructed by compacting soil in layers while stepping the face of the slope back at an angle. Subsequently, the face is protected from erosion by vegetation or other protective systems. Additional geosynthetic elements are incorporated into the reinforced steepened slopes to facilitate drainage, minimize ground water seepage and to assure the stability of the steepened slope and the erosion resistance of the facing (Fig. 3.52).

In the present-day geosynthetic engineering, there are various slope stabilization methods in practice. However, the specific application guidelines are described only for a few popular methods of slope stabilization along with their basic description.

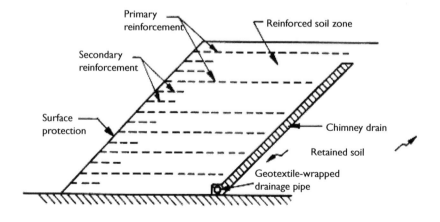

*Figure 3.52* Components of a reinforced steepened slope.

### Geotextile-wrapped drain (GWD) method

This method was proposed by Broms and Wong (1986) and was used successfully in Singapore to stabilize a steep slope in residual soil and weathered rock. By this method, the stability of existing unfailed soil slopes can be increased, failed slopes can be stabilized, or new steep slopes or high embankments can be constructed without exceeding the load-bearing capacity of soil. In these applications, the function of the geotextile, both as a tensile reinforcement and as a filter, is utilized.

In this method, the geotextile-wrapped drains consisting of granular materials are installed along the slopes as shown in Fig. 3.53(a). The drains reduce the pore water pressure within the slopes during the rainy season and thereby the shear strength is increased. The geotextile layer acts as a filter around the drains, which prevents the migration of soil (internal erosion) within the slope into the drains. It also reinforces the soil along the potential sliding zones or planes.

One additional advantage with this method is that the temporary decrease of the stability of the slope is only marginal during the construction of the deep trenches required for the drains. Here, only a limited width of the slope is affected. When concrete gravity or cantilever walls are used, the stability of the slope can be reduced considerably during the construction.

The required spacing of the drains wrapped in a geotextile, as well as dimensions of the drains, depend on the pore water pressures in the slope which can be evaluated by means of a flownet (Fig. 3.53(b)). The granular material in the drains is considered to be infinitely pervious in relation to the slope material. The pore water pressure in the slope is reduced considerably by the drains both above and between the drains as can be seen from the flownet. For general situations, 0.5 m wide and 1.0 m high drains spaced 3.0 m apart would be reasonable.

The drains should be located deep enough so that they intersect potential slip surfaces in the soil. The required depth of the drains depends on the difficulties of excavating trenches along the slopes. The maximum depth is about 4 m. For slopes in residual soils or weathered rocks, this depth is usually sufficient because most slope failures in these materials are shallow, having a maximum depth of failure surface less than 3–4 m.

The required tensile strength of the geotextile can be calculated by considering the force polygon for the sliding soil mass above possible sliding surfaces in the soil (Fig. 3.53(c)). The sliding surface is often located at the contact between the completely weathered and the underlying partially weathered material.

For a planar sliding surface, the orientation of the geotextile-wrapped drains should be perpendicular to the resultant of the normal reaction force and the force that corresponds to the mobilized shear strength along the potential failure surface, as shown in Fig. 3.53(c) in order to utilize the geotextile effectively.

The required number of layers $N$ of the geotextile in each drain can be determined as

$$N = \frac{FRs}{aT} \qquad (3.28)$$

where $R$ is the force per unit width (kN/m) to be resisted by the geotextile; $s$ is the drain spacing (m); $T$ is the tensile strength per unit width (kN/m) of the geotextile; $a$ is the effective perimeter of the drain (m); and $F$ is the factor of safety.

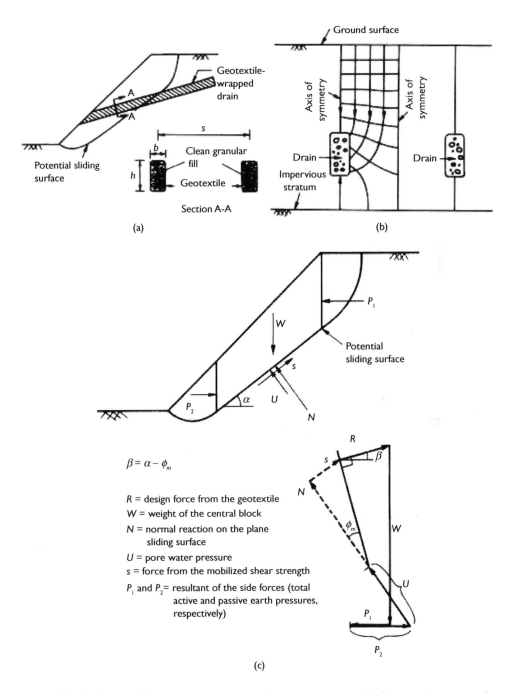

$\beta = \alpha - \phi_m$

R = design force from the geotextile
W = weight of the central block
N = normal reaction on the plane
    sliding surface
U = pore water pressure
s = force from the mobilized shear strength
$P_1$ and $P_2$ = resultant of the side forces (total
    active and passive earth pressures,
    respectively)

(c)

Figure 3.53 (a) Slope stabilization using the geotextile-wrapped drains; (b) flownet showing steady state seepage; (c) computation of design tensile reinforcement provided by the geotextile (after Broms and Wong, 1986).

The geotextiles available in the market generally require an elongation of 14–50%, before the ultimate tensile strength of the geotextile is mobilized. The strain required to mobilize the ultimate strength is much less for woven geotextiles than for nonwoven geotextiles. Only woven geotextiles should therefore be used. In view of the large strain required at failure, a factor of safety of at least three should be used in the design.

The length $L$ that is required to transfer the load in the geotextile to the surrounding soil can be calculated as follows:

$$L = \frac{Rs}{2(hK\sigma'_v + b\sigma'_v)\tan\phi'_a}$$

(3.29)

where $\sigma'_v$ is the vertical effective stress at mid-height (centre) of the drains; $K$ is the lateral earth pressure coefficient for the compacted granular material in the drains; $h$ is the height of the drains; $b$ is the width of the drains; and $\phi'_a$ is the friction angle between the geotextile and the soil.

The deformation $\delta$ of the geotextile to mobilize the required tensile force can be calculated from the following equation:

$$\delta = L \times \frac{e}{100}$$

(3.30)

where $e$ is the percent elongation needed to mobilize the required tensile resistance of the geotextile.

During the construction of the granular fill drains, it is important to compact the fill carefully. The compaction will increase the lateral earth pressure and therefore the friction between the geotextile and the soil results in the reduced transfer length $L$. For a well compacted fill, a value of $K$ equal to at least 1.0 can be used in the calculation of transfer length. The lateral earth pressure is highly dependent on the degree of compaction of the granular fill.

A second important point, with respect to the compaction of the granular fill drains, is that the compaction should be done in the downhill direction in order to pretension the geotextile. In this way, the elongation of the geotextile, which is necessary to mobilize the required tensile force as well as the required displacement of the slope, will be reduced.

### Anchored geosynthetic system (AGS) method

This method was suggested by Koerner (1984) and Koerner and Robins (1986) and is also known as the *anchored spider netting method*. It is an in-situ slope stabilization method in which a geosynthetic material (geotextile, geogrid or geonet) or other porous material is placed directly on the unstable or questionable slope and anchored to it with long steel rod nails at discretely reinforced nodes, 1–2 m apart. These nails must be long enough to penetrate up to, and beyond, the actual or potential failure surface. Figure 3.54 shows the idealized cross-section of a slope stabilized by this method along with its conceptualization. When the rods are properly fastened, they

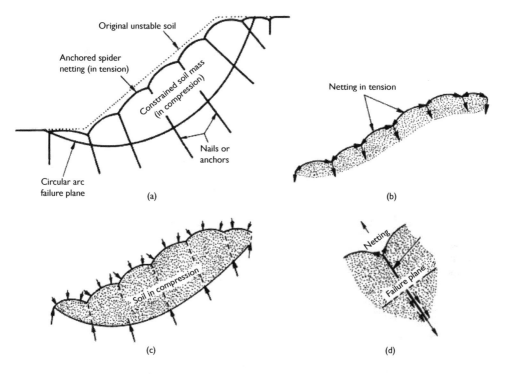

*Figure 5.54* (a) Idealized cross-section of the anchored spider netting in stabilizing a soil slope; (b) free-body diagram of netting; (c) free-body diagram of contained soil; (d) free-body diagram of anchor (after Koerner, 1984; Koerner and Robins, 1986).

begin pulling the surface netting into the soil, thus placing the net in tension and the contained soil in compression. When suitably deployed, this method offers a number of advantages in arresting slope failures, as listed below:

- the steel rods in penetrating the failure surface aid stability;
- the stress caused by netting at the ground surface aids stability;
- The surface netting stress mobilizes the normal stress at the base of the failure surface, which aids stability; and
- The entire system causes soil densification, which increases the shear strength parameters of soil.

It is important to recognize that the mechanism by which an AGS stabilizes a slope is different from that of reinforced earth or soil nailing. Both the reinforced earth ties and the soil nails are passive systems that rely on soil strains to mobilize pullout, bending and shear resistances of the inclusions. By contrast, the anchors in an AGS are actively tensioned during their installation. The increase in stability of the slope thus does not rely on soil movement to mobilize the soil-anchor interaction, but rather the increased stresses on the potential failure surfaces, imparted by the tensioned fabric, increase the stability of the slopes (Ghiassian *et al.*, 1996).

In recognition of the above, the analysis of a slope stabilized by anchored geosynthetics follows a traditional limiting equilibrium stability analysis for slopes. The effects of the AGS are considered as additional forces acting on a potential failure surface. Bending and shearing resistances of the anchors are disregarded for several reasons. First, the bending and the shearing resistances of the anchors are not mobilized. Second, the spacing of anchors is typically greater than the spacing in soil nailing and thus a coherent soil mass may not develop. A slope may fail by the erosion and the flow of soil around anchors; therefore, the bending and shearing resistances of the anchors may be irrelevant. Third, a variety of materials could be used for the anchors in an AGS, including cables with duck-billed anchors, which have essentially no bending resistance. Finally, the assumption is conservative.

### Reinforced soil structures (RSS) method

Slopes can be stabilized with the construction of geosynthetic-reinforced soil structures. If the reinforced soil structures have an inclination $\beta \leq 70°$, they are called the *reinforced soil slopes*; otherwise they are called the *reinforced soil retaining walls*, which have been addressed in Section 3.2. In such stabilization methods, the failed soil can be used as a backfill material to make them economical.

Rimoldi and Jaecklin (1996) summarized the construction methods for the green-faced reinforced soil walls and the steep slopes in four main schemes:

a   Straight reinforcement (Fig. 3.55(a)): This type of reinforcement, made of geosynthetics only, is mainly used for shallow slopes ($\beta \leq 50°$). Generally, the face is left exposed, or covered by a geomat or biomat to prevent erosion. Therefore, the reinforcing geosynthetic is installed just at the face, without any wraparound. The installation is very easy. The reinforcement is laid down horizontally and straight, and then the soil is spread and compacted to the required height, smoothing the face with a vibrating table or with the bucket of a backhoe.

b   Reinforcement wrapped around the face (Fig. 3.55(b)): In this scheme, the geosynthetic is used both for reinforcement of the fill and for the face protection from soil washing and progressive erosion, by wrapping it around the face of the slope. This 'wraparound' technique has been the most widely used construction method in Europe. The 'wraparound' installation procedure can be used with or without formworks. The use of formworks is particularly suggested when it is necessary to have a smooth and uniform face finishing, as discussed in Section 3.2.3.

The most simple construction method with the wraparound technique is without any formwork – it consists in placing a geogrid layer; in laying down, spreading and compacting the fill soil; in smoothing and leveling the face of the slope at the desired angle with a vibrating table or with the bucket of back-hoe; then the geogrid is wrapped around the face and fixed with a 'U' staple. This method provides a fast construction and affords good results if it is not necessary to obtain a perfectly smoothed face. In fact, bulging of the face often occurs, with unpleasant aesthetic effect. Wraparound can also be made using movable formworks, or straight steel mesh or steel mesh shaped as an 'L' or a 'C'. There can also be some other suitable means. If the steel meshes are used, they are left in place after

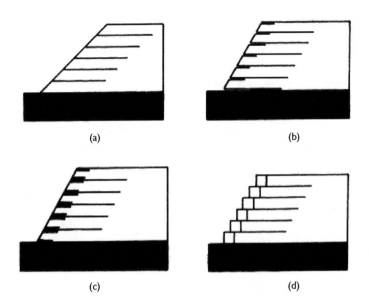

*Figure 3.55* The construction schemes for green-faced structures: (a) straight reinforcement; (b) wraparound reinforcement; (c) mixed scheme; (d) face blocks plus straight reinforcement (after Rimoldi and Jaecklin, 1996).

the construction is terminated, which saves a lot of time and hence allows a very fast construction rate. A typical team of 4–5 workers well equipped and experienced enough, can install about 50 m² of wall face in one working day, but in particular situations, 100 m² of face in one day can be achieved. The reinforcing geosynthetics can be connected to the steel meshes, but usually the two elements are independent.

c   Mixed scheme: straight reinforcement plus another geosynthetic wrapped around the face in 'C' shape (Fig. 3.55(c)): In this scheme, the two functions of reinforcement and face protection are played by two different geosynthetics. The reinforcing geosynthetic has high tensile strength and modulus, while the other one for the face protection is lighter and is engineered to support the growing vegetation and to retain the soil, preventing wash out and erosion.

d   Front blocks tied back by straight reinforcement (Fig. 3.55(d)): In this scheme, a front block is used both to support the facing during construction and for providing the final face finishing. Blocks are usually made of compacted soil, encased in containers, made either of gabion baskets or of geosynthetics wrapped all around. Blocks are mechanically connected to the straight reinforcing geosynthetics. This *front blocks method* has the advantage of not being dependent on the weather situation. The prefabrication does not disturb any traffic and can be near the site. A standard excavator is used to place the face blocks quickly and the same excavator is also used for backfilling. No hydro-seeding is needed because the grass seeds are already included inside the face blocks and grass starts growing immediately.

Over-steep geogrid reinforced slopes are usually associated with vegetation, and the facing of the slope is wrapped around by the geogrids or sometimes the facing is temporarily supported by a steel mesh allowing vegetation to grow through the mesh apertures. Hard facing is also in use with geogrid-reinforced soil walls. Hard facing as opposed to soft facing refers to large precast concrete panels or small modular concrete blocks (MCBs). The MCBs are laid dry (i.e. without mortar) and the geogrid reinforcements are placed between the block courses and connected by means of insert keys or pins or by only the frictional interface between the courses. The footings for the geogrid-reinforced modular concrete block wall systems (GRMCBWSs) can be constructed from granular compacted materials or from cast in place concrete. The walls are usually constructed with stepped facing, resulting in a batter ranging between 5° to 20°. The overall shape is equivalent to a steep slope as opposed to a vertical wall, and therefore the analysis can be carried out using the steep slope procedures. One advantage of GRMCBWSs is the simplicity of installation because the blocks are easily transportable. It is estimated that four persons can erect 30–40 m$^2$ of wall over an eight-hour working day. As for the cost comparison, it is estimated that the walls exceeding 1.0 m in height typically offer a 25–35% cost saving over the conventional cast-in-place concrete retaining walls.

The following points may be followed during the construction of reinforced soil slopes:

1   A level subgrade is prepared by clearing the site and removing all slide debris.
2   Geosynthetic reinforcement should be placed with the principal strength direction perpendicular to the slope face.
3   Lightweight compaction equipment should be used near the slope face to help maintain the face alignment
4   A face wrap may not be required for slopes up to 1 horizontal to 1 vertical, if the reinforcement is maintained at close spacing, not greater than 400 mm.
5   Drainage layers, if required, should be constructed directly behind or on the sides of the reinforced section.

### 3.5.4   Case studies

#### Case study 1

The *geotextile-wrapped drain (GWD) method*, as described in Section 3.5.3, was adopted for the stabilization of a landslide on the campus of the Nanyang Technological Institute (NTI) in Singapore. The landslide occurred in early 1984, during a period of heavy rainfall, on the NTI campus in the western part of Singapore. One student dormitory, Block E, was located at the toe of the slope. Two other dormitories were at the crest. An existing rubble wall, which had been constructed along the whole length of slope with height varying from 1.70 m to 3.50 m, failed during the landslide. The average height of the slope was about 7 m. The inclination of the slope to the horizontal was 37° prior to the failure. A scupper drain at the toe of the rubble wall was damaged and closed up as a result of the movement of the slope. The ground immediately in front of the displaced rubble wall heaved about

200 mm. The whole sliding mass continued to move at a slow rate during the rest of 1984. Large cracks appeared on the displaced rubble wall. The total displacement of the wall was approximately 700 mm in the end of 1984. The toe had moved about 300 mm. The slope was composed of residual soil and completely weathered sedimentary rocks.

The remedial stabilization works at the block E slope consisted of the installation of eight geotextile-wrapped crushed rock drains (Fig. 3.56). The drains, 0.5 m wide and 1.0 m high, were spaced 3.0 m apart. Based on the sliding surface and a residual friction angle of 18°, the required tensile force of the geotextile was 85 kN per metre of the slope or 255 kN per drain. For each drain, four layers of 3.4-m wide PET geotextile, with an ultimate tensile strength of 70 kN/m (238 kN per layer) at 14% elongation, were used. The geotextile was wrapped around the two sides and the bottom of each drain. The drains were connected to the crib wall at the lower end of the slope for drainage. The crib wall was filled with crushed rock to allow discharge of the water from the transverse drains. The horizontal layers of the geotextile were also placed in the slope between the ground surface and the transverse drains to increase the stability of the slope with respect to shallow slides above the transverse drains. The far end of each geotextile strip was anchored in the crushed rock drain. Another layer of the geotextile was placed along the drains between the horizontal strips as a filter to prevent the soil above from being washed/eroded into the drains. No further movements of the slope were observed after the installation of the drains.

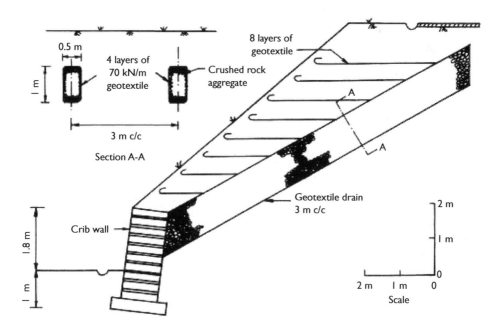

*Figure 3.56* Stabilization scheme – Nanyang Technological Institute block E slide, Singapore (after Broms and Wong, 1986).

## Case study 2

*The anchored geosynthetic system* (AGS) method, as described in Section 3.5.3., has been used to stabilize a 4.5-m high clayey silt slope at a uniform slope angle of 25° (Koerner and Robins, 1986). The slope was in an active state of failure. The slope was hand cleared of vegetation and graded so as not to have any concave depressions. A knitted geonet made of bitumen coated nylon was used as the netting. The anchors were 13-mm diameter steel rods in 1.2 m long sections which were threaded into one another during installation. Since the existing failure zone was a shallow slope failure the rods were only 2.4 m long. This length easily penetrated the failure plane as it drew the net into the surface soil. Two adjacent widths of netting were used on this slope (each being 5.6 m wide) along with a total of 73 steel rod anchors. Upon completion of the anchored spider netting, the slope was seeded with a rapid growing rye grass and mulched. The grass grew within two weeks and had completely hidden the netting. The slope was reported to be in a stable condition.

## Case study 3

Dixon (1993) reported the geogrid-reinforced soil repair of a slope failure in clay on the North Circular Road, London, U.K. Figure 3.57 shows the cross-section of the repair of the slip failure. The slope was a cut slope (side slope = 1V:2H, maximum height = 8 m) formed in London clay in 1975. Some seven years after construction, the slip failures began to occur along a 500-m length of cutting causing the damage to fence lines and spillage onto the carriageway. Main earthworks began with the excavation and removal from the site of a 35-m long strip of slipped soil in September 1985. The excavation extended beyond the failure plane with benched steps cut into the undisturbed clay. To control any seepage, a 300-mm thick granular drainage layer was included over the excavated surface on the north side, where the forest slopes towards the cutting. The general sequence then adopted was to reinstate the first strip using the fill excavated from an adjacent strip, thereby minimizing the double handling. The second strip was then reinstated using the fill excavated from a third strip and so on. The fill was tipped from a dump-truck, placed using the bulldozer and compacted to a maximum layer depth of 200 mm using the vibrating roller towed by the bulldozer. The 2.0 m width of 'Tensar' SS1 secondary reinforcements was obtained by cutting the standard 4.0-m wide rolls into half on site with a disc cutter. The 'Tensar' SR2 rolls were cut to the required length and laid perpendicular to the slope alignment. The adjacent rolls were butt jointed. The slope face was over filled and trimmed in the conventional manner. The earthworks were completed in early February 1986. No special site equipment or expertise was required for installation, which was carried out using the conventional plant and labour. The average construction time per 35 m long strip was about 3 days (typical strip quantities: excavation – 1200 m³, fill – 800 m³, gravel drain – 350 m³, geogrids – 2000 m²). After construction, no discernable movements were noted and the grass cover on the slope was reported to be in a good condition with a pleasing appearance.

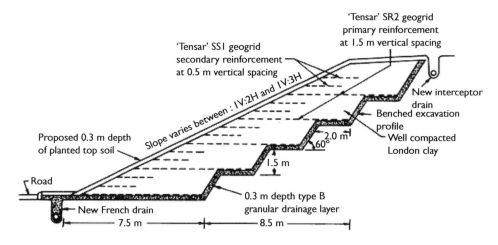

*Figure 3.57* Cross-section of the repair of the slip failure of a slope (after Dixon, 1993).

## Chapter summary

1   A geosynthetic-reinforced soil retaining wall has three basic components: back-fill (usually a granular soil), reinforcement layers (usually woven geotextiles or geogrids) and facing (geosynthetic wraps, segmental/modular concrete blocks, full-height precast concrete panels, welded wire panels, gabion baskets, etc.).

2   The limit equilibrium analysis is the most commonly used and simple design approach for designing the geosynthetic-reinforced soil retaining walls. The analysis and design involve the checks for internal stability, external stability and stability of the facing system. The facing design is affected by the level of compaction of backfill.

3   An embankment over the weak/soft foundation soil can be constructed by placing a geosynthetic (geotextile, geogrid or geocomposite) layer over the soft foundation soil as the basal geosynthetic layer, which serves as reinforcement, drainage, separation, filtration or a combination thereof. More than one geosynthetic layer may be required, if the foundation soil has voids or weak zones.

4   The use of a geosynthetic basal layer is generally attractive for low ratios between the foundation soil thickness and the embankment base width (say, less than 0.7). The analysis and design involve checks for overall slope stability, lateral spread, embankment settlement, overall bearing failure and pullout failure.

5   Use of geosynthetic layers (geotextile, geogrid, geocell or geocomposite) within a controlled granular fill beneath the footings and mat foundations provide improved load-bearing capacity and reduced settlements by distributing the imposed loads over a wider area of weak subsoil. The geosynthetic layer is considered to perform mainly reinforcement and separation functions. The improvement in bearing capacity is usually expressed in terms of the bearing capacity ratio (BCR).

6   The analysis and design of geosynthetic-reinforced soil foundations involve checks for bearing capacity failure of foundation soil above the uppermost geosynthetic layer, pullout of the geosynthetic layer, breaking of the geosynthetic layer and creep failure of the geosynthetic layer.

7   The studies have shown that the use of more than three geosynthetic layers in foundation soils are not beneficial, and the optimum size of the geosynthetic layer is about three times the width of the footing, $B$. For beneficial effects, the geosynthetic layers should be laid within a depth equal to the width of footing. The optimum vertical spacing of geosynthetic reinforcement layers is between $0.2B$ and $0.4B$. For a single layer reinforced soil, the optimum embedment depth is approximately $0.3B$.

8   The geosynthetic-reinforced soil slopes have provided innovative and cost-effective solutions to the slope stabilization problems, particularly after a slope failure has occurred or if a steeper than the safe unreinforced slope is desirable. Reinforced slopes are basically compacted fill embankments that incorporate geosynthetic tensile reinforcement arranged in horizontal planes.

9   The limit equilibrium method is most widely used to design geosynthetic-reinforced soil slopes. The analysis and design of geosynthetic-reinforced soil slopes involve checks for internal stability, global stability and surficial stability.

10   Geotextile-wrapped drain (GWD) and anchored geosynthetic system (AGS) methods are examples of stabilizing the slopes in the field.

## Questions for practice

(Select the most appropriate answer to the multiple-choice questions from Q 3.1 to Q 3.15.)

3.1   Which one of the following geosynthetics can be used as a reinforcement in the construction of reinforced soil retaining walls?
(a)   Nonwoven geotextile
(b)   Woven geotextile
(c)   Geonet
(d)   Geomembrane

3.2   The typical vertical spacing of reinforcement layers in the geotextile-reinforced soil walls varies between
(a)   0.1 m and 0.5 m
(b)   0.1 m and 1.0 m
(c)   1.0 m and 2.0 m
(d)   none of the above.

3.3   In the geotextile-reinforced soil retaining wall, the geotextile layer should be installed with its principal strength direction (warp direction)
(a)   inclined at 45° to the wall face
(b)   inclined at 60° to the wall face

(c) parallel to the wall face
(d) perpendicular to the wall face.

3.4 In the geotextile-reinforced soil retaining wall with a wraparound facing, the minimum lap length should generally be
(a) 0.2 m
(b) 0.5 m
(c) 1.0 m
(d) 1.5 m.

3.5 A nonwoven geotextile layer at the base of an embankment on the soft foundation soil
(a) acts principally as the reinforcement
(b) acts principally as the drainage layer and hence accelerates consolidation and subsequent gain in strength of the foundation soil
(c) acts principally as the separator
(d) causes compaction of the foundation soil.

3.6 The use of a geosynthetic basal layer is generally attractive, if the ratio between the foundation soil thickness and the embankment base width is
(a) lower than 0.7
(b) higher than 0.7
(c) extremely high
(d) any value.

3.7 Which one of the following is not the most critical failure mechanism for embankments on soft foundation soils?
(a) overall slope stability failure
(b) settlement
(c) lateral spreading
(d) overall bearing failure

3.8 The use of geosynthetic reinforcement layers within the foundation soil below the footing may
(a) decrease both the load-bearing capacity and the settlement of the foundation
(b) increase both the load-bearing capacity and the settlement of the foundation
(c) decrease the load-bearing capacity and increase the settlement of the foundation
(d) increase the load-bearing capacity and decrease the settlement of the foundation.

3.9 If $B$ is the footing width, then for a single layer geosynthetic-reinforced foundation soil, the optimum embedment depth of the geosynthetic layer is approximately
(a) $0.1B$
(b) $0.3B$
(c) $0.5B$
(d) $B$.

3.10   The number of geosynthetic reinforcement layers used to strengthen the foundation soil
  (a)  is 1
  (b)  is 2
  (c)  is 3
  (d)  can be greater than 1.

3.11   A cylindrical tank has to be supported on a geogrid-reinforced granular soil pad. If the uniaxial geogrid is used for the construction of the pad, then
  (a)  the main direction of the uniaxial geogrid should be laid radially towards the centre of the tank
  (b)  the main direction of the uniaxial geogrid should be laid perpendicular to the tank wall
  (c)  both (a) and (b)
  (d)  it may not cause any improvement in the strength and stiffness of the granular soil.

3.12   If the reinforced soil structures have an inclination to the horizontal $\beta > 70°$, they are called
  (a)  reinforced soil slopes
  (b)  reinforced soil retaining walls
  (c)  reinforced soil foundations
  (d)  both (a) and (b).

3.13   Within a soil slope, the geosynthetic sheets are usually placed in
  (a)  horizontal planes
  (b)  inclined planes towards the slope face
  (c)  inclined planes away from the slope face
  (d)  vertical planes.

3.14   The geosynthetic strain in slope stabilization applications can be allowed up to
  (a)  2% to 10%
  (b)  5% to 15%
  (c)  10% to 20%
  (d)  15% to 30%.

3.15   By the geotextile-wrapped drain (GWD) method of slope stabilization,
  (a)  stability of existing unfailed soil slopes can be increased
  (b)  failed slopes can be stabilized
  (c)  new steep soil slopes can be constructed
  (d)  all of the above.

3.16   Describe the basic components of a geosynthetic-reinforced soil retaining wall.

3.17   How can you make a geotextile wraparound facing UV-resistant?

3.18   Can you recommend fine-grained soils as a backfill material? Justify your answer.

3.19   What are the various failure modes of geosynthetic-reinforced soil retaining walls? Explain them briefly.

3.20   List the factors governing the lap length in wraparound facing of a geotextile-reinforced soil retaining wall.

3.21 Design a 10-m high geotextile-wrapped soil retaining wall with the following data:

*For the granular backfill*
Total unit weight, $\gamma_b = 18 \text{ kN/m}^3$
Angle of internal friction, $\phi'_b = 32°$

*For the geotextile*
Allowable tensile strength, $\sigma_{all} = 35 \text{ kN/m}$

*For the foundation soil*
Cohesion, $c = 30 \text{ kPa}$
Total unit weight, $\gamma = 16.8 \text{ kN/m}^3$
Angle of internal friction, $\phi = 15°$

*Foundation soil – geotextile interface shear parameters*
Friction angle, $\phi'_i = 0.95\phi$
Adhesion, $c_a = 0.9c$
Factor of safety against the geotextile rupture = 1.5
Factor of safety against the geotextile pullout = 1.5

3.22 What precautions should be taken during the construction of geotextile-reinforced soil retaining walls?

3.23 Describe a construction method for the permanent geosynthetic-reinforced soil retaining wall, widely used in Japan.

3.24 What are the special features of the case study, presented by Gourc and Risseeuw (1993), on a geotextile-reinforced wraparound faced wall built in late 1982 in Prapoutel, France.

3.25 What do you mean by a basal geosynthetic layer?

3.26 List the factors that may be of major concern when choosing the basal geosynthetic to function as a reinforcement.

3.27 What are the potential failure modes of an embankment on the soft foundation soil? Describe briefly.

3.28 A 3.5-m high and 7-m wide embankment is to be built on soft ground with a basal geotextile layer. Calculate the geotextile strength and modulus required in order to prevent block sliding on the geotextile. Assume that the embankment material has a unit weight of $17 \text{ kN/m}^3$ and angle of shearing resistance of $32°$ and that the geotextile-soil interface angle of shearing resistance is two-thirds of that value.

3.29 List the various alternatives with regard to the installation of a geosynthetic layer inside the embankment along with their advantages.

3.30 Draw a neat sketch to show the construction sequence for geosynthetic-reinforced embankments over soft foundation soils.

3.31 What is the purpose of placing narrow horizontal strips with wraparound along the side slopes of the embankment?

3.32 What should be the geosynthetic orientation in the linear embankments?

3.33 What were the challenging tasks during the construction of a long embankment through the tidal area of Deep Bay in the New Territories district of Hong Kong, as reported by Risseeuw and Voskamp (1993)?

3.34 Describe the ideal reinforcement pattern below a shallow footing. What are the difficulties in adopting this pattern in real-life projects? Can you suggest an effective practical reinforcement pattern?

3.35   What are the advantages of wraparound ends/edges of the geosynthetic reinforcement layers within the foundation soil?

3.36   Why are the improvements in load-bearing capacity of geosynthetic-reinforced
foundation soil not very significant at low deformations and considerably
better at high deformations? Can you suggest a method to improve the low-
deformation behaviour?

3.37   What are the different failure modes of geosynthetic-reinforced foundation
soils? Explain briefly.

3.38   Does the presence of a geosynthetic layer inside the granular fill modify the load
transmission mechanism through it?

3.39   List the parameters affecting the load-bearing capacity of a geosynthetic-reinforced foundation soil. Describe the effects of the most significant parameters.

3.40   What is BCR? Can it be lower than one?

3.41   Compare a typical load-settlement curve for a geosynthetic-reinforced soil with
that for an unreinforced soil.

3.42   What is the effect of prestressing the geosynthetic reinforcement on the settlement behaviour of a geosynthetic-reinforced soil?

3.43   Give a brief description of the geocell mattress foundation. How will you assess
its load-bearing capacity?

3.44   What are the prime modes of geosynthetic failure in a slope stabilization
application?

3.45   List the findings of the model test carried out by Das *et al.* (1996) on a geogrid-
reinforced clay slope.

3.46   Suggest a suitable reinforcement layout to prevent the shallow slips in the
shoulder of a 7.5-m high clay embankment with 1V:2H side slopes.

3.47   List the limitations of the limit equilibrium approach for designing geosynthetic-reinforced soil slopes.

3.48   Given a 12-m high embankment at a slope angle of $\beta = 42°$; the soil strength
parameters are $c = 20$ kPa and $\phi = 20°$ in both the embankment and the foundation sections. The total unit weight of soil, $\gamma = 17$ kN/m³. For a failure circle
located at coordinates (+2, +15) with respect to the toe at (0, 0) (see Fig. 3.50),
determine the factor of safety assuming a radius of 20 m. How many layers
of geotextiles spaced 250 mm apart and having an allowable tensile strength
of 70 kN/m placed at the interface of the foundation and the embankment
are required to increase the factor of safety by 30%? Developing a computer
programme, find the minimum factor of safety for both unreinforced and reinforced conditions.

3.49   Describe briefly the geotextile-wrapped drain (GWD) method of slope stabilization. What are the special features of this method?

3.50   What are the components of reinforced steepened slope?

3.51   Illustrate the various mechanisms that the soil nails in anchored spider netting
method provide in the soil slope stabilization.

3.52   In the anchored spider netting method of slope stabilization, the geotextile
is exposed on the slope surface. What advantages and disadvantages do you
observe?

# References

AASHTO (American Association of State Highway and Transportation Officials) (2007). *LRFD Bridge Design Specifications*, SI Units, Fourth Edition, American Association of State Highway and Transportation Officials, Washington, DC, USA.

Abramson, L.W., Lee, T.S., Sharma, S. and Boyce, G. (2002). *Slope Stability and Stabilization Methods*. John Wiley & Sons, Inc., New York.

Ali, F.H. and Tee, H.E. (1990). Reinforced slopes: field behaviour and prediction. *Proceedings of the 4th International Conference on Geotextiles, Geomembranes and Related Products*. The Hague, Netherlands, pp. 17–20.

Almeida, M.S.S., Britto, A.M. and Parry, R.H.G. (1986). Numerical modeling of a centrifuged embankment on soft clay. *Canadian Geotechnical Journal*, **23**, pp. 103–114.

Alston, C., Lowry, D.K. and Lister, A. (2015). Geogrid reinforced granular pad resting on loose and soft soils, Hamilton Harbour, Ontario. *International Journal of Geosynthetics and Ground Engineering*, **1**, 3, 21, pp. 1–11.

Anthoine, A. (1989). Mixed modeling of reinforced soils within the framework of the yield design theory. *Computers and Geotechnics*, **7**, 1 & 2, pp. 67–82.

Bassett, R.H. and Last, N.C. (1978). Reinforcing earth below footings and embankments. *Proceedings of the Symposium on Earth Reinforcement*, ASCE, New York, pp. 202–231.

Bathurst, R.J., Hatami, K. and Alfaro, M.C. (2012). Geosynthetic-reinforced soil walls and slopes – seismic aspects. Chapter 16, *Handbook of Geosynthetics Engineering*, Second Edition, Shukla, S.K., Editor, ICE Publishing, London, pp. 317–363.

Binquet, J. and Lee, K.L. (1975a). Bearing capacity analysis of reinforced earth slabs. *Journal of the Geotechnical Engineering Division*, ASCE, **101**, 12, pp. 1257–1276.

Binquet, J. and Lee, K.L. (1975b). Bearing capacity tests on reinforced earth slabs. *Journal of the Geotechnical Engineering Division*, ASCE, **101**, 12, pp. 1241–1255.

Bonaparte, R. and Christopher, B.R. (1987). Design and construction of reinforced embankments over weak foundations. Proceedings of the Symposium on Reinforced Layered Systems. *Transportation Research Record*, 1153, Transportation Research Board, Washington, D.C., pp. 26–39.

Broms, B.B. and Wong, I.H. (1986). Stabilization of slopes in residual soils with geofabric. *Proceedings of the 3rd International Conference on Geotextiles*. Vienna, Austria, pp. 295–300.

Broms, B.B. and Wong, K.S. (1990). Landslides. In: *Foundation Engineering Handbook* edited by H.Y. Fang, Van Nostrand Reinhold, New York.

BS 8006-1 (2010). Code of Practice for Strengthened/Reinforced Soils and Fills. British Standards Institution, London, UK.

Bush, D.I., Jenner, C.G. and Bassett, R.H. (1990). The design and construction of geocell foundation mattress supporting embankments over soft ground. *Geotextiles and Geomembranes*, **9**, 1, pp. 83–98.

CGS (Canadian Geotechnical Society) (2006). *Canadian Foundation Engineering Manual*. Fourth Edition, BiTech, Richmond, BC, Canada.

Chandaluri, V.K., Sawant, V.A. and Shukla, S.K. (2015). Seismic stability analysis of reinforced soil wall using horizontal slice method. *International Journal of Geosynthetics and Ground Engineering*, **1**, 3, pp. 23:1–23.10.

Christopher, B.R. and Holtz, R.D. (1985). *Geotextile Engineering Manual*. Report No. FHWA-TS-86/203, Federal Highway Administration, Washington, D.C.

Christopher, B.R., Gill, S.A., Giroud, J.P., Juran, I., Mitchell, J.K., Schlosser, F. and Dunnicliff, J.D. (1990). *Reinforced Soil Structures, Volume I: Design and Construction*. Federal Highway Administration, Washington, DC, Report FHWA-RO-89-043.

Christopher, B.R. and Leschinsky, D. (1991). Design of geosynthetically reinforced slopes. *Proceedings of the Geotechnical Engineering Congress*, Boulder, Colorado, pp. 988–1005.

Collin, J.G. (1996). Controlling surficial stability problems on reinforced steepened slopes. *Geotechnical Fabrics Report*, pp. 26–29.

Das, B.M., Omar, M.T. and Singh, G. (1996). Strip foundation on geogrid-reinforced clay. *Proceedings of the First European Geosynthetics Conference*. Eurogeo 1, Netherlands, pp. 419–426

de Buhan, P., Mangiavcchi, R., Nova, R., Pellegrini, G. and Salencon, J. (1989). Yield design of reinforced earth walls by homogenization method. *Geotechnique*, **39**, 2, pp. 189–201.

Dikran, S.S. and Rimoldi, P. (1996). Hard facing for steep reinforced slopes: A case history from the UK. *Proceedings of the First European Geosynthetics Conference*. Eurogeo 1, Netherlands, pp. 131–136.

Dixon, J.H. (1993). Geogrid reinforced soil repair of a slope failure in clay, North Circular Road, London, United Kingdom. In: *Geosynthetics Case Histories* edited by G.P. Raymond and J.P. Giroud on behalf of ISSMFE Technical Committee TC9, Geotextiles and Geosynthetics, pp. 168–169.

Fowler, J. and Koerner, R.M. (1987). Stabilization of very soft soils using geosynthetics. *Proceedings of the Geosynthetics '87*, New Orleans, Louisiana, IFAI, St Paul, MN, pp. 289–300.

Ghiassian, H., Hryciw, R.D. and Gray, D.H. (1996). Laboratory testing apparatus for slopes stabilized by anchored geosynthetics. *Geotechnical Testing Journal*, **19**, 1, pp. 65–73.

Gill, K.S., Choudhary, A.K., Jha, J.N. and Shukla, S.K. (2013). Experimental and numerical studies of loaded strip footing resting on reinforced fly ash slope. *Geosynthetics International*, **20**, 1, pp. 13–25.

Gourc J.P. and Risseeuw, P. (1993). Geotextile reinforced wrap around faced wall, Prapoutel, France. *Geosynthetics Case Histories* edited by G.P. Raymond and J.P. Giroud on behalf of ISSMFE Technical Committee TC9, Geotextiles and Geosynthetics, pp. 272–273.

Guido, V.A., Biesiadecki, G.L. and Sullivan, M.J. (1985). Bearing capacity of a geotextile-reinforced foundation. *Proceedings of the 11th International Conference on Soil Mechanics and Foundation Engineering*. San Francisco, Calif. pp. 1777–1780.

Haliburton, T.A., Douglas, P.A. and Fowler, J. (1977). Feasibility of Pinto Island as a long-term dredged material disposal site. Miscellaneous Paper, D-77-3, U.S. Army Waterways Experiment Station.

Ingold, T.S. (1982). An analytical study of geosynthetic reinforced embankments. *Proceedings of the 2nd International Conference on Geotextiles*. Las Vegas, USA, pp. 683–688.

IRC (Indian Roads Congress) (2013). IRC: 113-2013, *Guidelines for the Design and Construction of Geosynthetic Reinforced Embankments on Soft Subsoils*. Indian Roads Congress, New Delhi, India.

IRC (2014). IRC:SP:102 (2014), *Guidelines for Design and Construction of Reinforced Soil Walls*. Indian Roads Congress, New Delhi, India.

Jewell, R.A. (1990). Revised design charts for steep reinforced slopes. *Reinforced Embankments, Theory and Practice*, Thomas Telford, London, pp. 1–30.

Jewell, R.A. and Woods, R.I. (1984). Simplified design charts for steep reinforced slopes. *Proceedings of the Symposium on Reinforced Soil*, University of Mississipi.

Jewel, R.A., Paine, N. and Woods, R.I. (1984). Design methods for steep reinforced embankments. *Proceedings of the Symposium of Polymer grid reinforcement*, Institute of Civil Engineering, London, pp. 18–30.

Jha, J.N., Choudhary, A.K., Gill, K.S. and Shukla, S.K. (2013). Bearing capacity of a strip footing resting on reinforced fly ash slope: an analytical approach. *Indian Geotechnical Journal*, **43**, 4, pp. 354–366.

Jiang, G.L. and Magnan, J.P. (1997). Stability analysis of embankments: comparison of limit analysis with method of slices. *Geotechnique*, **47**, 4, pp. 857–872.

Kazi, M., Shukla, S.K. and Habibi, D. (2015). An improved method to increase the load-bearing capacity of strip footing resting on geotextile-reinforced sand bed. *Indian Geotechnical Journal*, India, **45**, 1, pp. 98–109.

Kazi, M., Shukla, S.K. and Habibi, D. (2015). Behaviour of embedded footing on geotextile-reinforced sand. *Ground Improvement*, DOI: 10.1680/grim.14.00022.

Koerner, R.M. (1984). In-situ soil stabilization using anchored nets. *Proceedings of the Conference on Low Cost and Energy Saving Construction Methods*. Rio de Janeiro, Brazil, pp. 465–478.

Koerner, R.M. and Robins, J.C. (1986). In-situ stabilization of soil slopes using nailed geosynthetics. *Proceedings of the 3nd International Conference on Geotextiles*. Vienna, Austria, pp. 395–400.

Koerner, R.M. (2005). *Designing with geosynthetics*. 5th edition, Prentice Hall, New Jersey, USA.

Leshchinsky, D. (1997). Software to facilitate design of geosynthetic-reinforced steep slopes. *Geotechnical Fabrics Report*, January-February 1997, pp. 40–46.

Leshchinsky, D. and Volk, J.C. (1985). Stability charts for geotextile reinforced walls. *Transportation research Record, No.* 1131, pp. 5–16.

Leshchinsky, D. and Volk, J.C. (1986). Predictive equation for the stability of geotextile reinforced earth structure. *Proceedings of the 3rd International Conference on Geotextiles*, Vienna, Austria, pp. 383–388.

Lovisa, J., Shukla, S.K. and Sivakugan, N. (2010). Behaviour of prestressed geotextile-reinforced sand bed supporting a loaded circular footing. *Geotextiles and Geomembranes*, 28, 1, pp. 23–32.

Michalowski, R.L. and Zhao, A. (1995). Continuum versus structural approach to stability of reinforced soil structures. *Journal of Geotechnical Engineering*, ASCE, 121, 2, pp. 152–162.

Murray, R. (1982). Fabric reinforcement of embankments and cuttings. *Proceedings of the 2nd International Conference on Geotextiles*, Las Vegas, USA, pp. 707–713.

NCMA (National Concrete Masonry Association) (1997). *Design Manual for Segmental Retaining Walls*. National Concrete Masonry Association, Herndon, VA.

Ochiai, H., Tsukamoto, Y., Hayashi, S., Otani, J. and Ju, J.W. (1994). Supporting capability of geogrid-mattress foundation. *Proceedings of the 5th International Conference on geotextiles, Geomembranes and related Products*, Singapore, pp. 321–326.

Paulson, J.N. (1987). Geosynthetic material and physical properties relevant to soil reinforcement applications. *Geotextiles and Geomembranes*, 6, 1–3, pp. 211–223.

Palmeira, E.M. (2012). Embankments. Chapter 5, *Geosynthetics and Their Applications*, Shukla, S.K., Editor, Thomas Telford, London, pp. 101–127.

Porbaha, A., Zhao, A., Kobayashi, M. and Ishida, T. (2000). Upper bound estimate of scaled reinforced soil retaining walls. *Geotextiles and Geomembranes*, 18, 6, pp. 403–413.

Rimoldi, P. and Jaecklin, F. (1996). Green faced reinforced soil walls and steep slopes: the state-of-the-art in Europe. *Proceedings of the First European Geosynthetics Conference*. Eurogeo 1, Netherlands, pp. 361–380.

Risseeuw, P. and Voskamp, W. (1993). Geotextile reinforced embankment over tidal mud, Deep Bay, New Territories, Hong Kong. *Geosynthetics Case Histories* edited by G.P. Raymond and J.P. Giroud on behalf of ISSMFE Technical Committee TC9, Geotextiles and Geosynthetics, pp. 230–231.

Rowe, R.K. and Soderman, K.L. (1985). An approximate method for estimating the stability of geotextile reinforced embankments. *Canadian Geotechnical Journal*, 22, 3, pp. 392–398.

Sawicki, A. (2000). *Mechanics of Reinforced Soil*. A.A. Balkema, Rotterdam

Sawicki, A. and Lesniewska, D. (1989). Limit analysis of cohesive slopes reinforced with geotextiles. *Computers and Geotechnics*, 7, 1–2, pp. 53–66.

Schmertmann, G.R., Chourery-Curtis, V.E., Johnson, R.D., Bonaparte, R. (1987). Design charts for geogrid reinforced soil slopes. *Proceedings of Geosynthetics '87*. New Orleans, pp. 108–120.

Selvadurai, A.P.S. (1979). *Elastic Analysis of Soil-Foundation Interaction*. Elsevier, Amsterdam.

Shukla, S.K. (1995). *Foundation model for reinforced granular fill – soft soil system and its settlement response*. PhD Thesis, Indian Institute of Technology Kanpur, India.

Shukla, S.K. (1997). A study on causes of landslides in Arunachal Pradesh. *Proc. of Indian Geotechnical Conference*, Vadodara, India, Vol. 1: pp. 613–616.

Shukla, S.K. (2002a). Shallow foundations. Chapter 5, *Geosynthetics and Their Applications*, Shukla, S.K., Editor, Thomas Telford, London, pp. 123–163.

Shukla, S.K. (2002b). *Geosynthetics and Their Applications*. Thomas Telford, London.

Shukla, S.K. (2012a). *Handbook of Geosynthetic Engineering*. Second Edition, ICE Publishing, London.

Shukla, S.K. (2012b). Shallow foundations. Chapter 6, *Handbook of Geosynthetics Engineering*, Second Edition, Shukla, S.K., Editor, ICE Publishing, London, pp. 129–161.

Shukla, S.K. (2014). *Core Principles of Soil Mechanics*. ICE Publishing, London.

Shukla, S.K. (2015). *Core Concepts of Geotechnical Engineering*. ICE Publishing, London.

Shukla, S.K. and Baishya, S. (1998). *Documentation and Identification of Geotechnical Parameters pertaining to Active Landslides on National Highway 52A between Banderdewa and Itanagar*. Report of the project sponsored by the Faculty of Natural Disaster Management, North Eastern Regional Institute of Science and Technology, Itanagar, Arunachal Pradesh, India.

Shukla, S.K. and Chandra, S. (1994a). The effect of prestressing on the settlement characteristics of geosynthetic-reinforced soil. *Geotextiles and Geomembranes*, 13, 8, pp. 531–543.

Shukla, S.K. and Chandra, S. (1994b). A generalized mechanical model for geosynthetic-reinforced foundation soil. *Geotextiles and Geomembranes*, 13, 12, pp. 813–825.

Shukla, S.K. and Chandra, S. (1994c). A study of settlement response of a geosynthetic-reinforced compressible granular fill – soft soil system. *Geotextiles and Geomembranes*, 13, 9, pp. 627–639.

Shukla, S.K. and Chandra, S. (1995). Modelling of geosynthetic-reinforced engineered granular fill on soft soil. *Geosynthetics International*, 2, 3, pp. 603–618.

Shukla, S.K. and Chandra, S. (1996a). Settlement of embankment on reinforced granular fill – soft soil system. *Proceedings of the International Symposium on Earth Reinforcement, Fukuoka, Japan*, pp. 671–674.

Shukla, S.K. and Chandra, S. (1996b). A study on a new mechanical model for foundations and its elastic settlement response. *International Journal for Numerical and Analytical Methods in Geomechanics*, 20, 8, pp. 595–604.

Shukla, S.K. and Chandra, S. (1997). Effect of densification on the settlement behaviour of reinforced granular fill – soft soil system. *Proceedings of the Indian Geotechnical Conference, Vadodara, India*, December 1997, Vol. 1, pp. 337–340.

Shukla, S.K. and Chandra, S. (1998). Time-dependent analysis of axi-symmetrically loaded reinforced granular fill on soft subgrade. *Indian Geotechnical Journal*, India, 28, 2, pp. 195–213.

Shukla, S.K. and Kumar, R. (2008). Overall slope stability of the prestressed geosynthetic-reinforced embankment on soft ground. *Geosynthetics International*, 15, 2, pp. 165–171.

Shukla, S.K. and Yin, J.H. (2003). Time-dependent settlement analysis of a geosynthetic-reinforced soil. *Geosynthetics International*, 10, 2, pp. 70–76.

Shukla, S.K. and Yin, J.H. (2006). *Fundamentals of Geosynthetic Engineering*. Taylor and Francis, London.

Simac, M.R. (1992). Reinforced slopes: a proven geotechnical innovation. *Geotechnical Fabrics Report*, pp. 13–25.

Steward, J., Williamson, R. and Nohney, J. (1977). Guidelines for use of fabrics in construction and maintenance of low-volume roads. Report No. FHWA-TS-78-205.

Tatsuoka, F., Tateyama, M., Uchimura, T. and Koseki, J. (1997). Geosynthetic-reinforced soil retaining wall as important permanent structures. *Geosynthetics International*, 4, 2, pp. 81–136.

Wayne, M. and Han, J. (1998). On-site soil usage with geogrid-reinforced segmental retaining wall (SRWs). *Geotechnical Fabrics Report*, pp. 20–25.

Wright, S.G. and Duncan, J.M. (1991). Limit equilibrium stability analyses for reinforced slopes. *Transportation Research Report*, No. 1330, pp. 40–46.

Wrigley, N.E. (1987). Durability and long-term performance of 'Tensar' polymer grids for soil reinforcement. *Materials Science and Technology*, Vol. 3, pp. 161–170.

Yin, J.H. (1997a). Modelling geosynthetic-reinforced granular fills over soft soil. *Geosynthetics International*, 4, 2, pp. 165–185.

Yin, J.H. (1997b). A nonlinear model of geosynthetic-reinforced granular fill over soft soil. *Geosynthetics International*, 4, 5, pp. 523–537.

Zanzinger, H. and Gartung, E. (2002). Performance of a geogrid reinforced railway embankment on piles. *Proceedings of the Seventh International Conference on Geosynthetics*, France, pp. 381–386.

Zhao, A. (1996). Limit analysis of geosynthetic-reinforced slopes. *Geosynthetics International*, 3, 6, pp. 721–740.

## Answers to selected questions

3.1   (b)
3.3   (d)
3.5   (b)
3.7   (c)
3.9   (b)
3.11  (c)
3.13  (a)
3.15  (d)

## Chapter 4

# Hydraulic and geoenvironmental applications of geosynthetics

## 4.1 INTRODUCTION

Filters and drains are key components of many hydraulic structures. Erosion/movement of soil by moving water and/or wind is a common problem in most soil structures. Containment facilities (landfills, ponds, reservoirs, canals and dams) are essentially required for the storage and/or disposal of materials in the form of liquids and solids. Seepage control in tunnels and underground structures requires an effective mechanism to keep them dry or leak-proof for their proper utilization. Geosynthetics are used in all these areas of hydraulic and geoenvironmental engineering. This chapter presents the use of geosynthetics in filters and drains, erosion control for slopes, containment facilities and tunnel linings. For each of these applications of geosynthetics, the details are described focusing on the basic description, the analysis and design concepts, the application guidelines, and some case studies.

## 4.2 FILTERS AND DRAINS

### 4.2.1 Basic description

The role of groundwater flow and good drainage is considered routinely in the stability of pavements, foundations, retaining walls, slopes, and waste-containment systems. Hence in the past several decades, geosynthetics have been in regular use either as filters, in the form of geotextiles (nonwovens and lightweight wovens), in conjunction with granular materials and/or pipes (Fig. 4.1(a)), or as both filters and drains in the form of geocomposites (Fig. 4.1(b)). The filters also form an essential part of many types of hydraulic structures. Thus, there are several application areas for filters and drains, including buried drains as pavement edge drains/underdrains, seepage water transmission systems in pavement base course layers and railway tracks, abutments and retaining wall drainage systems, slope drainage, erosion control systems, landfill leachate collection systems, drains to accelerate consolidation of soft foundation soils, drainage blankets to dissipate the excess pore pressure beneath the embankments and within the dams, and silt fences/barriers.

A filter consists of any porous material that has openings small enough to prevent movement of soil into the drain, and that is sufficiently pervious to offer little resistance to seepage. When a geosynthetic is used as a filter in drainage applications,

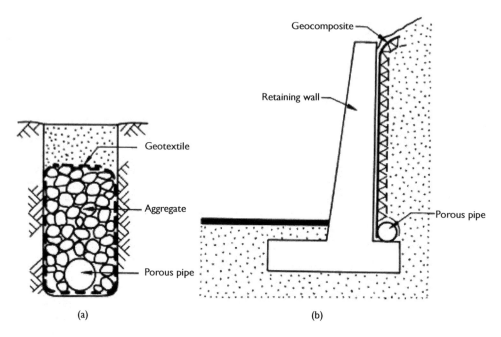

*Figure 4.1* (a) A use of the geotextile filter; (b) a use of the drainage geocomposite.

it prevents upstream soils from entering adjacent granular layers or subsurface drains. When properly designed, the geosynthetic filter promotes the unimpeded flow of water by preventing the unacceptable movement of fines into the drain, which can reduce the performance of the drain. Geosynthetic filters are being used successfully to replace conventional graded granular filters in several drainage applications. In fact, the filter structures can be realized by using granular materials (i.e. crushed stone) or geotextiles or a combination of these materials (Fig. 4.2). The choice between a graded granular filter or a geotextile filter depends on several factors. In general, the geotextile filters provide easier and more economical placement/installation, and continuity of the filter medium is assured whether the construction is below or above the water level. In addition, quality control can be ensured more easily for the geotextile-filter systems. Table 4.1 provides a comparison of granular and geotextile filters.

When using the riprap/geotextile filter, it is recommended that a layer of aggregate should be placed between the geotextile and the riprap, for the following reasons (Giroud, 1992):

- to prevent the damage of the geotextile by large rock pieces;
- to prevent the geotextile degradation by light passing between large rocks;
- to apply a uniform pressure on the geotextile, thereby ensuring a close contact between the geotextile filter and the sloping ground, which is necessary to ensure a proper filtration;
- to prevent the geotextile movement between the rocks because of wave action, thereby ensuring permanent contact between the geotextile filter and the sloping ground, which is also necessary to ensure a proper filtration.

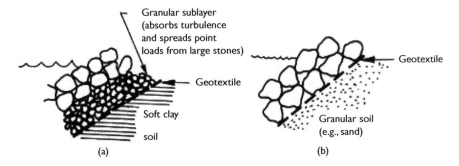

*Figure 4.2* Filter layers using geotextile.

*Table 4.1* Comparison of granular and geotextile filters (modified from Pilarczyk, 2000).

| Objective | Granular filter | Geotextile filter |
|---|---|---|
| **Similarities** | | |
| • Complex structure and distribution of open area | | |
| • Sensitive in respect to changing permeability | | |
| **Differences** | | |
| Determination of characteristic opening size | By particle-size analysis | By pore-size analysis |
| Thickness (filtration length) | Long | Very short |
| Porosity | 25–40% | 75–95% |
| Compactibility | Low | High for needle-punched nonwoven geotextiles |
| Uniformity | Natural variation in grading and density | Greater uniformity due to industrial processing and control |
| Transmissivity | Independent of stresses | Often dependent on exerted stresses |
| Internal stability | Can be unstable | Stable |
| Durability | High | High but not yet properly defined |
| Placement and execution | Proper quality control necessary, more excavation required | Quick and relatively easy placement, less excavation required |

In sediment control applications like the silt fences/barriers used to remove soil from runoff, the filter performance of geosynthetics are evaluated in terms of the *filtering efficiency (FE)*. This term is defined as the percentage of sediment removed from the sediment-laden water by a geosynthetic over a specified period of time.

It is a misconception that a geosynthetic can replace the function of the granular filter completely. A granular filter serves also other functions related to its thickness and weight. It can often be needed to damp (i.e. to reduce) the hydraulic loadings (internal gradients) to an acceptable level at the soil interface. After which a geotextile can be applied to fulfil the filtration function.

Geotextiles with high in-plane drainage ability and several geocomposites are nowadays commercially available for use as drains themselves, thus replacing the traditional granular drains. The drainage geocomposites consist of drainage cores of extruded and fluted sheets, three dimensional meshes and mats, random fibres and geonets, which are covered by a geotextile on one or both sides to act as a filter. The cores are usually produced using polyethylene, polypropylene or polyamide (nylon). Geocomposite drains may be prefabricated or fabricated on the site. They offer readily available material with known filtration and hydraulic flow properties, easy installation and therefore construction economies, and protection of any waterproofing applied to the structure's exterior (Hunt, 1982).

Vertical strip drains (also known as the prefabricated vertical band drains (PVD) or wick drains) are geocomposites used for land reclamation or for stabilization of soft ground. They accelerate the consolidation process by reducing the time required for the dissipation of excess pore water pressure. The efficiency of drains is partly controlled by the transmissivity, that is, the discharge capacity can be measured, using the drain tester, to check their short-term and long-term performances. The discharge capacity of drains is affected by several factors such as confining pressure, hydraulic gradient, length of specimen, stiffness of filter, and the duration of loading. The experimental study, conducted in the laboratory by Broms *et al.* (1994), suggests that the effect of length of the drains and duration of loading on the discharge capacity of drains is small, whereas the stiffness of the drain filter can have a considerable effect. The discharge capacity of the drain decreases with decreasing stiffness of the filter.

Presently, the drainage geocomposites are designed for structures requiring vertical drainage, such as bridge abutments, building walls and retaining walls. The composite normally consists of a spacer sandwiched between two geotextile sheets. This construction combines in a single flexible sheet.

The primary function of the geosynthetic in subsurface drainage applications is filtration. The successful use of a geotextile in a filtration application is dependent on a thorough knowledge of the soil to be retained. The essential properties to be determined are particle-size distribution, permeability, plasticity index and dispersiveness. It is crucial to adequately characterize the soil to be retained in order to ensure its compatibility with the chosen geotextile. In certain applications, such as use of geotextile filters below waste deposits, the nature of the leachate is of crucial importance because the bacterial growth process may render the geotextile impermeable. Therefore, the leachate parameters such as total suspended solids, chemical-oxygen demand and biological-oxygen demand may require to be determined (Fourie, 1998).

When a geotextile is placed adjacent to a base soil (the soil to be filtered), a discontinuity arises between the original soil structure and the structure of the geotextile, as discussed in Section 1.6.3 (see Chapter 1). This discontinuity allows some soil particles, particularly the particles closest to the geotextile filter and having diameters smaller than the filter opening size, to migrate through the geotextile under the influence of seepage flows. This condition is shown in an idealized manner in Fig. 4.3(a). For a geotextile to act as a filter, it is essential that a condition of equilibrium is established at the soil/geotextile interface as soon as possible, after installation, to prevent the soil particles from being piped indefinitely through the geotextile; if this were to happen, the drain would eventually become blocked. As the fines are washed out from the base soil, the coarser particles located at the filter interface will maintain their positions, and a natural filtration zone will be formed immediately above the soil/geotextile interface. These larger particles will, in turn, stop smaller particles,

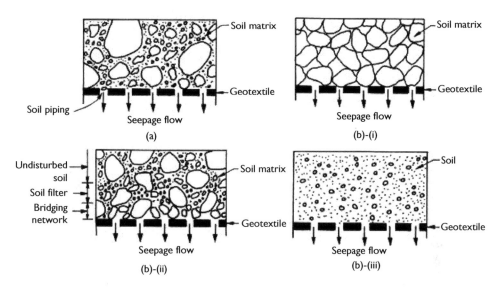

*Figure 4.3* (a) Idealized soil-geotextile interface conditions immediately following the geotextile installation; (b) idealized interface conditions at equilibrium between three different soil types
and geotextile filter − (i) single-sized soil and geotextile filter, (ii) well-graded soil and
geotextile filter, and (iii) cohesive soil and geotextile filter (Courtesy of Terram Ltd., UK).

which then stop even finer particles. As a consequence, the coarse particles at the filter
interface cause a filtration phenomenon within the soil itself, stopping the migration
of fines. At equilibrium, which may take normally between 1 and 4 months to occur
in practice, the soil adjacent to the filter becomes more permeable.

The structure, or stratification, of the soil immediately adjacent to the geotextile at
the onset of equilibrium conditions dictates the filtering efficiency of the system. The
stratification is dependent on the type of soil being filtered, the size and frequency of
the pores of geotextile, and the magnitude of the seepage forces present. Figure 4.3(b)
shows typical stratification occurring with three different soil types – single-sized soil,
well-graded soil and cohesive soil. When the soil is well graded, considerable rearrangement of the soil takes place. At equilibrium, three zones may be identified: the
undisturbed soil, a 'soil filter' layer which consists of progressively smaller particles
as the distance from the geotextile increases and a bridging layer which is a porous,
open structure. Once the stratification process is complete, it is actually the soil filter
layer which actively filters the soil. If the geotextile is chosen correctly, it is possible
for the soil filter layer to be more permeable than the undisturbed soil. The function
of the geotextile is to ensure that the soil remains in an undisturbed state without any
soil piping as shown in Fig. 4.4. Note that the time taken to reach 'constant system
permeability conditions' should coincide with the time taken for 'zero soil piping' for
effective use of geosynthetic filter. Also note that for most soils (we can refer to them
as stable soils; more discussions can be found in Section 4.2.2), it is not necessary for
a geosynthetic filter to screen out all the particles in the soil. Instead, a geosynthetic
filter needs only to restrain the coarse fraction of the various particle sizes present.

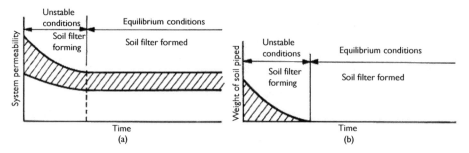

*Figure 4.4* Overall requirements for optimal filter performance (after Lawson, 1986).

In filter applications, the design must be prepared so as to avoid, throughout the design life, the following three phenomena causing a decrease of the permeability of the geotextile filter in course of time:

1   blocking
2   blinding
3   clogging

When a geotextile is selected to retain the particles of low-concentrated suspensions or whenever there is a lack of direct contact between the soil and the geotextile, the coarse particles of a size equal to or larger than the pore sizes of the geotextile may migrate and locate themselves permanently at the entrance of the pores of the geotextile, as shown in Fig. 4.5(a). This phenomenon that develops at the soil-geotextile interface resulting in a decrease of geotextile permeability is called *blocking*. *Blinding* is a phenomenon similar to blocking and is used to describe the mechanism occurring when the coarse particles retained by the geotextile, or geotextile fibres, intercept the fine particles migrating from the soil in such a way that a low-permeable layer (often called *soil cake*) is formed very quickly at the interface with the geotextile, thereby reducing the hydraulic conductivity of the system (Fig. 4.5(b)). The phenomenon of accumulation of soil particles within the openings (voids) of a geotextile, thereby reducing its hydraulic conductivity, is called *clogging* (Fig. 4.5(c)). This phenomenon may result in a complete shut off of water flow through the geotextile filter.

Note that the clogging, in general, takes place very slowly, and the blinding of the filter is far more detrimental than clogging. The geotextiles with a tortuous surface in contact with the soil, such as needle-punched nonwoven geotextiles, do not favour the development of a continuous cake of the fine soil particles, whereas the geotextile filters with a smooth surface may favour the development of such a cake (Giroud, 1994). Furthermore, the geotextiles with a tortuous surface do not favour the mechanism of blocking, because they do not have the individual openings.

## 4.2.2   Analysis and design concepts

Application of geotextiles as the filters is the most common application of geosynthetics in civil engineering constructions. The ability of a geotextile to allow sufficient water flow without migration of soil particles is a critical design requirement for filtration

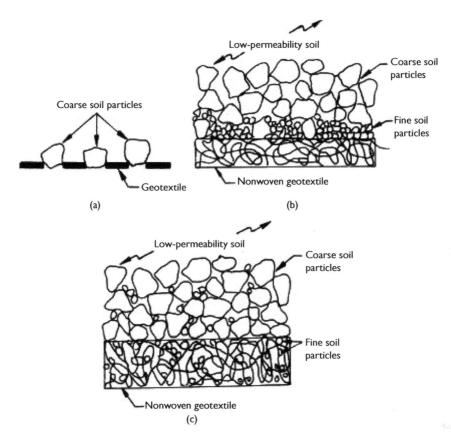

*Figure 4.5* Schematic views of (a) blocking; (b) blinding; (c) clogging mechanisms (after Palmeira and Fannin, 2002).

and drainage applications. Design approaches for the geotextile filters are based largely on experience and are wholly empirical in nature. Proper geotextile performance is required for the long-term serviceability of the structure. Various elements of filtration system (soil/waste, filter, drain, water/leachate) must be considered along with external conditions such as unidirectional or bidirectional flow, construction equipment and survivability, static and/or dynamic loading, and long-term durability. To achieve the satisfactory filter performance by geosynthetics, especially geotextiles, the following functions must be fulfilled during the design life of application under consideration:

1   Maintain an adequate permeability (or hydraulic conductivity)/permittivity to allow flow of water from the soil layer without significant flow impedance so as not to build up excess hydrostatic pore water pressure behind the geosynthetic *(permeability/permittivity criterion)*.
2   Prevent significant wash out of soil particles, that is, the soil piping *(retention or soil-tightness or piping resistance criterion)*.
3   Avoid the accumulation of soil particles within the geosynthetic structure, called the clogging, resulting in a complete shut off of water flow *(anti-clogging criterion)*.

4   Survive the installation stresses and any other long-term mechanical, biological or chemical degradation impacts for the lifetime of the structure to perform effectively (*survivability and durability criterion*).

Note that the permeability criterion places a lower limit on the pore size of the geotextile, whereas the retention criterion places an upper limit on the pore size of the geotextile. In other words, the permeability criterion requires a large pore size because the permeability of a geosynthetic filter increases with its increasing pore size; on the other hand the retention criterion requires a reduction of the pore size to restrict the migration of soil particles. These two criteria are, in principle to some extent, contradictory if they have to be fulfilled simultaneously. However, in the majority of cases, it is possible to find a filter that meets both the permeability criterion and the retention criterion. Several different geosynthetic filter criteria have been developed (Giroud, 1982, 1996; Lawson, 1982, 1986; Hoare, 1982; Wang, 1994) largely based on the conventional granular filter criteria, which were first formulated by Terzaghi and Peck (1948). All of these criteria are applicable for specific filter applications. These criteria use soil permeability and compare it with the geotextile permeability for establishing the permeability criterion, whereas they compare soil particle-size distribution with the geotextile pore-size distribution for establishing the retention criterion.

While establishing the geosynthetic filter criteria for the drainage applications, the following basic filtration concepts are kept in mind:

1   If the largest pore size in the geotextile filter is smaller than the larger particles of soil, then the filter will retain the soil.
2   If the smaller openings in the geotextile are sufficiently large enough to allow smaller particles of soil to pass through the filter, then the geotextile will not blind or clog.
3   A large number of openings should be present in the geotextile so that proper flow can be maintained even if some of the openings later become clogged.

It must be noted that the filter criteria and the design method for field application of the filters should be developed on the basis of data obtained from the detailed soil-geotextile performance testing in the field and/or the laboratory. However, in the absence of such real data, the criteria discussed in the current section can be considered.

It is a general misconception that the pore sizes of a filter should be smaller than the smallest particle size of the soil to be protected, because it would lead to using quasi-impermeable filters (which, of course, would not meet the permeability criterion). In some cases, the filter openings can be larger than the largest soil particles and the filter will still retain the soil (Giroud, 1984). In fact, soil retention does not require that the migration of all soil particles be prevented. Soil retention simply requires that the soil behind the filter should remain stable; in other words, some small particles may migrate into and/or through the filter provided that this migration does not affect the soil structure, that is, does not cause any further movement of the soil mass. At the same time, the filter and the drainage medium located downstream of the filter should be such that they can accommodate the migrating particles without clogging.

The filtration mechanism as explained in the previous section shows that the geotextile filter acts essentially as the catalyst, which induces the formation of the natural filter in the soil. It is basically the soil filter zone which most significantly controls the water flows. The sooner the natural filter is established, the smaller the number of particles that will migrate. For the ideal geotextile filter performance, the permeability (or the hydraulic conductivity) of the particle network at the soil-filter interface, as well as of the geotextile filter itself, should always be equal to or greater than the permeability of the parent soil. It is important that after an initial period of instability during the formation of the soil filter, the permeability of the soil-filter system should remain relatively constant over time.

The permeability criteria of geotextile filters, commonly suggested, are in the following form:

$$k_n \geq A k_s \tag{4.1}$$

where $k_n$ is the coefficient of cross-plane permeability of geotextile; $k_s$ is the coefficient of permeability of the protected soil; and $A$ is a dimensionless factor varying over a wide range, say 0.1 to 100.

The permeability criterion, $k_n \geq k_s$, has long been advocated by many researchers on the assumption that the geotextile needs to be no more permeable than the protected soil. Christopher and Holtz (1985) recommend the criterion, $k_n \geq 10k_s$, for critical soil and hydraulic conditions in which the clogging has been shown to cause roughly an order of magnitude decrease in the geotextile permeability. The criterion, $k_n \geq 0.1k_s$, was proposed by Giroud (1982) on the premise that a geotextile with only 10% of the permeability of the soil would still have a much greater flow capacity than the soil because the length of the flow path is directly related to the flow rate through a porous medium.

The presence of a filter, even when very permeable, disturbs the flow in the soil located immediately upstream. The selected filter should have permeability such that the disturbance to the flow – for example, the pore water pressure and the flow rate – is small and acceptable. For the geotextile filters, the typical permeability criteria for some specific applications are as follows (Giroud, 1996):

For a standard drainage trench:

$$k_n > 10k_s \tag{4.2}$$

For a typical dam-toe drain:

$$k_n > 20k_s \tag{4.3}$$

For the dam clay cores:

$$k_n > 100k_s \tag{4.4}$$

Note that the critical applications may require the design of even higher $k_n/k_s$ ratios, due to the high gradient that can occur in the filter vicinities.

Federal Highway Administration (FHWA) also established the following permittivity requirements for subsurface drainage applications (Holtz *et al.*, 1997):
For < 15% passing 75 μm:

$$\psi \geq 0.5 \; s^{-1} \tag{4.5}$$

For 15–50% passing 75 μm:

$$\psi \geq 0.2 \; s^{-1} \tag{4.6}$$

For > 50% passing 75 μm:

$$\psi \geq 0.1 \; s^{-1} \tag{4.7}$$

The retention criteria govern the upper (piping) filtration limit for filters and ensure that the soil to be protected is not continually piped through the geotextile filter and into the drainage medium. Failure to adopt appropriate retention criteria for filter design can have costly and potentially catastrophic consequences. The retention criteria of geotextile filter commonly suggested are in the following form:

$$O_f \leq BD_s \tag{4.8}$$

where $O_f$ is a certain characteristic opening size of geotextile filter; $D_s$ is a certain characteristic particle diameter of the soil to be protected; it indicates the particle diameter, such that $s\%$, by weight, of the soil particles are smaller than $D_s$; and $B$ is a dimensionless factor varying over a certain range.

The magnitude of $B$ depends on a number of factors, including soil types, hydraulic gradient, allowable amount of soil to be initially piped, the test method to determine $O_f$ and $D_s$, and the state of loading (confined and unconfined) (Faure and Mlynarek, 1998; Lawson, 1998). It is commonly determined by permeameter testing, which has the advantage of allowing near-field conditions to be modelled. Figure 4.6(a) shows the procedure used to determine the retention criteria for a specific soil type based on such testing. The relationship between the weight of soil passing through the geotextile filter is plotted against $O_f/D_s$, say $O_{95}/D_{85}$ ratios tested for various times. The piping limit conventionally is established as the maximum stable ratio of $O_{95}/D_{85}$ below which the soil is not continually piped through the geotextile filter. Having derived the appropriate value of $B$ from the permeameter evaluation, an appropriate retention criterion can be presented for this individual soil type in terms of the format shown in Equation (4.8). Figure 4.8(b) shows a series of test results for a specific tropical residual soil of essentially granular structure.

To overcome the time element associated with the permeameter testing of individual soil types, the standard retention criteria have been developed over a wide range of groups in the past. In general, for filtering the granular soils, the values of $B$ range from 0.5 to 1.0. For filtering the fine-grained soils with a plasticity index less than 10%, the values of $B$ range from 2 to 3. To filter the cohesive soils with a plasticity index greater than 10%, the required geotextile AOS is normally independent of soil-particle size (cohesive soils do not behave as individual particles) and consequently, $O_{95} < 0.2$ mm normally would suffice (Lawson, 1998).

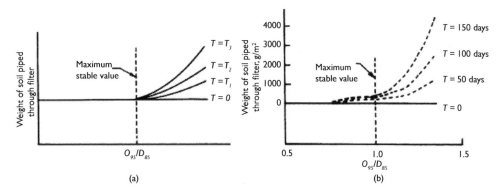

*Figure 4.6* (a) General procedure for determining the piping limit; (b) determination of piping limit for Hong Kong completely decomposed granite (CDG) soils (after Lawson, 1998).

All the existing retention criteria for geotextile filters reported in the literature are functions of various opening sizes of the geotextile such as $O_{95}$, $O_{90}$, $O_{50}$ and $O_{15}$, and diameter of soil particles such as $D_{90}$, $D_{85}$, $D_{50}$ and $D_{15}$ depending mostly on the uniformity coefficient of the soil, $C_u (= D_{60}/D_{10})$. Most of the criteria are given in the form of the $O_f/D_s$ ratio, called the *soil tightness number*, not exceeding a certain value or a range. The typical ranges of variation of $O_{95}/D_{50}$, $O_{95}/D_{85}$ and $O_{95}/D_{90}$ are, respectively, 1–6, 1–3 and 1–2.

For geotextile filters, the retention criteria as per FHWA guidelines developed by Christopher and Holtz (1985) are as follows:

*Steady state flow conditions*

$$O_{95} \leq BD_{85} \tag{4.9}$$

where, for a conservative design, $B = 1$, or for a less conservative design, for $D_{50} > 75\ \mu m$:

$B = 1$ for $C_u \leq 2$ or $\geq 8$ (4.10a)

$B = 0.5C_u$ for $2 \leq C_u \leq 4$ (4.10b)

$B = 8/C_u$ for $4 \leq C_u \leq 8$ (4.10c)

and for $D_{50} \leq 75\ \mu m$:

$B = 1$ for woven geotextiles (4.10d)
$B = 1.8$ for nonwoven geotextiles (4.10e)

For cohesive soils (Plasticity Index, PI > 7):

$$O_{95} \leq 0.3\ mm \tag{4.11}$$

*Dynamic, pulsating and cyclic flow (if the geotextile can move)*

$$O_{95} \leq 0.5\ D_{85} \tag{4.12}$$

Lawson (1998) has pointed out that if the geotextile filters that fall outside the boundaries indicated by appropriate retention criteria are used, immediate catastrophic failures do not occur. Over time, with continual soil piping, a loss of serviceability in the immediate filter area may arise. This may be evidenced by undue deformations, structural cracking, etc. It is only if these serviceability problems are allowed to persist without maintenance that subsequent collapse can occur.

A more comprehensive approach to the soil retention criteria has been suggested by Luettich *et al.* (1992). It has been proposed by Lafleur *et al.* (1993) that the filter opening size must fall within a narrow range. If it is too large, erosion will take place; if it is too small, blocking or clogging can occur near the interface, resulting in a decreased system discharge capacity. For all applications where the geotextile can move, and when it is used as sandbags, it is recommended that samples of site soils should be washed through the geotextile to determine its particle-retention capabilities (Holtz *et al.*, 1997).

*Long-term flow* capability of geosynthetics (generally geotextiles) with respect to the hydraulic load coming from the upstream soil is of significant practical interest. Filtration tests, such as the gradient ratio test for cohesionless soils or hydraulic conductivity ratio test for cohesive soils, as stated in Section 2.5 (see Chapter 2), must be performed to recommend the anti-clogging criterion, especially for critical/severe applications. In these tests, GR ≤ 3.0 or HCR ≥ 0.3 should generally be satisfied as the anti-clogging criterion in order to ensure satisfactory filter performance in the field. In the absence of such real data, particularly for less critical applications, the selected geotextile filter should satisfy the following anti-clogging criteria (Christopher and Holtz, 1985):

$$O_{95} \geq 3D_{15} \text{ for soil with } C_u > 3 \tag{4.13}$$

For soil with $C_u \leq 3$, select the geotextile with maximum opening size possible from the retention criteria.

*Other qualifiers*:
For soils with percentage passing 75 μm > 5%:

$$\text{Porosity, } n > 50\% \tag{4.14a}$$

for nonwoven geotextile filters, and

$$\text{POA} \geq 4\% \tag{4.14b}$$

for woven geotextile filters.

For soils with percentage passing 75 μm < 5%:

$$\text{Porosity, } n > 70\% \tag{4.15a}$$

for nonwoven geotextile filters, and

$$\text{POA} \geq 10\% \tag{4.15b}$$

for woven geotextile filters

In order to avoid the possibility of clogging of the geotextile filter, its opening size or percentage open area cannot be too small. If the base soil, that is the soil to be protected, is internally stable, then there is less possibility of occurrence of clogging. A soil is said to be internally stable (or self-filtering) if its own fine particles do not move through the interconnected pores of its coarser fraction. The internal stability has been found to depend on the shape of the gradation curve for cohesionless soils and on the dispersive ability for cohesive soils (Kenney and Lau, 1985; Mlynarek and Fannin, 1998). Typically, the plastic soils, the uniformly graded granular soils (coefficient of uniformity, $C_u$ less than approximately 3), and the well-graded soils ($C_u > 4$, and coefficient of curvature, $C_c > 1$) behave as stable soils (Fig. 4.7(a)). Unstable soils cannot perform self-filtration, that is, they have the potential to pipe internally.

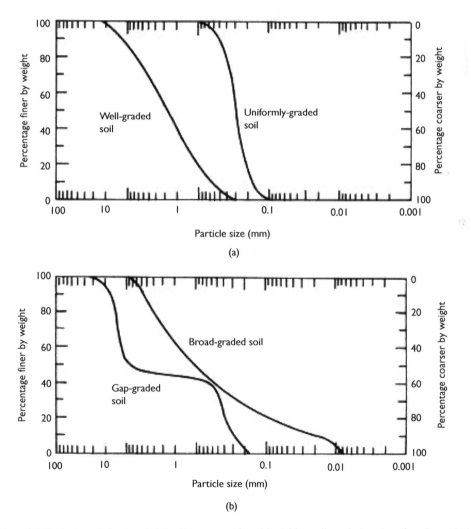

*Figure 4.7* Typical particle-size distribution curves for: (a) stable well-graded and uniformly-graded granular soils; (b) potentially unstable soils (after Richardson and Christopher, 1997).

Such soils may include gap-graded soils, non-plastic broadly-graded soils, dispersive soils (sugar sands, and rock flour), the wind-blown silt deposits (i.e., loess-type soils) with interbedded sand seams, and other highly erodible soils. The gap-graded soils have coarse and fine fractions, but very little medium fraction is present (Fig. 4.7(b)). If there is an insufficient quantity of soil particles in the medium fraction, the fine soil particles pipe through the coarser fraction and a soil filter bridges behind the geotextile. In broadly-graded soils with concave upward particle-size distributions and having uniformity coefficient $C_u > 20$), the gradation is distributed over a very wide range of particle sizes, such that the fine soil tends to pipe through the coarser particles. The dispersive soils are fine-grained natural soils that deflocculate in the presence of water and, therefore, are highly susceptible to erosion and piping (Sherard *et al.*, 1972).

Situations such as those involving internally unstable, high hydraulic gradients and very high alkalinity groundwater have been identified to create severe clogging problems. Iron, carbonate, and some organic deposits can chemically clog the geotextiles. Under such situations, one should avoid the use of geotextile filters and should use a granular filter, or should open up the geotextile to the point where some soil loss will occur, if the downstream conditions permit such a soil loss. In fact, the unstable soils require a more rigorous geotextile evaluation, if one wants to use it as a filter.

Certain filtration and drainage applications, such as in landfills, may expose the geotextile to chemical or biological activity that can drastically influence its durability as well as its filtration and drainage properties. If biological clogging is a concern, a higher porosity geotextile may be used. It is also better to have an inspection and maintenance program to flush the drainage system in the drain design and operation. Thus, it is important to note that the properties of the soil and the geosynthetic as well as the characteristics of the fluid passing through the filter influence the hydraulic characteristics of the soil-geosynthetic systems.

For filtration and drainage applications, the geotextile should also meet certain minimum strength and endurance properties to survive the installation stresses as well as the long-term degradation impacts. The limits must be established on the basis of site-specific evaluation, testing, and design. However, for routine projects, the users can have some specific guidelines in selecting geotextiles from the available guidelines in standards/codes of practice/manuals, such as AASHTO M 288-06 (2011) geotextile specifications (AASHTO, 2013), as described in Section 7.18 (see Chapter 7). These survivability requirements are the default criteria that can be used in the absence of site-specific information. Note that the data on static puncture are necessary for the filtration function. When the site loading conditions are such that there is a potential risk of static puncture of the filter, the data on static puncture should be obtained as a priority.

For designing a geotextile filter, one should identify the conditions under which it will be required to perform. These can include the project conditions (critical nature of the project) and the physical conditions (soil, hydraulic and stress conditions). The critical nature of the project will help determine the level of design effort, and the physical conditions will establish the geotextile requirements (Christopher, 1998). For the critical projects, the design should be based on a thorough analysis, including the performance test results.

Flow directions and hydraulic gradients can significantly influence the behaviour and stability of an engineered filter. Flow can be classified as either steady-state or dynamic. The steady state flow is usually present in trench drains used to lower down the groundwater level beneath roads and parking lots, behind retaining walls, under foundations, and below recreation fields. The steady-state flow suggests that the water movement occurs in one principal direction; this is the simplest application for a geo-textile filter. The dynamic pulsating flow is usually encountered in edge drains used to remove the surface infiltration water from roads; dynamic cyclic flow is encountered in permanent erosion-control applications for shorelines, stream banks, and canals. In contrast to steady-state flow, dynamic flow may occur in more than one direction. If the geotextile is not properly weighted down and in intimate contact with the soil to be protected, or if dynamic, cyclic, or pulsating loading conditions produce high localized hydraulic gradients, then the soil particles can continuously move without formation of any natural soil filter behind the geotextile, severely taxing its ability to perform (Christopher, 1998). While recommending the geotextile filters for wave attack applications, or any other situation in which turbulent or two-dimensional flow conditions can occur – for example, erosion control systems, one should be very careful.

While designing with geotextiles in filtration applications, the basic concepts are essentially the same as when designing with granular filters. The geotextile must allow the free passage of water and prevent the erosion and migration of soil particles into the drainage system or into the armour of the revetment depending on the type of application throughout the design life of the structure. The simplified design proce-dures for a geotextile filter for stable soils in subsurface drainage systems can be sum-marized in the following steps:

*Step 1:* Evaluate the soil to be filtered (the retained soil). As a minimum, this should include:

- visual classification
- consistency limits
- particle-size distribution analysis

*Step 2:* Determine the minimum survivability requirements.
*Step 3:* Determine the minimum permeability using the permeability criterion.
*Step 4:* Determine the maximum opening size using the retention criterion as well as the clogging criterion.
*Step 5:* Select the geotextile in accordance with Steps 2, 3 and 4.
*Step 6:* Perform the filtration test to meet the requirements of retention and anti-clogging criteria, if the application is critical.

For unstable soils, one should consult the subject expert and plan for performing soil-specific laboratory testing.

For designing the drainage system, the maximum seepage flow into the system must be estimated. In the case of in-plane drainage with thick geotextile or geocom-posite, the flow rate per unit width of the geosynthetic should be compared with the flow rate per unit width requirement of the drainage system. Since the in-plane flow capacity for geosynthetic drains reduces significantly under compression as well as with time due to creep, the final design must be based on the performance test under the anticipated design loading conditions with a safety factor for the design life of the project. Note that the objective of design is to ensure stability throughout the design life by a reduction in pore water pressure or depth of water table. The steps required in a typical design can be summarized as follows:

*Step 1:* Keeping the critical nature and the site conditions of application in view, define the stability requirement and design life.
*Step 2:* Determine the particle-size distribution curve and the coefficient of permeability of soil samples from the site.
*Step 3:* Select the drainage aggregate, if it has to be used along with the geotextile filter.
*Step 4:* Assess the reduction in pore water pressure or reduction in water table depending on the requirement in a particular application, and estimate the water flow into and through the drainage system based on the hydraulic gradient and the permeability of the soil.
*Step 5:* Determine the type and dimensions of the drainage system.
*Step 6:* Determine the geosynthetic requirements considering the permeability criterion, retention criterion, anti-clogging criterion and survivability criterion, as discussed above, and then select the proper geosynthetic accordingly.
Step 7: Monitor the installation during construction and observe the drainage system during and after the storm periods.

If the geocomposite drains are being used for drainage applications, then their design must satisfy the following criteria (Corbet, 1992):

1   The core must resist the applied normal and shear loads without collapsing.
2   Under the sustained load, the core must not reduce significantly in thickness (compression creep).
3   The core must allow the expected water flow to reach the discharge point without the buildup of water pressure in the core.
4   The core must support the geotextile filter.

Geosynthetic drains in the form of band drains as described in Section 1.5 (see Chapter 1) are nowadays frequently installed within the saturated soil mass to provide a vertical drainage, which can be obtained in the conventional method by constructing sand drains of appropriate diameter. In such specific applications, the complete drainage design of band drains as per the radial consolidation theory requires an estimation of equivalent sand drain diameter, $D_e$, which can be calculated using the following expression:

$$D_e = \frac{2(B + \Delta x)}{\pi} \tag{4.16}$$

where $B$ is the width and $\Delta x$ is the thickness of band drain. The above expression is derived based on the fact that the effectiveness of a drain depends, to a great extent, upon the circumference of its cross-section but very little upon its cross-sectional area (Kjellman, 1948; Shukla, 2015).

The vertical compression test for the geocomposite pavement panel drains may be conducted to simulate vertical, horizontal, and eccentric loading resulting from an applied vertical load under various backfill conditions. The results of the test may be used to evaluate the vertical strain of the panels and the core area change for a given load.

The design, specification and construction of any drainage or filter system should recognize that the backfill conditions and materials must be selected, placed, and compacted so that the geosynthetic product and soil act in concert to carry the applied loads without excessive strains, either vertical, horizontal, or at any load angle. Construction forces and in-service static and dynamic load-induced compression must be considered properly. Appropriate filter gradation criteria must be followed in the selection of granular backfill material to minimize the migration of soil fines into the voids of backfill material in the presence of hydraulic gradients. The backfill material selection and the placement method should be based primarily on achieving adequate compaction without damaging the drainage and filter materials, while also achieving intimate contact with the soil face. Permeability of the backfill material must also be considered in its selection to promote a higher ground water flow to the drainage system. To enhance the placement, especially around the geocomposite drains and to prevent the damage to these structures, the aggregate size should not generally exceed 19 mm.

## EXAMPLE 4.1

A geotextile-wrapped trench drain is to be constructed to drain a soil mass. Determine the appropriate hydraulic properties of the geotextile to function as a filter in a critical application with the following soil properties:

$D_{10} = 0.14$ mm
$D_{15} = 0.18$ mm
$D_{60} = 0.65$ mm
$D_{85} = 1.1$ mm
$k_s = 2 \times 10^{-4}$ m/s
Percentage passing 75 μm <5%

## SOLUTION

From Equation (4.2), the cross-plane permeability $k_n$ of the geotextile should meet the following requirement:

$$k_n > 10 \, k_s$$

or

$$k_n > (10 \times 2 \times 10^{-4}) \, \text{m/s} = 2 \times 10^{-3} \, \text{m/s}$$

Since the soil has less than 15% passing 75 μm, therefore from Equation (4.5),

$$\psi \geq 0.5 \, s^{-1}$$

Coefficient of uniformity,

$$C_u = \frac{D_{60}}{D_{10}} = \frac{0.65\,\text{mm}}{0.14\,\text{mm}} = 4.6$$

Since $C_u$ lies between 4 and 8, therefore the value of factor $B$ for its use in Equation (4.9) can be calculated from Equation (4.10c) as

$$B = \frac{8}{C_u} = \frac{8}{4.6} = 1.7$$

From Equation (4.9),

$$O_{95} \le BD_{85}$$

or

$$O_{95} \le (1.7 \times 1.1)\,\text{mm} = 1.87\,\text{mm}$$

Since $C_u > 3$, the geotextile should meet the following anti-clogging criterion (Equation (4.13)).

$$O_{95} \ge 3D_{15}$$

or

$$O_{95} \ge (3 \times 0.18)\,\text{mm} = 0.54\,\text{mm}$$

Also, from Equations (4.15a) and (4.15b),

Porosity, $n > 70\%$

for nonwoven geotextile filters, and

POA $\ge 10\%$

for woven geotextile filters.

Thus, the geotextile filter should have the following hydraulic properties:

$k_n > 2 \times 10^{-3}\,\text{m/s}$

$\psi > 0.5\,\text{s}^{-1}$

$0.54\,\text{mm} \le \text{AOS} \le 1.87\,\text{mm}$

$n > 70\%$ for nonwoven geotextile filters
POA $> 10\%$ for woven geotextile filters

## EXAMPLE 4.2

A geosynthetic has to be selected to provide the drainage behind a 10-m high retaining wall with a vertical backface as shown in Fig. E4.2. The coefficient of permeability of the soil backfill is $1 \times 10^{-5}$ m/s. Determine the required transmissivity of the geosynthetic to function as a drain. Would an ordinary single layer of nonwoven geotextile be adequate?

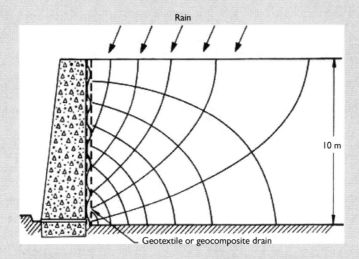

Figure E4.2  Example 4.2

## SOLUTION

The flow net shown in Fig. E4.2 can be used to estimate the rate of flow $q$ per unit length of the wall as (Shukla, 2014)

$$q = k_s H \frac{n_f}{n_d}$$

where $k_s$ is the coefficient of permeability of soil backfill, $H$ is the total head loss, $n_f$ is the number of flow channels, and $n_d$ is the number of potential drops. In the present problem, $k_s = 1 \times 10^{-5}$ m/s, $H = 10$ m, $n_f = 6$ and $n_d = 8$. Therefore,

$$q = (1 \times 10^{-5})(10)\left(\frac{6}{8}\right) \text{ m}^2/\text{s} = 7.5 \times 10^{-5} \text{ m}^2/\text{s}$$

Hydraulic gradient,

$$i = \frac{10 \text{ m}}{10 \text{ m}} = 1.0$$

Transmissivity, $\theta$, can be calculated using the following expression (Equation (2.14)):

$$Q_p = \theta i B$$

or

$$Q_p/B = \theta i$$

or

$$q = \theta i$$

or

$$\theta = \frac{q}{i} = \frac{7.5 \times 10^{-5}}{1.0} = 7.5 \times 10^{-5} \text{ m}^2/\text{s}$$

The typical values of transmissivity for the most common nonwoven geotextiles fall into the range of $10^{-4}$–$10^{-6}$ m²/s depending greatly on the normal stress acting on the geotextile. We can compare the required transmissivity to the actual value, obtaining a factor of safety as follows:

$$FS = \frac{\theta_{\text{allowable}}}{\theta_{\text{required}}} = \frac{10^{-4} \text{ to } 10^{-6}}{7.5 \times 10^{-5}} = 1.33 \text{ to } 0.013, \text{ which is not adequate.}$$

Therefore, **a single layer of geotextile is not suited** for drainage application behind the retaining wall. In fact, a geocomposite having much greater in-plane flow capacity should be used.

### 4.2.3  Application guidelines

Geosynthetics are used as a filter and/or a drain in many applications; the guidelines described below can be useful in most of such applications.

1    The geosynthetic filters and drains should be well protected to prevent any degradation, and care should be taken to avoid their contamination.
2    The geosynthetic, particularly a woven geotextile, should be placed with the machine direction following the direction of water flow.
3    The geotextile filters should be placed in intimate contact with the soil to be retained. An intimate contact between filter and soil is one of the major demands for all geotextile filters. If it is lacking, the adjacent soil can become suspended as water percolates towards the filter. Filtering a suspension flow is more difficult for geotextiles, because this condition hinders the forming of a secondary filter. Usually the geotextile acts as a kind of 'catalyst' for developing filtration in the adjacent soil. Suspended soils cannot build a secondary filter, so clogging and/or blinding will result. If the construction process makes it difficult to establish a

close contact, a sand fill can be used between the geotextile and the base soil to provide a granular filter, which otherwise would be created in the soil itself.

4   The ends of subsequent rolls and parallel rolls of the geotextile should be over-lapped a minimum of 0.3–0.6 m. The overlaps may be increased for high hydraulic flow conditions and heavy construction. Joints (see Section 7.8 and Section 7.9, Chapter 7), if provided, must prevent the infiltration of soil particles.

5   The overlying stones should be placed without damaging the geotextile. A cushion layer may be provided over the geotextile. This will reduce the impact (drop and abrasion) of large elements on the geotextile. The thickness of such a cushion layer should be approximately equal to the diameter of the stones to be placed on it, up to about 0.4 m (Heibaum, 1998). During the initial period after installation, the large fill elements will agitate due to traffic, dynamic hydraulic loads or deforma-tions of the subsoil caused by the new load until an equilibrium is found. Thin and very light geotextiles may not sustain this load; hence the geotextile has to be cho-sen properly. When in doubt, some thickness should be added to the cushion layer.

6   Construction procedure, suggested by Holtz et al. (1997), can be adopted for the geotextile-wrapped underdrains. The major construction steps are shown in Fig. 4.8.

7   In the drainage system of retaining walls, the geosynthetic drain should be located away from the wall in an inclined orientation so that it can intercept seepage before it impinges on the back of the wall. Placement of a thin vertical drain directly against the retaining wall may actually increase seepage forces on the wall due to rainwater infiltration (Terzaghi and Peck, 1948; Cedergren, 1989).

8   Backfill materials should be placed by methods that will not disturb or damage the geosynthetic product. Backfill should be placed in a maximum of 150 mm lifts and compacted to a minimum of 90% standard Proctor density. Excessive com-paction of backfill directly against the drain or filter should be avoided; otherwise loading during compaction may cause damage or deformation to the filter or the drain materials and their joints, if any. Coarse-grained, clean materials such as crushed stone, gravel and sand are more readily compacted using the vibratory

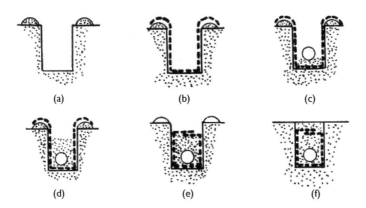

*Figure 4.8* Construction procedure for the geotextile-wrapped underdrains: (a) excavate the trench; (b) place the geotextile; (c) add the bedding and the pipe; (d) place/compact the drainage material; (e) wrap the geotextile over top; (f) compact the backfill.

equipment. Fine materials with a high plasticity should not be used as a backfill material.

9  For installation of geocomposite pavement edge drains, the trenches should be excavated with stable, straight and smooth sidewalls. If the sidewall of the pavement side trench is not reasonably straight and smooth, or if undercutting or sidewall sloughing occurs, the geocomposite drain material should not be installed against the sidewall of the pavement side trench. In such cases, the product may be installed in the centre of the trench or against the shoulder sidewall. If installed in the centre of the trench, the product must be supported during installation and the backfill should be placed in such a way as to keep it straight, vertical and stable. Also the trench width may be increased to a minimum of 150 mm plus the product thickness. Fig. 4.9 shows a typical type and arrangement of prefabricated geocomposite drain. The trench depth and width should be sufficient to carry the design flow below the base course – subgrade interface such that water will not be retained in the structural pavement section. The trench width is also governed by the required space for adequate compaction of the backfill using the compaction equipment without damaging the geocomposite drainage panel. Geocomposite pavement drains should not be laid in standing or flowing water.

Prefabricated geosynthetic drains should be properly tied-in to the collector and the outlet drainage systems. Outlet pipe, generally laid at a slope of 3%, should have sufficient diameter to remove the collected water from the geocomposite drain at a rate equal to or greater than the flow capacity of the geocomposite drain.

10 Since the function of the geotextile in a silt fence, as shown in Fig. 4.10, is to filter and allow the settlement of soil particles from the sediment-laden overland water flow, the geotextile at the bottom of the fence should be buried in a 'J' configuration to a minimum depth of 150 mm in a trench so that no flow can pass under the silt fence. The trench should be backfilled and the soil compacted over the geotextile. The geotextile should be spliced together with a sewn seam only at a support post, or two sections of fence may be overlapped instead (AASHTO, 2013).

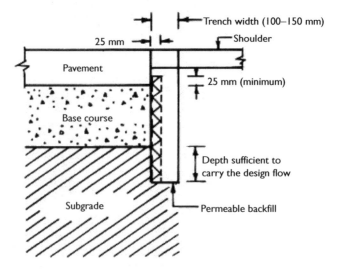

*Figure 4.9* A typical type and arrangement of geocomposite buried drain.

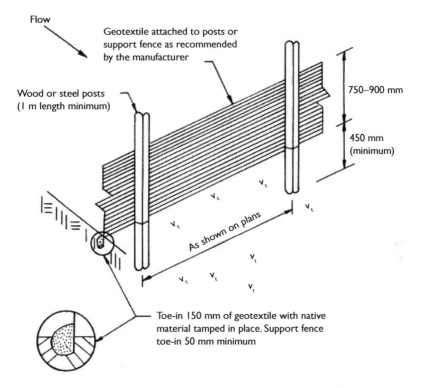

Flow

Geotextile attached to posts or
support fence as recommended
by the manufacturer

Wood or steel posts
(1 m length minimum)

750–900 mm

450 mm
(minimum)

V$_t$

V$_t$

V$_t$

V$_t$

V$_t$

V$_t$

V$_t$

V$_t$

V$_t$

V$_t$

As shown on plans

Toe-in 150 mm of geotextile with native
material tamped in place. Support fence
toe-in 50 mm minimum

*Figure 4.10* Typical silt fence details (after AASHTO, 2013).

## 4.3  SLOPES – EROSION CONTROL

### 4.3.1  Basic description

The problem of soil movement due to erosive forces by moving water and/or wind as well as by seeping water is called *soil erosion*. Gravity is also one of the prime agents of soil erosion, particularly on steep slopes. Soil erosion is associated with negative economic and environmental consequences in many areas, such as agriculture, river and coastal engineering, highway engineering, slope engineering, and several sections of civil engineering. Construction sites with unvegitated steep slopes are prime targets for soil erosion.

Soil erosion by moving water is caused by two mechanisms: (1) detachment of particles due to raindrop impact, and (2) movement of particles from surface water flow (Wu and Austin, 1992). The dislodged particles carry with them seeds and soil nutrients. Natural growth of vegetation on the exposed soil slope surface is thus hindered. High velocity runoffs can cause not only surface soil movement downslope, but their scouring effects can cause total undermining of slopes. Rain erosion can act upon a land surface of any degree of slope; however, the severity of rain erosion increases with increasing slope steepness and slope length (Ingold, 2002).

The exposed denuded slopes become increasingly vulnerable to erosion agents and are ultimately destabilized. To control erosion is to curb or restrain (not to stop completely) the gradual or sudden wearing away of soils by wind and moving water. The goal of any erosion control project should be to stabilize soils and manage erosion in an economical manner. Since the surface water flow cannot be eliminated, the most feasible solution to erosion problems is the slope protection. The slope protection serves two functions: (1) it slows down the surface water flow, and (2) it holds the soil particles, grass or seedlings in place. If an element (natural or synthetic) is incorporated into the soil to prevent the detachment and transportation of soil particles, then the slope would be able to withstand greater forces.

The solutions of soil erosion problems typically involve the use of basic erosion control techniques such as soil cover and soil retention. The use of revetments is very common in civil engineering practice for erosion control (Fig. 4.11). A cover layer (called the armour) of a revetment can be permeable or impermeable. An open cover layer substantially reduces the uplift pressures, which can be induced in the sublayers and provides protection against the external loads. Riprap, blocks and block mats, grouted stones, gabions and mattresses, and concrete and asphalt slabs are most commonly used revetment armours.

Riprap consists of stone dumped in place on a filter blanket (4.11(a)) or prepared slope to form a well-graded mass with a minimum of voids. The stone used for riprap is hard, dense, durable, angular in shape, resistant to weathering and to water action, and free from overburden, spoil, shale and organic material. The riprap material can also be placed on the gravel bedding layer and/or a woven monofilament or nonwoven geotextile filter (Fig. 4.11(b)).

Concrete block systems consist of prefabricated concrete panels of various geometries, which may be attached to and laid upon a properly designed woven monofilament or nonwoven geotextile filter. Gabions and mattresses are compartmented rectangular containers made of galvanized steel hexagonal wire mesh or rectangular plastic mesh and filled with hand-sized stone. Compared with rigid structures, the advantages of gabions include flexibility, durability, strength, permeability and economy. The growth of native plants is promoted as the gabions collect sediment in the stone fill. A high percentage of installations is underlaid by woven monofilament and nonwoven geotextiles to reduce hydrostatic pressure, facilitate sediment capture and prevent washout from behind the structure (Theisen, 1992).

The basic function of a revetment is to protect the slopes (coastal shorelines, river/stream banks, canal slopes, hill slopes and embankment slopes) against the hydraulic

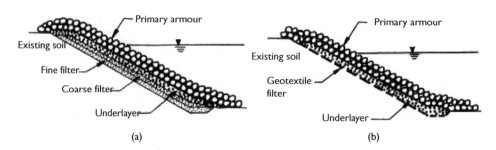

*Figure 4.11* Revetment systems: (a) conventional revetment system consisting wholly of granular materials; (b) revetment system containing a geotextile filter.

loadings (forces by water waves and currents as well as seepage forces). The resistance of the erosion control is derived mainly from friction, weight of different elements, interlocking and mechanical strength. As a result of difference in strength properties, the critical loading conditions are also different. Maximum velocities and impacts will be determinants for grass mats and riprap, as they cause displacement of the material. Uplift pressures and impacts, however, are of prime importance for paved revetments and slabs, as they tend to lift the revetment. As these phenomena vary both in space and time, the critical loading conditions vary both with respect to position along the slope and the time during the passage of water wave, particularly along the coastal shorelines. Cement concrete and bituminous blocks will mainly respond to uplift forces as the maximum loads are distributed more evenly over a layer area, thus causing a higher resistance against uplift, compared with loose block revetment (Pilarczyk, 2000).

Fundamentally, the function of armouring systems is to protect and hold in place the filtering system, which is nowadays almost always a geotextile. The first line of defense against the soil erosion is the geotextile, since it is in intimate contact with the soil. If the block systems do not have enough mass and frictional characteristics, this intimate contact between the soil and the filter may be lost, which usually results in a rapid degradation of the slope beneath the armouring system.

Scarcity of land and often-limited financial means are forcing engineers to become more innovative and to utilize new products. Geosynthetics have already earned a recognized position in the area of erosion control. Gotextiles and geonets are nowadays used as a replacement for graded granular filters typically used beneath the revetment armour to keep erodible soil in place, and they have been found very effective in erosion control. The basic objective in using the geotextile filter is to effectively protect the subsoil from being washed away by hydraulic loads.

Geosynthetic nets and meshes available in various forms have proven to be successful in both temporary and long-term erosion control projects. These products protect the soil surface from water and wind erosion while accelerating the vegetative development. Nettings or meshes may contain UV stabilizers for controlled degradation. Perhaps most advantageous to the environment, these meshes and nettings may ultimately become biodegradable.

Three-dimensional erosion control geosynthetic mats and geocells that are nowadays commercially available with various dimensions can be used in permanent erosion control systems. As was mentioned in Section 1.2 (see Chapter 1), the geocells are three-dimensional honeycomb structures that have a unique cellular confinement system formed by a series of self-containing cells up to 20 cm deep. They have the ability to physically confine the soil placed inside the cells (Fig. 4.12). They retain the soil, the moisture and the seed, and thus create situations for growth of vegetative mats on slopes where the vegetation may be difficult to establish. The vegetative mats provide reinforcement and the system's cells increase the natural resistance of these mats to erosive forces and protect the root zone from soil loss. At the same time, the cellular confinement system facilitates the slope drainage.

The long-term nondegradable (permanent) rolled erosion control three-dimensional products are also manufactured commercially as a *turf reinforcement mat (TRM)*. The turf reinforcement is in fact a method or system by which the natural ability of vegetation to protect soil from erosion is enhanced through the use

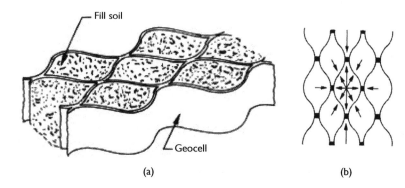

*Figure 4.12* (a) Physical confinement of soil in a geocell; (b) confinement forces generated by resistance of the cell walls and by the passive resistance of the soil in the adjacent cells.

of geosynthetic materials. A flexible three-dimensional matrix retains seeds and soil, stimulates seed germination, accelerates seedling development and, most importantly, synergistically meshes with developing plant roots and shoots to permanently anchor the matting to the soil surface. TRMs provide more than twice the erosion protection of unreinforced vegetation. In fact, these systems are capable of withstanding the short-term high velocity flows without erosion; thus they are most suitable where heavy runoff or channel scouring is anticipated. The higher resistance to flow has resulted in the widespread practice of using turf reinforcement as an alternative to riprap, concrete and other armour systems in the protection of open channels, drainage ditches, detention basins and steepened slopes.

A grassed clay dike revetment is also one of the types of revetments used mostly in agricultural applications aimed at preventing erosion of the dikes by hydraulic forces. The development of a strong grass revetment is a matter of time. Also its success depends on the good maintenance.

The most common and natural element used for erosion control is vegetation. The roots of grasses protect the slope surface from erosion. The deeper roots of plants, shrubs and trees tend to reinforce and stabilize the deeper soils. The application of vegetation as bank protection is preferred rather than the application of conventional materials such as riprap, concrete blocks, etc. If necessary, vegetation and appropriate geosynthetics (geomats, geonets, geocells, etc.) can be applied in combination (Fig. 4.13). The selection of vegetation must be done on the basis of soil and climatic conditions of the specific area of application. The vegetation will on the one hand stabilize the body of the channel, consolidate the soil mass of the slope and bed, and reduce erosion; on the other hand, the presence of vegetation results in extra turbulence and retardation of flow. Geotextiles and other perforated geosynthetics and open blocks provide additional strength to the root mat and can reduce much of the direct mechanical disturbance to the plants and the soil.

In erosion control applications, where vegetation is considered to be a long-term solution from an environmental point of view, short-term erosion control is technically performed excellently in a diverse set of environments and soil conditions by the jute products (*geojutes*). It has been observed that the geojute nets may have a life of about 1 to 2 years in the contact with soil. A jute net may absorb water up to four

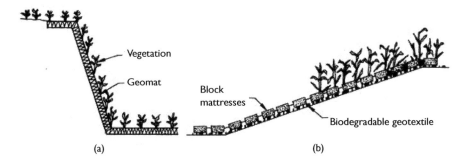

*Figure 4.13* Erosion control using a geosynthetic mat and geotextile along with vegetation.

times its dry weight and transfers it to the soil it is in contact with, thereby giving the full benefit of moisture for growth of vegetation when used for erosion control. Geocoir nets are another type of biodegradable material which can be effectively used in a manner similar to the geojute nets. The geocoir nets degrade much slower than the geojute nets; the expected field life is about 2 to 3 years and thus may provide protection to the slopes for a longer time than the geojute nets. Geocoir nets are also resistant to saline water, but compared to the jute nets, their water absorption capability is lower (IRC, 2011). The low cost (despite significant costs of transportation) and the inherent variability of soil application well accommodate a natural fibre product. An additional advantage of these biodegradable products is that on biodegradation, they improve the quality of the soil for quick vegetation growth. However, a geojute has drawbacks: its open weave construction leaves the soil exposed, the organic material tends to shrink and swell under changing moisture, and it is extremely flammable. Note that the temporary erosion protection is important, but the long-term goal of any vegetated erosion control technique is to provide a permanent erosion protection through permanent vegetation and/or subsequent root reinforcement.

Geotextiles are also used in toe and bed protection, which consists of the armouring of the beach or bottom surface in front of a structure to prevent scouring and undercutting by water waves and currents (Fig. 4.14). The stability of toe is essential because its failure will generally lead to a failure of the entire structure.

In many cases, the geotextile is used to wrap a fill material (sand, gravel, asphalt or mortar), creating *geobags*, *geotubes* or *geomats*, known collectively as the *geocontainers*, which are used in hydraulic and coastal engineering (Fig. 4.15). The geotextile cover has to act as a filter towards the fill, which is permeable, as is the container. Sometimes the container also is used as a filter element with respect to the subsoil, that is, when used as a scour fill or a scour prevention layer. The volume of geocontainers actually used varies from 100 to 1000 m$^3$. Geocontainers are suitable for slope, toe and bed protection but the main application is construction of groynes, perched beaches and offshore breakwaters. They can also be used to store and isolate contaminated materials obtained from harbour dredging and/or as bunds for the reclamation works. Geocontainers offer the advantages of simplicity in placement and constructability, cost effectiveness, and minimum impact on environment. Thus, they can be a good alternative for traditional materials and erosion control systems and hence they deserve to be applied on large scales.

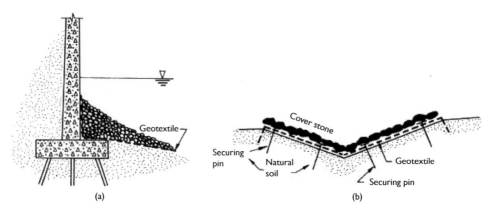

*Figure 4.14* (a) Scour protection for an abutment; (b) erosion control of a ditch.

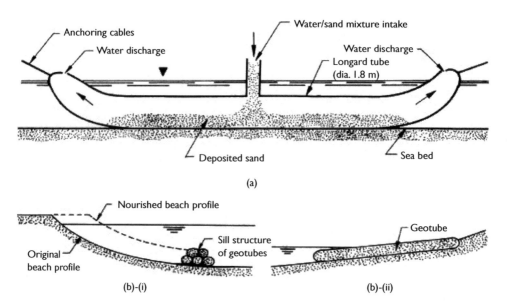

*Figure 4.15* Geocontainer: (a) procedure for filling a geotube; (b) erosion control applications (after Pilarczyk, 2000).

In the past, numerous geosynthetic erosion control systems, including rolled erosion control products (RECPs), which are manufactured or fabricated into rolls, have been developed to provide an aesthetically appealing and maintenance-free option. Some of these geosynthetic systems are intended to provide temporary protection to the top soil cover and seeds against raindrop impact and sheet erosion until the vegetation can grow, after which they biodegrade or photodegrade. Others are intended to remain intact and to control erosion even in the absence of vegetation. These systems are usually preformed mats, meshes, blankets and cells. For these systems, the degradation of the geosynthetic material constitutes the failure of the system. Theisen and Richardson (1998) presents an excellent and comprehensive overview of the various existing short-term and long-term erosion control systems, as summarized in Tables 4.2 and 4.3, respectively. These systems may offer cost advantages and improved aesthetics over more traditional designs.

*Table 4.2* Short-term erosion control systems (after Theisen and Richardson, 1998).

| Type | Description | Flow velocity range, fps |
|---|---|---|
| Hay/straw/hydraulic mulches | Typically machine applied over newly seeded sites | 1–3 |
| BOP (Biaxially oriented process nets) | Polypropylene or polyethylene nets used to anchor loose fibre mulches, such as straw, or as a component of erosion control blankets | 2–5[1] |
| ECM (woven erosion control meshes) | Twisted fibers of polypropylene, jute, or coir, woven into a dimensionally stable blanket. Excellent for bioengineering or sod reinforcement | 3–6[1] |
| ECB (erosion control blankets) | BOP stitched or glued on one or both sides of a biodegradable fibre blanket composed of straw, wood, excelsior, coconut, etc. | 3–6[1] |

Note:
[1] Depending on vegetation, composition, and density.

*Table 4.3* Long-term erosion control systems (after Theisen and Richardson, 1998).

| Type | | Description | Flow velocity, fps[1] |
|---|---|---|---|
| Soft armoring systems | GCS (vegetated geocellular containment) | Polymeric honeycomb-shaped three-dimensional cell systems filled with soil and vegetation | 4–6[2] |
| | FRS (UV-stabilized fibre roving systems) | Strands of polypropylene or fiberglass fibre blown on the ground surface, then anchored in place using emulsified asphalt | 6–9 |
| | TRM (turf-reinforcement mats) | A three-dimensional matrix of polypropylene, polyethylene, or nylon fibres or yarns, mechanically stitched woven or thermally bonded. Designed to be seeded, and then filled with soil. | 10–25 |
| | CBS (vegetated concrete block systems) | Articulate or hand-placed concrete blocks filled with soil, then vegetated | 10–25[2] |
| Hard armoring systems | GCS (geocellular containment systems) | Polymeric honeycomb cells filled with gravel or concrete | 6–25[2] |
| | FFR (fabric-formed revetments) | Geotextiles filled with grout or slurry | 15–25 |
| | CBS (concrete block systems) | Articulating or hand-placed concrete blocks | 15–25 |
| | Gabions | Rock-filled wire baskets | 15–25 |
| | Riprap | Quarried rock of sufficient density | 6–30 (depends on mean diameter) |

Notes:
[1] Some systems with greater mass and/or ground cover may exceed these limits.
[2] Depending on infill material.

Based on an experimental study, Ahn *et al.* (2002) pointed out that mulching mats can be used effectively to provide good plant growth and adequately stabilized soil slopes. The mulching mat can consist of geosynthetics and jute nets with a needle-punched structure (Fig. 4.16). It should be manufactured to have the following properties:

1   biodegradable
2   very light and portable
3   capable of holding water but remaining substantially unaffected, and
4   capable of holding seeds and fertilizer to prevent them from being washed away by rain and wind.

A mulching mat with the above properties has several advantages. First, since seeds and fertilizers are tied into the mat, the seeds are not washed or blown away from the soil slope. Second, the mulching mat prevents evaporation from the soil slope and helps plants to grow more successfully. Third, the mulching mat maintains the ground temperature and reduces damage.

The choice and suitability of a particular geosynthetic system for controlling soil erosion depends on whether the system is intended to provide long-term or short-term protection, the degree of protection it can provide under different climatic, topographic and physiographic conditions and the cost-protection efficiency measure (Rustom and Weggel, 1993). Selection of an appropriate rolled erosion control products to protect the disturbed soil slopes depends on many factors, including expected project life, down-slope length, soil type, vegetative class, local climatic conditions, slope angle, slope orientation, drainage patterns and available experience. Most manufacturers of rolled erosion control products provide extensive case histories, field and laboratory test data and design software to make the job easier.

Figure 4.17 indicates the prototype data on stability of revetments subject to water current attack. It is noted that the concrete systems are excellent erosion control measures, whereas the grass mats work most effectively if the water current continues for a short period of time.

It is important to note the results of experimental studies, conducted by Cancelli *et al.* (1990), to separate out the performance of several erosion control products based on rain splash and runoff in controlling the soil erosion on a 1 vertical (V) to 2 horizontal (H) slope of silty fine sand (Fig. 4.18). It was reported that under rain

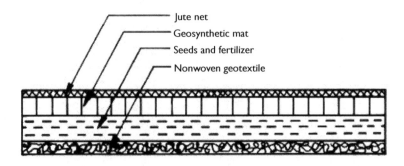

*Figure 4.16* Geosynthetic mulching mat (after Ahn *et al.*, 2002).

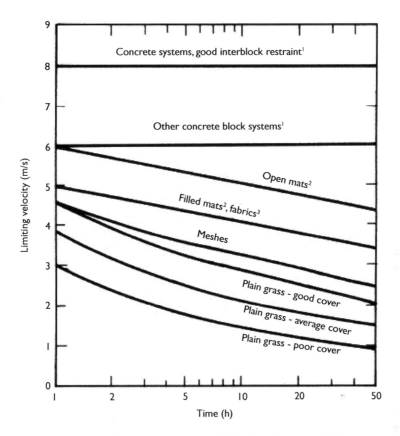

*Figure 4.17* Limiting velocities for erosion resistance (after Hewlett *et al.*, 1987).
Notes:
[1]*Minimum superficial mass 135 kg/m²;*
[2]*Minimum nominal thickness 20 mm;*
[3]*Installed with 20 mm of soil surface or in conjunction with a surface mesh;* all reinforced grass values
  assume well-established good grass cover.

splash, the jute returned by far the highest efficiency with a soil loss of approximately 3 g/l of rainfall (Fig. 4.18(a)); however, under runoff the jute returned by far the worst result (Fig. 4.18(b)). These results suggest that a geocomposite based on jute and synthetic products will have a better efficiency in terms of soil loss under a typical combination of rain splash and runoff.

### 4.3.2   Analysis and design concepts

Revetment systems are very effective in erosion control of slopes, including coastal shore-lines, stream banks, canal banks, hill slopes, and embankment slopes. In the conventional systems, the graded granular layers are used as the filters beneath ripraps and other revetment systems. In the current construction practice, the geotextile layers are commonly used as a replacement of graded granular filters in the riprap erosion control systems.

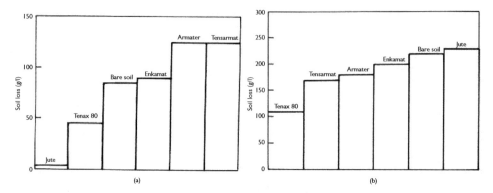

*Figure 4.18* Erodibility of erosion control products: (a) based on rain splash; (b) based on runoff (after Cancelli *et al.*, 1990).

To evaluate the stability of the revetment (cover layer and sublayers), information is required about the hydraulic conditions, the structural properties and the possible failure mechanisms. For designing the revetments, the designer should note that the geotextile filter is only one of the structural components involved, and there are a few more components, as shown in Fig. 4.19(a), to be designed. A failure of any one component may cause the failure of the entire revetment structure. To achieve a perfect erosion control system for a slope, the following aspects must be taken into account in the design process:

1   stability of cover layer, sublayers, subsoil considering the whole system as well as the individual element;
2   flexibility, that is, the ability to follow settlement;
3   durability of cover layer and geotextile filter;
4   possibility of inspection of failure;
5   easy placement and repair;
6   low construction and maintenance cost;
7   overall performance.

The design of revetments, like other hydraulic structures, must be based on an integral approach of the interaction between the structure and the subsoil. The main geotechnical limits that should be evaluated in the design of the revetments on a sloping ground are the following:

1   overall stability of slopes;
2   settlements and horizontal deformations due to the weight of the structure;
3   seepage of groundwater;
4   piping or internal erosion due to seepage flow;
5   liquefaction caused by the cyclic loading of water due to wave actions or earthquakes.

A complete analytical approach to the design of revetments incorporating geotextiles does not currently exist. While certain aspects, particularly in the hydraulic field, can be relatively accurately predicted, the effect of various forces (Fig. 4.19(b))

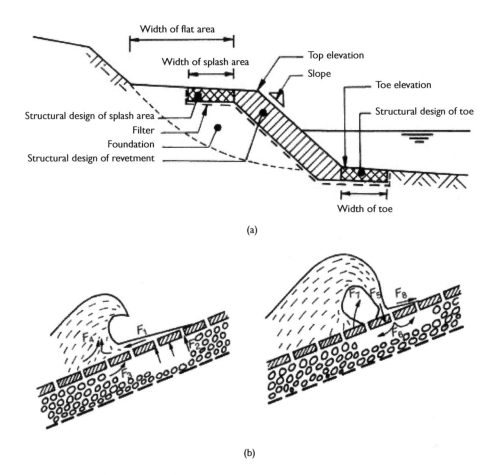

*Figure 4.19* (a) Design components of a typical revetment structure (after Pilarczyk, 2000); (b) various forces due to water waves that may act on the revetment system.

Notes: $F_1$ – forces due to down-rush; $F_2$ – uplift forces due to water in filter; $F_3$ – uplift forces due to approaching wave front; $F_4$ – forces due to change in velocity field; $F_5$ – wave impact; $F_6$ – uplift forces due to mass of water falling on slope; $F_7$ – force caused by low pressures on slope due to air entrainment; $F_8$ – forces due to up-rush.

on the revetment cannot be represented with confidence in a mathematical form for all possible configurations and systems. Therefore, the designer must make use of empirical rules or past experiences. Using this approach, it is likely that the design will be conservative. Since there is a great variety of possible composition of erosion control systems, it is not possible to describe the complete designs of all these systems in this section. However, the design of a geotextile filter applicable to all the systems is being discussed. The readers can find more details, in the works of Fuller (1992), on the design, particularly of typical articulating block system (ABS) revetments with a geotextile filter in coastal conditions.

Since filtration is a primary function of geotextiles, the design steps for the geotextile layer remain essentially the same as the design for geotextile filters in subsurface drainage systems discussed in Section 4.2.2. However, while designing the geotextile filters for erosion control systems, the following special considerations should be given:

1   As the riprap stones or concrete blocks may cover some portions of the geotextile filter, it is essential to evaluate the flow rate required through the open area of the system, and select a geotextile that meets those flow requirements.
2   For < 15% passing 75 μm, $\psi \geq 0.7\ s^{-1}$.
3   The largest opening in the geotextile should be small enough to retain even the smaller particles of the base soil. It means the value of $B$ in the retention criterion should be reduced to 0.5 or less. Usually, no transport of soil particles should be allowed and thus the washout of soil particles should be completely prevented, independent of the level of hydraulic loading, because settlement and loss of stability it could result in (see Fig. 4.20). In this situation, the geosynthetic filter is called the *geometrically tight filter*. However, a very limited washout is sometimes acceptable; in that case the filter is called the *geometrically open filter* which has the openings larger than the size of certain soil particles.
4   Where the geotextile can move, an intermediate layer of gravel-sized particles may be placed over the geotextile and riprap of sufficient weight should be placed to prevent the dynamic flow action from moving either riprap stone or geotextile.
5   Keeping in view the severe hydraulic conditions caused by continual or even reversing dynamic flows, the soil-geotextile filtration tests in accordance with ASTM D5101-12 for cohesionless soils (ASTM, 2012a) and ASTM D5567-94 for cohesive soils (ASTM, 2011) should always be performed with site soil samples to enable appropriate selection of geotextiles.
6   Since the placement of riprap is generally more severe than the placement of drainage aggregate, Class 2 classification for monofilament, and Class 1 for all others should be considered to meet the survivability requirements, as discussed in Section 7.18 (see Chapter 7), in the absence of any site-specific evaluation, testing and design.

A number of geosynthetic manufacturers have developed their own design manuals. However, a proper basic knowledge of the background of the design methodology must be used to verify the real value of the design procedure of a particular product

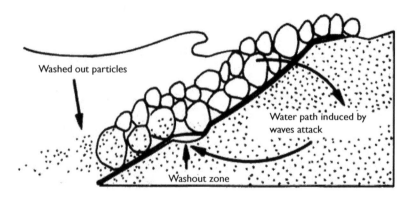

*Figure 4.20* Development of filter failure resulting from the washout of fines (after Mlynarek and Fannin, 1998).

before its use in field applications. Full-scale prototype testing is a good method of verifying designs but the costs may limit application to only major projects. On smaller projects, a physical modelling of the cover layer under hydraulic attack can be carried out to verify a design or to refine the results of mathematical modelling.

## EXAMPLE 4.3

Consider a revetment system as shown in Fig. E4.3 with the following parameters:

$\alpha$ = angle of slope in degrees
$\gamma_w$ = unit weight of water in kN/m³
$W_c'$ = submerged weight per unit area of protective covering material in kN/m²
$\Delta h$ = head loss across the geotextile in m

Under what condition, will there be no geotextile uplift?

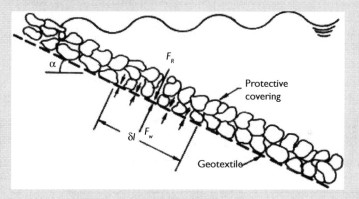

*Figure E4.3* Example 4.3.

## SOLUTION

There will be no geotextile uplift if the force, $F_R$, from the riprap perpendicular to the slope exceeds the force, $F_w$, from the water pressure beneath the geotextile. Mathematically, for no geotextile uplift

$$F_R > F_w$$

or

$$(W_c' \cos \alpha)\Delta l > (\gamma_w \Delta h)\Delta l \text{ , where } \Delta l \text{ is the length of any part of the slope}$$

or

$$\Delta h < (W_c' \cos \alpha)/\gamma_w$$

For routine field applications, the stability of the covering material in waterway revetments incorporating geotextiles can be assessed using an analytical approach based on a stability number, $S_N$, defined as below (Pilarczyk, 1984a):

$$S_N = \frac{H}{\gamma'_R D} \tag{4.17}$$

where $H$ is the wave height (m), $D$ is the depth of protective covering material (m), and $\gamma'_R$ is the submerged relative unit weight of the covering material (dimensionless) as defined below:

$$\gamma'_R = \frac{\gamma_c - \gamma_w}{\gamma_w} \tag{4.18}$$

where $\gamma_c$ is the unit weight of protective covering material (kN/m$^3$), and $\gamma_w$ is the unit weight of water. Usually $\gamma'_R$ varies from 1.24 to 1.38.

The minimum depth of the protective covering material required to withstand the wave action can be determined from the table of required stability numbers (Pilarczyk, 1984a,b; Tutuarima and Wijk, 1984) listed in Table 4.4. It should be noted that in each case, it is assumed that the permeability of the geotextile exceeds that of the soil. If the geotextile permeability is only equal to that of the soil, then the above required stability numbers should be reduced by 40% or to 2.0, whichever gives the higher value (Tutuarima and Wijk, 1984).

Note that the design principle for waterway revetments based on the stability number approach as described here can also be applied to the coastal erosion control. Since the waves of larger wave heights are usually encountered in coastal environment, heavier revetments will be required to control the coastal erosion. It has been reported that for very high waves, the stability number approach seriously underestimates the stability of riprap armour stones. In such a situation, the following formula should be used to determine the appropriate stone weight, $W$ (Hudson, 1959):

$$W = \frac{\gamma_s H^3 \tan \alpha}{\lambda_D (G_s - 1)^3} \tag{4.19}$$

Table 4.4 Required stability numbers for waterway revetment systems.

| Protective covering | Required stability number |
| --- | --- |
| Unbonded riprap | < 2 |
| Free blocks | < 2 |
| Asphalt grouted open aggregate | < 4.3 |
| Sand-filled mattresses | < 5 |
| Articulated blocks | < 5.7 |
| Grouted articulated blocks | < 8 |

where $H$ is the wave height, $\gamma_s$ is the unit weight of solid stones, $G_s$ is the specific gravity of the stones, $\lambda_D$ is the damage coefficient, and $\alpha$ is the slope angle. For no damage and no overtopping of the revetment, $\lambda_D = 3.2$.

The stone weight found from Equation (4.19) can be converted into an average stone diameter, $D_{50}$, using the following expression:

$$D_{50} = \sqrt[3]{0.699W} \qquad (4.20)$$

where $W$ is in tonnes and $D_{50}$ is in metres.

The size of the stone obtained from Equation (4.20) is likely to be too large for direct placement on a geotextile filter sheet. In such a situation, the intermediate layer or layers of smaller stones of suitable grading that will not cause any damage to the geotextile should be provided between the large stone armour having a minimum thickness of $2D_{50}$ and the geotextile.

## EXAMPLE 4.4

Consider a waterway revetment system with the following parameters:

> Wave height, $H = 1.2$ m
> Unit weight of protective covering material (unbonded riprap), $\gamma_c = 23$ kN/m³

Assume that the permeability of the geotextile is greater than the permeability of soil to be protected. Determine the minimum depth of the protective covering. Take the unit weight of water as $\gamma_w = 10$ kN/m³

## SOLUTION

From Equation (4.18), the relative submerged unit weight of the covering material is

$$\gamma_R' = \frac{23 - 10}{10} = 1.3$$

From Eq. (4.17), the stability number is

$$S_N = \frac{1.2}{1.3D}$$

From Table 4.4, for unbonded riprap,

$$S_N < 2$$

or

$$\frac{1.2}{1.3D} < 2$$

or

$$D > \frac{(1.2)}{(2)(1.3)}$$

or

$$D > 0.462 \text{ m}$$

Thus, the minimum depth of protective covering is **0.462 m.**

## EXAMPLE 4.5

Determine the average size of the armour stone in a revetment system with the geotextile filter required to protect a 25° slope from the waves up to 2.5 m high assuming that no overtopping of the revetment occurs. Take the specific gravity of stone as $G_s = 2.70$.

## SOLUTION

As $G_s = 2.70$, $\gamma_s = G_s \gamma_w = (2.70)(1) = 2.70 \text{ t/m}^3$

From Equation (4.19), the stone weight is

$$W = \frac{\gamma_s H^3 \tan \alpha}{\lambda_D (G_s - 1)^3} = \frac{(2.70)(2.5)^3 (\tan 25°)}{(3.2)(2.70 - 1)^3} = 1.25t$$

From Equation (4.20), the average size of the armour stone is

$$D_{50} = \sqrt[3]{0.699W} = \sqrt[3]{(0.699)(1.25)} = 0.956 \text{ m}$$

If the erosion control systems are required using bags, tubes or mats, the geotextile must be permeable but soil tight, usually $O_{90}$ for the geotextile should be about 0.1 to 0.2 mm. Large bags (>1 m³) are fabricated usually of strong polyester geotextile with mass per unit area larger than 500 g/m² and tensile strength > 10 kN/m. These bags must be filled in place using a pumped sand slurry or concrete. Their large size makes them more resistant to movement under water wave attacks. A schematic overview of the failure mechanisms of a stacked sand-filled geotube structure is presented in Fig. 4.21. The designer for each particular project should carefully examine all these

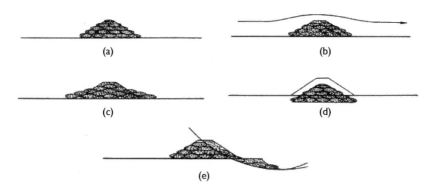

*Figure 4.21* Possible failure modes of geosystem structures: (a) theoretical cross-section; (b) hydraulic stability; (c) internal stack stability; (d) stability against squeezing/settlement; (e) geotechnical slope stability (after ACZ, 1990).

failure modes. The material durability and the long-term behaviour of geosystems need special attention. Systematic monitoring of realized projects, including failure cases, and evaluation of the prototype and laboratory data may provide useful information for verification purposes, as well as for design.

Rolled erosion control products (RECPs) are initially selected on the basis of a combination of somewhat arbitrary factors. By understanding the predictive methods for soil loss, one can predict the soil loss rate and understand the role of these rolled erosion control products and field techniques in limiting the process. For this purpose, the Universal Soil Loss Equation (USLE), developed by the US Soil Conservation Service in the 1930s, can be used. This equation can be stated as follows (Ingold, 2002):

$$X = R \times K \times S \times L \times C \times P \tag{4.21}$$

where $X$ is the annual soil loss per unit area (metric tons per ha); $R$ is the rainfall erosion index that reflects the erosion potential from regional precipitation; $K$ is the soil-erodibility factor; $S$ is the slope-steepness factor; $L$ is the slope length factor; $C$ is the cover and management factor (protected and unprotected conditions); and $P$ is the erosion control practice factor that reflects the maintenance activities of the facility. For more details about all these factors, readers can refer to the contribution by Ingold (2002).

Well-established vegetation is an effective and attractive form of protection for slopes exposed to mild and moderate surface erosion. When designing grassed slopes and waterways, it is required to take into account immediate and long-term flow resistance based upon longevity of the erosion control products being used. Two design concepts are used to evaluate and define a channel configuration that will perform within acceptable limits of stability. These methods are defined as the *permissible velocity approach* and the *permissible tractive force (boundary shear stress) approach*. Under the permissible velocity approach, the channel is assumed stable if the adopted velocity is lower than the maximum permissible velocity. The tractive force approach focuses on stresses developed at the interface between flowing water and the materials forming the channel boundary (Chen and Cotton, 1988; Theisen, 1992).

The permissible velocity approach uses Manning's equation, in which, for the given depth of flow, $D$, the mean velocity, $V$, in the cross section may be calculated as

$$V = \frac{1.49}{n} R^{2/3} S^{1/2} \tag{4.22}$$

where $n$ is Manning's roughness coefficient; $R$ is the hydraulic radius, equal to the cross-sectional area, $A$, divided by the wetted perimeter, $P$; and $S$ is the slope of the channel, approximated by the average bed slope for uniform flow conditions.

The tractive force approach uses a simplified shear stress analysis which is as follows:

$$\tau = \gamma DS \tag{4.23}$$

where $\tau$ is the tractive force; $\gamma$ is the unit weight of water; $D$ is the maximum depth of flow; and $S$ is the average bed slope or energy slope.

A design based on the flow velocity may be limited because the maximum velocities vary widely with channel length ($L$), shape ($R$) and roughness coefficients ($n$). In reality, the force developed by the flow, not the flow velocity itself, challenges the performance of erosion control systems. The maximum shear stress criterion is thus necessary to ensure properly engineered design of channel lining erosion control systems. Note that the velocity and the tractive force are not directly proportional. Under certain conditions, a decrease in velocity may increase the depth of flow, thereby increasing the shear stress. Flow duration is another significant parameter affecting the design of erosion control systems, particularly grassed systems. As the duration of flow progresses, the resistance of grassed surface reduces. The short-term performance of fully vegetated surfaces is impressive at nearly 4 m/s (Theisen, 1992).

As discussed in Section 4.3.1, the jute geotextile is capable of reducing the erosive effects of rain drops and controlling the migration of soil particles of the exposed surface. On biodegradation, the jute geotextile forms a mulch and fosters quick vegetative growth. The choice of the right type of jute geotextile and plant species is critical for effective results. The species of vegetation needs to be selected carefully considering the local soil and the climatic conditions. The Indian Standard IS: 14986 suggests the names of plants useful for stabilization of bunds, terrace fences and steep slopes and gullies (BIS, 2001). The choice of jute geotextile basically depends on the type of soil to be protected. It must be ensured primarily that the slope to be protected from rain water erosion is geotechnically stable. The selection of jute geotextile is also required to be done in consideration of the extreme rainfall in a limited time-span at that location as the intensity of rainfall is more important than the average annual rainfall at a place for assessing the erosion index ($R$) and deciding on the choice of a particular type of jute geotextile. IS:14986 provides broad guidelines to the choice of the jute geotextile type for users (BIS, 2001).

### 4.3.3    Application guidelines

Erosion control systems such as the riprap revetments with geosynthetic filters and the mattress revetments may require the attention of the following guidelines:

1    Depressions in the slope should be filled to avoid geotextile bridging and possible tearing when the cover materials are placed. A well-compacted slope is important in order to produce a smooth surface and thus ensure that there is a good connection between the revetment and the subsurface.

2    The geotextile sheet should be placed with warp/machine direction, in case of a woven geotextile, in the direction of water flow, which is normally parallel to the slope for erosion control runoff and wave action, and parallel to the stream or channel in the case of streambank and channel protection. It should be placed in intimate contact with the subsoil without wrinkles or folds. There should be a minimum offset of 1.5 m between the adjacent ends.

3    Jointing systems, which are without strain, can be made with an overlap of 0.5 to 1.0 m or with a lapped ('J') seam (see Section 7.8 and Section 7.9, Chapter 7). Joints under stress must be avoided as much as possible. In particular cases where heavy forces occur in the main direction as well as at right angles to the main direction, it is usual to apply two layers, one in each direction. When overlapping, successive sheets of the geotextile should be overlapped upstream over downstream, and/or upslope over downslope. In cases where the wave action or the multidirectional flow is anticipated, all seams perpendicular to the direction of flow should be sewn. Overlaps should be made along the slope parallel to the direction of water flow. Overlapped seams of roll ends should be a minimum of 300 mm except where placed under water. In such instances, the overlap should be a minimum of 1 m. Overlaps of adjacent rolls should be a minimum of 300 mm in all instances (AASHTO, 2013)). Geotextile sheets can be held in position by ballasting with sandbags or by pinning loosely with large-headed polymer pins (Fig. 4.22).

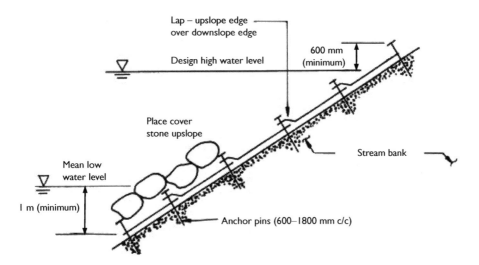

*Figure 4.22* Cross-section of the stream-bank slopes with revetment (after AASHTO, 2013).

4    Place the armour (cover layer) over the geotextile as quickly as possible, prefer-
     ably within 14 days. In underwater applications, the geotextile and the cover
     layer should be placed the same day.

5    The armour system placement should begin at the toe and proceed up the slope.
     Riprap and heavy stone filling should not be dropped from a height of more
     than 300 mm. Stone pieces with a mass of more than about 100 kg should not
     be allowed to roll down the slope. The placing process should avoid any damage
     or stretching to the geotextile. It is a good practice to insist that the installation
     contractor demonstrates that the chosen placing method does not result in dam-
     age to the geotextile. Alternatively, the use of a cushion layer (granular sublayer)
     between the armour and the geotextile spreads the load and reduces the contact
     stress.

6    The geotextile must be checked for its temperature resistance at 130–140 °C
     when used in conjunction with a bituminous armour placed in situ.

7    Once in place, the individual mattresses, if used, should be joined so that the
     edges cannot be lifted up under the action of water waves and currents. In addi-
     tion, the top and bottom edges of the revetment, including the geotextile filter,
     should be anchored as shown in Fig. 4.23. In such a case, a toe structure may not
     be needed to stop revetments sliding. Keying the geotextile into the crest of the
     slope can be avoided, provided the geotextile can significantly be extended above
     the anticipated maximum high water level.

8    Bags for the revetments should be filled and stacked against a prepared stable
     slope with their long axes parallel to the shoreline. While a stacked-bag revetment
     can be placed on a steeper slope, it should not exceed 1 vertical to 1.5 horizontal.
     A stacked-bag revetment should preferably be two bags thick, for example, with
     the outside layer concrete-filled and the interior bags sand-filled. Concrete-filled
     bags can be stabilized by steel rods driven through the bags (Fig. 4.24).

9    When the turf reinforcement mats (TRMs) are used, they should be installed first,
     then seeded and filled with soil. High strength TRMs provide sufficient thick-
     ness and void space to permit the soil filling/retention and the development of

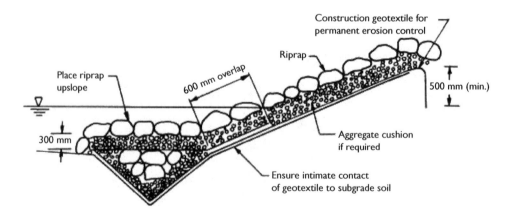

*Figure 4.23* Key details at top and toe of slope for the geotextiles for permanent erosion control
(after AASHTO, 2013).

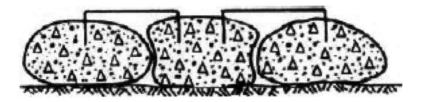

*Figure 4.24* Reinforcement of concrete bags.

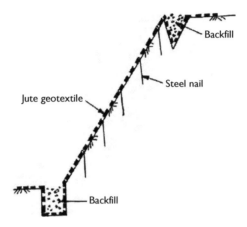

*Figure 4.25* Use of geonatural (jute geotextile) for erosion control of sloping ground aiming to speed up vegetation growth.

vegetation within the matrix. Many erosion control revegetation mats (ECRMs) are generally installed prior to seeding.

10 If geonaturals (jute/coir geotextiles) are used aiming to speed up the vegetation growth, then straw or hay mulch must be placed beneath them to achieve optimum results. Anchoring trenches (450-mm deep and 300-mm wide) should be excavated at the top and the toe of the slope along the length of the slope. Overlaps should be a minimum of 150 mm at sides and ends. The jute geotextile at the higher level on the slope should be placed over the portion to its next at a lower level. Side overlaps of the geonatural piece should be placed over its next piece on one side and under the next piece on the other. It should be fixed in position by steel staples (usually of 11 gauge dia) as shown in Fig. 4.25 or by split bamboo pegs. Stapling should be done normally at an interval of 1.5 m both in longitudinal and transverse directions. Seeds of vegetation or saplings of the appropriate plant species may be spread or planted at suitable intervals through the openings of the geonatural. Installation should be completed preferably before the monsoon to take advantage of the rains for quick germination of seeds. Watering/maintenance of the vegetation should be carried out as per the specialist advice of agronomist/botanist.

11 Installation of erosion control mats and blankets can follow the manufacturer's recommended procedures. Fig. 4.26 illustrates one such general installation procedure.

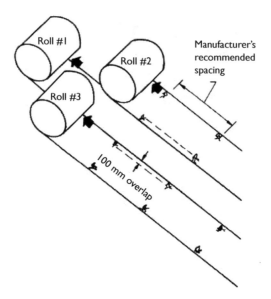

*Figure 4.26* Typical installation instruction for erosion control mats and blankets (after Agnew, 1991).

## 4.3.4 Case studies

### Case study 1

The research and applications of geonaturals, especially geojutes and geocoirs, have been directed towards developing these materials as alternatives to geosynthetics. Due to their biodegradability and short life, the majority of target use areas have been their applications as soil saver for supporting vegetation growth and as temporary filter and drainage layers. The design life of geonaturals may end when the vegetation grows and the soil layers consolidate to some extent.

Erosion of railway and road embankments and hill slopes is caused principally by rain and wind. Erosion of the top soil gradually destabilizes the earthen embankments. Denuded hill slopes are always vulnerable to erosive forces of rains, particularly during the monsoon. The jute geotextile when applied on an exposed soil surface acts as miniature check dams or micro terraces, reduces the kinetic energy of rain splashes, diminishes the intensity of surface runoff, prevents the detachment and the migration of soil particles and ultimately helps in the quick growth of vegetation by formation of mulch. The jute geotextile therefore helps in controlling erosion in road and railway embankments and hill slopes naturally. In India, the jute geotextiles have been used in several erosion control projects in the past. Some of them are described very briefly below (BIS, 2001):

- sand dune stabilization (5000 m²) with 500 g/m² jute geotextile in Digha Sea Beach, Midnapur, Forest Department, Government of West Bengal, India in 1988; 80% covered by vegetation after 6 months;

- control of top soil erosion (10000 m²) with 300 g/m² and 400 g/m² jute geotextiles in Arcuttipur, T.E. Cachar, Assam, India in July 1995; 93–97% reduction in soil loss;
- erosion control in embankment (3000 m²) in Valuka, Maldah Irrigation Department, Government of West Bengal, India in August 1996; no damage by rains in 1996 and 1997;
- afforestation and erosion control (2000 m²) with 25 g/m² jute geotextile in Hijli and Porapara, Midnapore Forest Department, Government of West Bengal, India in August 1997; growth of the trees in the treated area significantly higher; no sign of erosion.

## Case study 2

The erosion of riverbanks and coastlines is counteracted by protection of the surface to resist the forces generated by the flow and waves. The method widely used is to install a layer of stone pitching on the bank to stop the loss of soil. The rise and fall of the water level as well as the wave action causes water to flow into the pitched bank and then drain away. This two-way flow, known as dynamic flow, is capable of dislodging and carrying away soil lying below the stone protection and ultimately causing the revetment to fail. Installing a filter between the stone layer and the soil may reduce the soil erosion. Traditionally, a granular layer is used as a filter, which allows the water to pass through freely, but not the soil particles. The design and choice of a suitable granular material for this filter can be done, but it is not an easy task to achieve the function of filter accurately. The use of geosynthetic filters in such cases has proven to be an attractive alternative.

The Kolkata Port Trust authorities used jute geotextile as a revetment filter for the riverbank protection at Nayachor Island, Haldia, India during 1992 (Sivaramakrishnan, 1994). The eroded site was first prepared to form a uniform slope 1

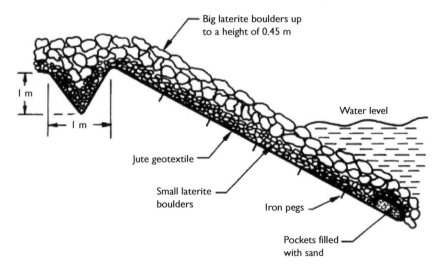

*Figure 4.27* Riverbank protection at Nayachor Island, Haldia, India (after Sivaramakrishnan, 1994).

vertical to 1 horizontal. The bare jute geotextile was unrolled over the slope of the embankment, starting from the top of the bank. The geotextile was anchored at the top in a trench 1 m × 1 m and similarly at the sides. The overlapping portions were nailed with 10-inch long iron pegs at an interval of 1 m. The bottom portion of the jute geotextiles was fabricated in such a manner that it had multiple pockets to fill sand in them. This was done to anchor the geotextile in its place and protect the erosion by reverse current and eddies. After the entire area was covered with the jute geotextile, small laterite boulders were placed over the jute geotextile. The small laterite boulders were laid to provide a cushion effect to the geotextile. On the top of the small laterite boulders, big laterite boulders, weighing approximately 15 to 20 kg, were pitched to a height of 1.5 feet as shown in Fig. 4.27. The entire operation was carried out during the low tide. A good amount of siltation, up to a height of 600 mm, was observed after a period of eight months, which indicates that the jute geotextile was effective in protecting the slope against its erosion.

## 4.4 CONTAINMENT FACILITIES

Containment facilities in various forms are constructed to meet the varying needs of society. These containment facilities can be categorized as the following three types (Giroud and Bonaparte, 1989):

1    facilities containing solids such as landfills, waste piles, and ore leach pads;
2    facilities containing liquids such as dams, canals, reservoirs (to which a variety of names are given such as ponds, lagoons, surface impoundments, and liquid impoundments); and
3    facilities containing mostly liquids at the beginning of operations and mostly solids at the end, such as settling ponds, evaporation ponds and sludge ponds.

Geosynthetics are generally used in the construction of all these containment facilities to perform various functions: fluid barrier, drainage, filtration, separation, protection and reinforcement. Out of these functions, a fluid barrier is generally the required function to be performed by the geosynthetic in almost all the containment facilities. In other words, in most of the containment facilities, the fluid barrier is the primary function of the geosynthetic. The present section describes some of the containment applications with more attention to the fluid barrier function of the geosynthetics.

### 4.4.1 Basic description

#### 4.4.1.1 Landfills

Our activities create several types of waste, such as municipal solid waste (MSW), industrial waste and hazardous waste. We should always attempt to minimize the amount of waste by designing and implementing programmes focussed on waste *reuse*, *recycling*, and *reduction*, may be called the *RRR – concept*. The remaining waste has to be disposed off by suitable disposal methods, such as incineration, deep well injection, surface impoundments, composting, and shallow/deep burial in soil and rock. Incineration is not a viable method of disposal for a wide variety of wastes, and furthermore, it may lead to air pollution. It also creates an ash residue that still must be landfilled. In fact, the need for landfilling of solid wastes will continue indefinitely for a number of reasons.

An engineered landfill is a controlled method of waste disposal. It is not an open dump. It has a carefully designed and constructed envelope that encapsulates the waste and that prevents escape of *leachate* (the mobile portion of the solid waste as contaminated water) into the environment. Leachate is generated as the liquid squeezed out of the waste itself (*primary leachate*) and by water that infiltrates into the landfill and percolates through the waste (*secondary leachate*). It consists of a carrier liquid (solvent) and dissolved substances (solutes).

There are basically two major types of landfill: the *MSW landfill* (also called the *sanitary landfill*) to keep commercial and household solid wastes, and the *hazardous waste landfill* for the deposition of hazardous waste materials. The MSW landfill is the most common type of landfill. The geometrical configurations of this landfill commonly include *area fill, trench fill, above and below ground fill* and *valley (or canyon) fill*, as shown in Fig. 4.28. The area fill type of landfill is used in areas with a high groundwater table or where the ground is unsuitable for excavation. The trench fill is generally used only for small waste quantities. The depth of trench excavation normally depends on the depth of the natural clay layer and the groundwater table. The above-and-below type of landfill is like a combination of the area fill and the trench fill. If the solid waste is kept between hills, that is, in the valley, it is called a valley fill type of landfill.

The process of selecting a landfill site is complex and sometimes costly; however a proper siting can be economical to the extent it contributes to the reduction in design and/or construction costs, as well as in long-term expenses with operation and

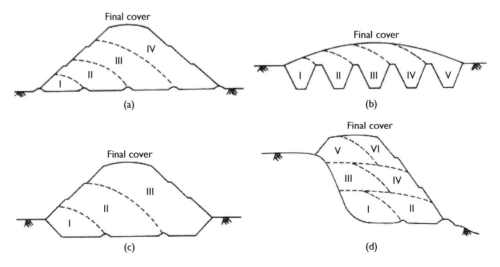

*Figure 4.28* Types of solid waste landfill geometry: (a) area fill; (b) trench fill; (c) above and below ground fill; (d) valley fill (after Repetto, 1995).

maintenance. Factors that must be considered in evaluating the potential sites for the long-term disposal of solid waste include haul distance, location restrictions, available land area, site access, soil conditions, topography, climatological conditions, surface water hydrology, geologic and hydrogeologic conditions, local environmental conditions, and potential end uses of the completed site (Tchobanoglous *et al.*, 1993). The geology of the site is an important barrier to migration of harmful substances. The ground should have a low hydraulic conductivity and a high capacity for the adsorption of toxic materials; it must be sufficiently stable and should not undergo excessive settlements under the load of the landfill body.

Engineered landfills, particularly MSW landfills, consist primarily of the following elements or systems (Fig. 4.29):

1    *Liner system:* This system consists of multiple barrier and drainage layers and is placed on the bottom and lateral slopes of a landfill to act as a barrier system against the leachate transport, thus preventing the contamination of the surrounding soil and groundwater.

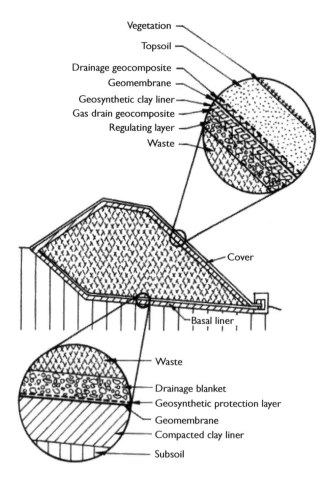

*Figure 4.29* Schematic diagram of a municipal solid waste landfill containment system.

2  *Leachate collection and removal system:* This is used to collect the leachate produced in a landfill and to drain it to a wastewater treatment plant for treatment and disposal. The materials used to construct this system are high-permeability materials, including the following:

- soils such as sands and gravels, often combined with pipes;
- geosynthetic drainage materials such as thick needle-punched nonwoven geotextiles, geonets, geomats, and geocomposites

3  *Gas collection and control system:* This is used to collect the gases (generally methane and carbon dioxide) that are generated during decomposition of the organic components of the solid waste. One can use the landfill gases to produce a useful form of energy.

4  *Final cover (top cap) system:* This system consists of barrier and drainage layers to minimize water infiltration into the landfill so that the amount of leachate generated after closure can be reduced.

Note that out of four major components of a landfill, the liner system is the single most important component. The barrier in a landfill liner or cover system may consist of a compacted clay layer (CCL), geomembrane (GM), geosynthetic clay layer (GCL), and/or a combination thereof. Figure 4.30 shows some examples of lining systems. If the combination of a geomembrane layer and an underlying layer of low-permeability soil (clayey soil), placed in a good hydraulic contact (often called *intimate contact*), is used as a liner, then this combined system is called the *composite liner*. It is important to note that the terms 'liner' and 'lining system' are not synonymous.

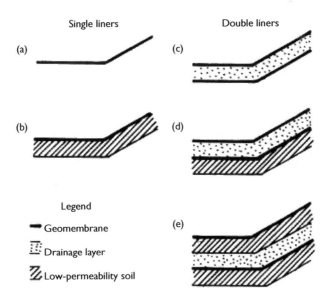

*Figure 4.30* Some examples of lining systems: (a) single geomembrane liner, (b) single composite liner, (c) double geomembrane liner; (d) double liner with geomembrane top (or primary) liner and composite bottom (or secondary) liner; (e) double composite liner (after Giroud and Bonaparte, 1989).

The liner refers to only the low-permeability barrier that impedes the flow of fluids towards the ground; whereas lining system includes the combination of the low-permeability barriers and the drainage layers in a containment facility.

Double composite liners with both primary and secondary leachate collection systems are essential for hazardous waste landfills. A leakage, if any, through a primary liner, can be collected and properly disposed of in the case of a double lined system. In case of single-lined systems, downstream monitoring wells are required for post-construction leak detection. The cost of design and construction of such monitoring wells may exceed the cost of additional liner/leak detection layers of a double lined system. With a leak detection layer, not only can the quantity of liquid be monitored over time, but the liquid can also be remediated in place because the secondary liner is in place to prevent the leachate from leaving the site.

Most major types of geosynthetics, namely geomembranes, geotextiles, geonets, geosynthetic clay liners and geogrids, are used in landfill engineering to perform various functions: fluid barrier, drainage, filtration, protection and reinforcement. In landfills, the geotextiles replace the conventional granular soil layers resulting in decreased weight and reduction in landfill settlement. They also prevent the puncture damage to the geomembrane liner by acting as a cushion in addition to functioning as a filter in leachate collection systems. Thus, the geotextiles can be used to augment and/or replace the conventional protective soil layers for geomembrane liners. Even with a geotextile, the minimum protective soil layer should be 300 mm thick. The geotextile will have a better puncture protection, if it has a higher mass per unit area.

In the case of sanitary landfills, it is always recommended to provide a cover for working faces at the end of the working day to control the disease vectors, odours, fires, etc., thus eliminating the threat to human health and environment. A 15-cm thick soil layer is traditionally used as the daily cover material. Geosynthetics are more effectively used as an alternative daily cover material in reusable or nonreusable forms. The reusable geosynthetic is placed over the working face at the end of the day and retrieved prior to the start of the next operating day. Tyres, sandbags, or ballast soils are placed along the edges to anchor the geosynthetic cover.

### 4.4.1.2 Ponds, reservoirs and canals

Liquid containment and conveyance facilities, such as ponds, reservoirs and canals are required in several areas including hydraulic, irrigation and environmental engineering. Unlined ponds, reservoirs and canals can lose 20–50% of their water to seepage. Traditionally, soil, cement, concrete, masonry or other stiff materials have been used for lining ponds, reservoirs and canals. The effectiveness and longevity of such materials are generally limited due to cracking, settlement and erosion. Sometimes the traditional materials may be unavailable or unsuitable due to construction site limitations, and they may also be costly.

Flexible geosynthetic lining materials, such as geomembranes, have been gaining popularity as the most cost-effective lining solution alone or in combination with conventional lining material for a number of applications, including irrigation and potable water. Figure 4.31 shows typical schematics of liquid containment and conveyance facilities (ponds, reservoirs and canals) involving application of geosynthetics in addition to conventional materials.

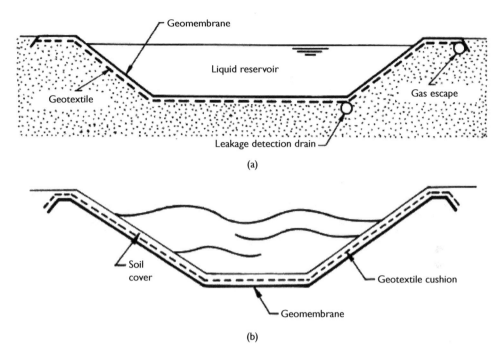

*Figure 4.31* Typical cross-sections: (a) liquid pond/reservoir; and (b) canal.

Geosynthetic liner/barrier materials can be classified as GMBs, GCLs, thin-film geotextile composites or asphalt cement-impregnated geotextiles. The selection of lining material is governed by the location and the environmental factors. Placement, handling and soil covering operations can also affect geosynthetic selection. When GMBs are used as the lining material, the geotextiles can be used with geomembranes for their protection against the puncture by the granular protective layer, which may also be required to prevent UV- and infrared-induced ageing of geosynthetics, as well as any effects of vandalism and burrowing animals. A geotextile, if used below the GMB liner, can function as a protection layer as well as a drainage medium for the rapid removal of leaked water, if any. For economic reasons, the GMB liner may be left uncovered.

### 4.4.1.3  Earth dams

Earth dams are water impounding massive structures and are normally constructed using locally available soils and rocks. One of the principal advantages of earth dams is that their construction is very economical compared to the construction costs of concrete dams. Apart from the conventional materials used in an earth dam, geosynthetics are also employed in the recent times for new dam constructions, and for the rehabilitation of older dams. Properly designed and correctly installed geosynthetics, in an earth dam, contribute to increase in its safety which corresponds to a positive environmental impact on dam structures (Singh and Shukla, 2002). The reasons for which the geosynthetics are used extensively in earth dam construction and rehabilitation are the following:

- The use of geosynthetics in earth dams may serve several functions: water barrier, drainage, filtration, protection and reinforcement.
- The geosynthetics are soft and flexible – therefore, they can endure some elasto-plastic deformations resulting from the subsidence, expansion, landslide and seepage of soil.
- The geosynthetics (geotextiles and geogrids) possess certain mechanical strength, which is favourable for selection of dam-filling materials.
- The permeability of geomembranes is much lower than that of clay or concrete.

The long-term performance of various components of an earth dam is critical to the performance of the dam as a whole. If a geotextile is to be used as a filter, a careful assessment of the properties, extensive testing, and monitoring are required to ensure its suitability. The locations in earth dams where the geotextile filters may be used are in the downstream chimney drain and in the downstream drainage blanket (see Fig. 4.32). If the dam is subjected to a rapid drawdown, then the drainage systems using geosynthetics may also be installed on the upstream side of the core. In the past, the geotextile filters, mostly nonwovens, have been used for the construction or the rehabilitation of numerous embankment dams (i.e. earth or earth and rockfill dams) in various parts of the world.

The chimney drain concept can also be used for rehabilitation purposes. In the case of embankment dams, that exhibit seepage through their downstream slope, the construction of a drainage system in the downstream zone is required. A geocomposite drain, placed on the entire downstream slope or only on the lower portion of it and covered with backfill, also performs well. The geocomposite drain must be connected with the new toe of the dam with outlet pipes or with a drainage blanket. This technique has been used at *Reeves Lake dam* in the USA which is a 13 m high dam repaired in 1990 by placing the geocomposite drain (including a PE geonet core between two PP thermally-bonded nonwoven geotextile filters) on the downstream slope (Wilson, 1992).

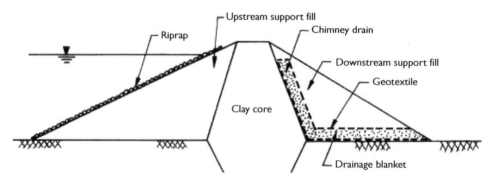

*Figure 4.32* A typical cross-section of an earth dam with geotextile filters.

To reduce the rate of seepage through the dam embankment, a geomembrane sheet may be installed on the upstream face of the embankment. The geomembrane sheet acts as a barrier to water flow. A thick geotextile must be placed on one or both sides of the geomembrane, to protect it from the potential damage by adjacent materials, typically the granular layer underneath and the external cover layer. The lower geotextile is generally bonded to the geomembrane in the factory; while the upper geotextile is independently placed between the geomembrane and the cast-in-place concrete cover layer (see Fig. 4.33).

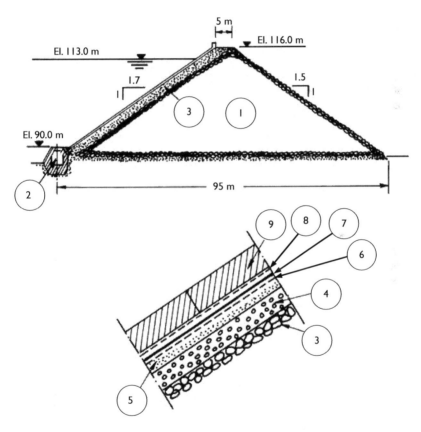

*Figure 4.33* Codole rockfill dam, constructed in 1983 in France with the use of geomembrane barrier and geotextile protective layers (after ICOLD, 1991).

Notes:
1  Rockfill (up to 1 m size)
2  Inspection and drainage gallery
3  Sand and gravel layer (2 m thick, 25–120 mm particle size)
4  Gravel layer (0.15 m thick, 25–50 mm particle size)
5  Cold premix layer (50 mm thick, 6–12 mm particle size)
6  Geotextile (mass per unit area = 400 g/m²) bonded to geomembrane
7  PVC geomembrane (thickness = 2 mm)
8  Geotextile (mass per unit area = 400 g/m²)
9  Concrete slabs (0.14 m thick, 4.5 m × 5 m size)

Geomembranes can be used for the lining of concrete or masonry dams. In this application, a thick needle-punched nonwoven geotextile is used as a cushion and drainage layer between the geomembrane and the dam. The geotextile is connected to a collector pipe at the toe of the dam. Because of the geotextile layer, the concrete, generally saturated with water, is allowed to drain, thereby slowing the mechanisms of concrete deterioration caused by the presence of water. This method can also be used to control seepage through the wall.

Geosynthetics are also used in dams as a reinforcement. If steep slopes are required, then the geogrids or the woven geotextiles are generally used for reinforcement purposes. For rehabilitation purposes, like increasing the height of the dam itself to increase the storage capacity or the free board, the use of geosynthetic is ideal, and intensive research is recommended in this direction. The first application of geosynthetic reinforcement in dam construction occurred in 1976 when Maraval Dam, which is 8 m high, was constructed in Pierrefeu, France (Giroud, 1992).

Geosynthetics are also used to control the surficial erosion of earth dams, both for new construction and rehabilitation purposes. The upstream slope of dams can be eroded by wave action and the downstream slope by runoff from precipitation or overtopping in case of flood. In all these cases, solutions incorporating the geosynthetics have been used. In the case of erosion caused by rain water, the entire downstream slope and the upper portion of the upstream slope is protected using typical techniques adopted for the river bank revetments, as a rip-rap, in which the geotextiles perform as a filter or, in other solutions (see ICOLD, 1993), as soil-cement blankets, concrete slabs, bituminous concrete layers and so on, in which the geosynthetics could be incorporated with a separation or even a reinforcement function. The products commonly used to control the surficial erosion due to atmospheric agents are mainly geomats and geocells. Sometimes biotechnical mats are also adopted, particularly when the biodegradation is desirable as in the case of a temporary role to be played during the vegetation growth. This is a common practice to solve the problems induced by erosion due to rainfall and consequent run off as described in Section 4.3.

The most challenging application of geosynthetics is related to the protection against overtopping, which represents a very crucial aspect of dam engineering. Many failures of embankment dams have been induced in the past by overtopping. Articulate concrete blocks linked by cables, and resting on a geotextile, can be used in order to protect the crest and the downstream slope against overtopping. For this purpose, the woven geotextiles are adopted mainly to perform as a filter material. However, the opening size of the geotextile can be selected not only to satisfy the filter criteria but also to allow penetration by grass roots. In fact, the articulate blocks are covered by the grassed topsoil layer to give it a natural appearance and additional anchorage for articulate concrete blocks. This system was used in the Blue Ridge Parkways dams in the USA (see Fig. 4.34). Geosynthetics, associated with earth materials and vegetation, can thus form a stable solution to resist overtopping phenomena in earth embankment dams. Gabions and mattresses can also be used to protect the upstream and downstream faces of earth dams.

The use of geosynthetics is associated with a reduction of natural earth materials which are to be exploited and placed on the dam sites. This construction practice shows a positive environmental impact.

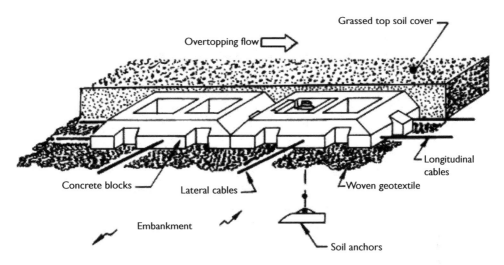

*Figure 4.34* Articulate concrete blocks with a geotextile filter and a grassed topsoil cover for the downstream face protection against overtopping of Blue Ridge Parkways dams in the USA, 8.5–12 m high (after Wooten *et al.*, 1990).

## 4.4.2  Analysis and design concepts

### 4.4.2.1  *Landfills*

The most important requirement of a landfill is that it should not pollute or degrade its environment, that is, it should always be environmentally friendly. So, the primary engineering assignment in designing, constructing and operating landfills is to provide efficient barriers against contamination. This requirement is achieved by both careful siting and adopting proper design/construction methods. The site of the landfill must be geologically, hydrologically and environmentally suitable. A detailed design of several landfill elements is necessary. The proper functioning of each of these elements is essential to construct and maintain a landfill in an environmentally sound manner.

Since water is the most important transporting agent for pollutants, the infiltration of water into and the extraction of water out of the waste body must be controlled by reliable technical means, such as leachate collection and removal systems. The drainage facilities must maintain the minimum gradients to facilitate the gravitational flow. The physical and biological properties of the waste, as well as the availability of construction materials, are important parameters for the design of liners and covers. Aiming at a justifiable degree of safety with respect to the environment, nationally or regionally responsible authorities issue minimum requirements and some basic rules for the design of landfill liners and covers. These rules differ from one country to another, and sometimes even within one country. The readers can find the details of German practice on liners and landfill covers in the contribution by Zanzinger and Gartung (2002).

Special attention is paid to the properties and the placement of the waste material. The waste body is considered a barrier by itself. The refuse should be in such a condition, that the stability of the landfill is granted, there is little or no tendency for harmful material to being dissolved and transported with seeping water, and the deformation due to settlement should be predictable and small. So the integrity of the cover would not be impaired in the long-term. In summary, the landfill structure forms a multibarrier system. Each of the barriers has to meet certain minimum technical requirements, independent of the performance of other barriers.

A municipal solid waste (MSW) landfill must be designed and constructed to accept a highly variable waste system. It must be able to prevent groundwater pollution, collect leachate, permit gas venting, and provide for groundwater and gas monitoring. Most, if not all, of the design and construction principles for MSW landfills apply equally to the hazardous waste landfills.

The landfill project must consider various regulations, formed by the responsible agencies such as the United States Environmental Protection Agency (USEPA), governing landfill siting, design, construction, operation, groundwater and gas monitoring, landscaping plan, closure monitoring and maintenance for the design life that may be 30 years.

The properties, advantages, limitations, and design requirements for compacted clay liner, geomembrane liner, and geosynthetic clay liner are summarized in the following paragraphs.

### Compacted clay liner (CCL)

Compacted clay liner for the landfill must have a low permeability to prevent or minimize the leachate leakage, adequate shear strength for stability, and minimal shrinkage potential to prevent desiccation cracking. Since compaction on the wet-side of optimum water content minimizes hydraulic conductivity (Lambe, 1958; Mitchel *et al.*, 1965; Boynton and Daniel, 1985; Shukla, 2014), the clay soil liners must be designed for the wet-side. The range typically varies from 0–4% points wet of standard or modified Proctor optimum. Typically, the clay liners must have a hydraulic conductivity, $k \leq 1.0 \times 10^{-9}$ m/s. The water content and the dry unit weight must be established in a range so that the compacted clay liner will also have adequate strength to withstand high overburden pressures and shear stresses depending on the height of the landfill that may be up to 75 m high. Note that the practice of wet-side compaction may cause problems in arid regions where the near-surface clays may desiccate during the periods of drought. It is a difficult task to find a way to compact clay soil with both low hydraulic conductivity and low shrinkage potential. However, solutions such as using soils rich in sand, placing the soil at the lowest practical water content or avoiding the use of highly plastic soils can be adopted to meet the design requirements appropriately. In designing the compacted clay layer, the causes of failure such as subsidence, desiccation cracking, and freeze/thaw cycling must be considered. The required minimum thickness of the compacted clay liner is normally 0.5 m. For hazardous waste landfills, the thickness may be increased up to 1.5 m placed in lifts of 0.25 m each.

### Geomembrane liner

The most widely used geomembrane in landfill engineering is high-density polyethylene (HDPE), because this offers excellent performance for landfill liners and covers. If greater flexibility than HDPE is required, then linear low-density polyethylene (LLDPE) geomembranes can be used. The geomembranes with textured surfaces on one side or both sides can be used for improving the stability on slopes. The introduction of textured geomembranes significantly helps, allowing for 3(H):1(V) side slopes and even steeper ones in some cases. The soil-geomembrane and geomembrane-geotextile interface friction values should be correctly estimated because these values are critical for the proper design of geomembrane-lined side slopes of landfills. Cover soils can slide over the surface of the geomembrane. Alternatively, the geomembrane can also fail by pulling out of the anchor trench and slipping downhill.

For an intact geomembrane, the transfer of moisture or gas transmission across the membrane occurs by diffusion and the rates are very low. For example, the water vapour transmission rate through 30 mil (0.75 mm) thick HDPE geomembrane is 0.02 g/m²/day, whereas the methane gas transmission rate through 24 mil (0.6 mm) thick HDPE geomembrane is 1.3 ml/m²-day-atm (USEPA, 1988). In order to warrant a sufficient robustness of the geomembrane in handling, the specified minimum thickness of approved geomembranes is 2.5 mm. This thickness also happens to be very satisfactory with respect to the sealing function. However, the HDPE-geomembranes of 2.5 mm are not very flexible. The minimum width of the geomembrane is 5 m in order to minimise the amount of field seaming needed to create large waterproof sheets (Zanzinger and Gartung, 2002).

The tensile stresses, developed due to the unbalanced friction forces (Fig. 4.35(a)) and/or due to localized subsidence (Fig. 4.35(b)), must be properly analyzed. The former situation arises when a material with high interface friction (like sand or gravel) is placed above the geomembrane and a material with low interface friction (like high water content clay) is placed beneath the geomembrane. The geomembrane goes into a state of pure shear and carries a tensile force. The factor of safety, $FS_T$, for geomembrane against the tensile failure is expressed as

$$FS_T = \frac{T_a}{T_r} \tag{4.24}$$

where $T_a$ is the allowable tensile force per unit width in the geomembrane, and $T_r$ is the required tensile force per unit width in the geomembrane.

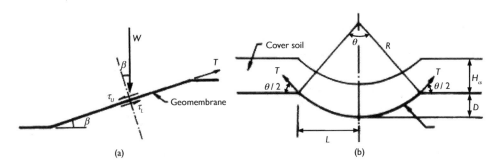

*Figure 4.35* (a) Shear and tensile stresses acting on a geomembrane due to unbalanced forces; (b) tensile stresses in a geomembrane mobilized by cover soil and caused by subsidence.

The second situation arises whenever the localized subsidence occurs beneath a geomembrane that supports a cover soil, often happening in the landfill closure situation where the underlying waste has been poorly and nonuniformly compacted. The out-of-plane forces from the overburden cause some induced tensile stresses in the geomembrane, depending upon the dimensions of the subsidence zone and the cover soil properties (Koerner and Hwu, 1991). The calculated value of the required tensile strength of geomembrane for the specific site situation must be compared with an appropriate laboratory simulation test (e.g. three-dimensional axisymmetric tension test) for allowable tensile force in the geomembrane.

Geotextiles can be recommended with geomembranes to function as a cushion layer in order to enhance its puncture resistance during installation and in-service in containment systems including landfills. Design and selection of a geotextile for specific geomembrane types and thickness consist of evaluating the local stress conditions. Determining the puncture force, $P$, as per the site conditions, the design puncture point load, $P_L$, is calculated as follows:

$$P_L = FS \times P \tag{4.25}$$

where $FS$ is the factor of safety for long-term loading conditions, generally greater than 10. Based on the value of $P_L$, mass per unit area required for each type of geotextile can be considered. Figure 4.36 provides an example design chart for continuous filament PP needle-punched nonwoven geotextiles on HDPE geomembranes. A typical example of the protective layer is an HDPE needle-punched nonwoven geotextile of 1200 g/m² plus 100–150 mm of sand or crushed stone of maximum 8 mm particle size.

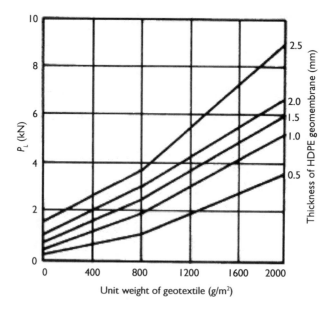

*Figure 4.36* Example design chart for continuous filament polypropylene needle-punched nonwoven geotextiles on HDPE geomembranes (after Werner *et al.*, 1990).

The landfill cover should be designed with the same degree of attention and care that is applied to the soil liner. Note that unlike a liner, a soil cover acts as a hydraulic barrier against infiltrating water from outside only; it is not required to act as a barrier against leakage of leachate solutes under combined *advection* and *diffusion*. The advection refers to the process by which the solutes are transported simultaneously along with the flowing fluid/solvent in a porous medium under a hydraulic gradient, whereas the diffusion refers to the movement of solutes/dissolved substances under a chemical and concentration gradient. The solutes can diffuse in the same direction as the advective movement, or they can diffuse in an opposite or counter direction. Fig. 4.37 shows a schematic diagram of conventional landfill design showing advective and diffusive flows acting in the same direction. The relative importance of diffusion as a leakage pathway increases as the hydraulic conductivity of the barrier decreases (Qian *et al.*, 2002).

Geomembrane-lined soil slopes (Fig. 4.38) require proper stability checks. The stability of the overlying materials (soil/drainage geosynthetic) as well as the tensile stresses that may be induced in the underlying geomembranes should always be performed. The interface friction values between the geomembrane and the overlying materials, generally evaluated from simulated direct shear tests, play a great role in the stability analysis. Both the stability of the overlying soil materials and the reduction of tensile stresses in the geomembrane can be accommodated by reinforcing the cover soil with either the geogrids or the geotextiles (Koerner and Hwu, 1991).

Recommendations must be made for the use of geomembrane without any hole or opening. If there is a hole in the geomembrane liner, the leachate will move easily through the hole and the seepage will take place through the soil subgrade (Fig. 4.39(a)). With a clay soil liner alone, the seepage takes place over the entire area of the liner (Fig. 4.39(b)). With a composite liner, only a limited amount of leachate will pass through any hole in the geomembrane, but it will then encounter the low-permeability clay soil, which will impede further the migration of the limited amount of leachate passing through the hole (Fig. 4.39(c)). Thus, a composite liner (i.e. the geomembrane on low permeability soil) is more effective in reducing the rate of leakage through the liner than either a geomembrane alone or a soil liner alone (Giroud and Bonaparte, 1989). The designer must keep in mind that the geomembrane should not be separated from the clay liner with permeable materials, such as a bed of sand or a geotextile, because this would jeopardize the intimate contact (Fig. 4.40).

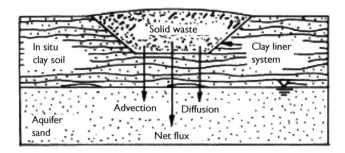

*Figure 4.37* Schematic diagram of conventional landfill design showing advective and diffusive flows acting in the same direction (after Qian et al., 2002).

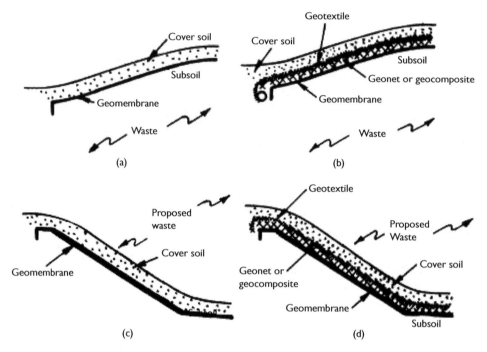

Figure 4.38 Geomembrane-lined slopes: (a) landfill cover with soil above the geomembrane; (b) landfill cover with drainage geosynthetic above the geomembrane; (c) landfill liner with soil above the geomembrane; (d) landfill liner with drainage geosynthetic above the geomembrane (after Koerner and Hwu, 1991).

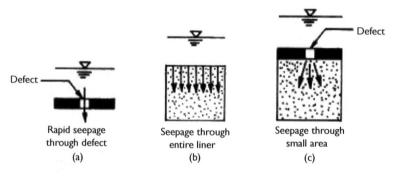

Figure 4.39 Seepage patterns through: (a) geomembrane liner; (b) clay soil liner, and (c) composite soil (after Qian et al., 2002).

For designing a leak detection monitoring system, the leakage assessment should be based on the analytical approaches supported by empirical data from other existing operational facilities of similar design. Note that the leakage is significantly affected by the performance of the geomembrane liner controlled by the defects or penetrations of the liner, including imperfect seams, punctures or pinholes caused by construction defects.

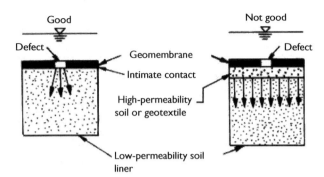

*Figure 4.40* Proper design of composite liner for intimate hydraulic contact between the geomembrane and the compacted soil (Daniel, 1993).

### Geosynthetic clay liner (GCL)

Geosynthetic clay liners are used as a substitute for the compacted clay liners in cover systems and composite bottom liners. They are installed on horizontal surfaces as well as on the slopes by unrolling and overlapping the edges and ends of the panels. Overlaps self-seal when the bentonite comes in contact with water, that is, hydrates. When a geosynthetic clay liner hydrates, the bentonite swells in the pores, thereby forming a watertight sheet that also offers a protection to the overlying geomembrane liner. It has many advantages over a CCL, including the following (USEPA, 1993; Snow *et al.*, 1994):

- easily shipped to any site and thus can be made easily available at any site;
- simple and rapid installation;
- no requirement of heavy equipment for installation;
- less requirement of vehicular traffic and less energy use for installation;
- lower thickness, approximately 0.5 in. (13 mm), and hence conservation of landfill space;
- material quality (consistency and uniformity) maintained in a controlled environment;
- lower consumption of construction water, dust generation, and vehicular traffic during installation;
- lower susceptibility to desiccation cracking;
- self-healing capabilities if punctured;
- better resistance to freeze/thaw and wet/dry cycles;
- availability of tensile strength developed by the geotextiles or geomembranes;
- can tolerate significantly more differential settlement;
- relatively simple, straightforward, common-sense procedures for quality assurance, thus making the assurance system economical.

Note that the compacted clay liners have also some advantages over the geosynthetic clay liners, such as the large thickness (approximately 2–3 ft (600–900 mm)) makes them virtually puncture proof, and greatly increases the breakthrough time by diffusion, and at the same time there is a long history of use of compacted clay liners. In fact, the substitution of a GCL for a CCL should be decided based on the evaluation of contaminant transport equivalency between them. This evaluation should be based on comparing not only advective mass fluxes through the liner, but also diffusive mass flux during the lifetime of the landfill. The diffusive mass flux decreases and the advective mass flux increases with time for both the GCL and the CCL. Thus, the contaminant transport by diffusion is important during the early stages. However, the contaminant mass transported by diffusion is relatively small and the effect of diffusion can be ignored during later stages (Qian *et al.*, 2002).

There are several factors that affect the hydraulic conductivity of GCLs, such as type of permeants and confining stress. The effects of wet-dry cycling and freeze-thaw cycling also must be considered when selecting the GCLs in the bottom liner and the final cover systems for landfills. With increase in confining stresses, the hydraulic conductivity of a GCL generally decreases significantly, mainly because of lower void ratio of bentonite resulting from the higher confining stresses. Alternate wetting and drying may occur in a GCL in the final cover systems of landfills and the site remediation projects. When exposed to freeze-thaw cycling, the hydraulic conductivity of a GCL does not get any significant changes, although the CCLs generally undergo large increases in hydraulic conductivity when exposed to freeze-thaw cycling. The designers must note that in general, GCLs fall between CCLs and GMBs in terms of ability to maintain their hydraulic integrity during distortion such as that induced by differential settlement in landfill final covers. The hydraulic conductivity of typical GCLs is generally equal to or less than $1.0 \times 10^{-11}$ m/s to $5.0 \times 10^{-11}$ m/s.

The stability of GCLs is an important design consideration because of the low shear strength of the bentonite after hydration. For the higher shear strength applications, the reinforced geosynthetic clay liners (e.g. geotextile-encased, stitch-bonded or geotextile-encased, needle-punched) should be recommended. The design should consider the possibility of the shearing failure involving a GCL at the following three locations (Daniel *et al.*, 1998):

1   the external interface between the top of the GCL and the overlying material (soil or geosynthetic);
2   internally within the GCL;
3   the external interface between the bottom of the GCL and the underlying material (soil or geosynthetic).

The design values of internal shear strength of GCLs should be measured on a product-specific basis from the laboratory direct shear tests under conditions closely simulating those expected in the field (Fox *et al.*, 1998). The reduction of internal or interface shear strength from peak to residual is dependent on the reinforcement type of the GCLs or contact materials at the interface. As a general guideline, unreinforced GCLs are not recommended for slopes steeper than 10(H):1(V) (Frobel, 1996; Richardson, 1997). In fact, one should not design the slopes that exceed the safe slope angle for the GCLs or their respective interfaces within the systems. Stitch-bonded and needle-punched GCLs probably are suited equally for applications involving a low normal stress (e.g., pond and lagoon liners and cover systems), whereas the needle-punched GCLs are probably the better choice for applications where a high normal stress is applied (e.g., landfill bottom liners).

All landfills have at their base a leachate drainage layer consisting of a natural soil (sands and gravels) and/or a geosynthetic drainage material (i.e. geocomposite, such as geotextile bonded to one or both surfaces of a geonet). Landfills with a double-composite liner system have both primary and secondary leachate drainage layers, called the leachate collection and leak detection layers, respectively. The most essential requirement for a landfill leachate drainage layer is that it should have adequate drainage capacity to handle the maximum leachate flow produced during the landfill operations. The leachate head buildup in the drainage layer should generally be less than 12 in. (0.3 m) (Qian et al., 2002). Leachate pipes are generally installed in trenches that are filled with gravel. The trenches are lined with geotextile to minimize the entry of fines from the liner into the trench and eventually into the leachate collection pipe. Typical trench details are shown in Fig. 4.41. Usually the design shown in Fig. 4.41(a) is used in landfills in which the liner material is clay and the design shown in Fig. 4.41(b) is used in landfills in which the primary liner material is geomembrane. It is essential to have a deeper excavation below the collection trench so that the liner has the same minimum design thickness even below the trench. The geotextile, which acts as a filter, should be folded over the gravel. Alternatively, a graded sand filter may be designed to minimize the infiltration of fines into the trench from waste. The design of geotextile filters and drains has already been discussed in Section 4.2. A leachate pipe may fail due to clogging, crushing or faulty design. The design and maintenance of leachate pipes for each of these situations must be considered properly.

Starting at the bottom, a typical double composite liner system consists of a minimum 2 ft (0.6 m) thick CCL (or GCL that can be equivalent to a 2 ft (0.6 m) thick CCL), followed by a secondary GMB liner, secondary leachate collection (or leak detection) layer, a minimum 2 ft (0.6 m) thick primary CCL (or GCL that can be equivalent to a 2 ft (0.6 m) thick CCL), a primary GMB, and a primary leachate collection system. A 2-ft (0.6 m) thick protective sand blanket tops off this sequence. The leachate collection system consists of a layer of geonet and geotextile. The former provides a good in-plane drainage conveyance and the latter good cross-plane drainage together with the ability to exclude (filter out) the fines. The geomembranes must be at least 1.5 mm (60 mils) thick if HDPE, or 0.75 (30 mils) thick if made from other polymers. The permeability of the subbase and CCL must not exceed $1.0 \times 10^{-9}$ m/s.

Proper closure is essential to complete a filled waste landfill, particularly of hazardous type. The cover system must be protected from burrowing animals, wind

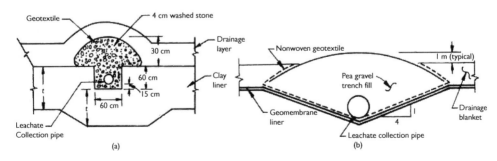

*Figure 4.41* Typical leachate collection trench details for: (a) clay liner; and (b) geomembrane liner (after Bagchi, 1994).

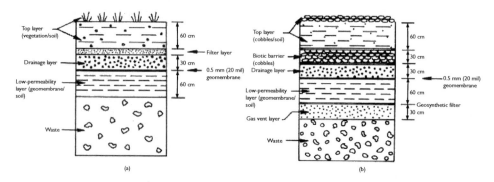

*Figure 4.42* USEPA-recommended landfill cover designs (after Landreth and Carson, 1991).

and water erosion, wet-dry cycles, and freeze-thaw cycles. In fact, it should be devised at the time the site is selected, and the plan and the design of the landfill containment structure are chosen. The location, the availability of low-permeability of soil, the stockpiling of good topsoil, the availability and use of geosynthetics to improve performance of the cover system, the height restrictions to provide stable slopes, and the use of the site after the post-closure care period are typical considerations. The design goals of the cover system are that further maintenance is minimized and that human health and the environment are protected. For hazardous waste facilities, a final cover with minimum requirements (Fig. 4.42(a)) consists of the following from bottom to top:

1   a 60 cm (24 in.) layer of compacted natural or amended soil with a hydraulic conductivity of $1 \times 10^{-9}$ m/s in intimate contact with a minimum 0.5 mm (20 mil) geomembrane liner;
2   a drainage layer: a minimum 30 cm soil layer having a minimum hydraulic conductivity of $1 \times 10^{-4}$ m/s, or a layer of geosynthetic materials with the same performance characteristics;
3   a top, vegetation/soil layer: a top layer with vegetation (or an armoured top surface) and a minimum of 60 cm of soil graded at a slope between 3% and 5% to prevent erosion and to promote drainage from the area.

Where the type of waste may create gases, vent structures (either soil or geosynthetic) must be included in the cover (Fig. 4.42(b)). Plant roots or burrowing animals, collectively called the bio-intruders, may disrupt the drainage and the low-permeability layers to interfere with the drainage capability of the layers. A 90-cm (3 ft) biotic barrier of cobbles directly beneath the top vegetation layer (Fig. 4.42(b)) may stop the penetration of some deep-rooted plants and may stop the invasion of burrowing animals. Settlement and subsidence should be evaluated for all covers and designed into the final cover plans. The cover design process should consider the stability of all the waste layers and their intermediate soil covers, the soil and foundation materials beneath the landfill site, all the liner and leachate collection systems, and all the final cover components. When a significant amount of settlement and subsidence is expected within a few years (2–5 years) of closure, an interim cover might be proposed – one that protects human health and environment. When settlement/subsidence is essentially complete, the interim cover should be replaced or incorporated into the final cover.

The complete design of a landfill requires a variety of calculations during the design process in order to demonstrate regulatory compliance and ensure proper design. For this purpose, available computer software can be used by the designers. However, they should always keep in mind the limitations of the software being used.

### 4.4.2.2 Ponds, reservoirs and canals

Ponds, reservoirs and canals require lining systems for their effective performance. The primary design consideration for any lining project is the loss of contained liquid throughout the intended service life. The intrinsic permeability of less than $1.0 \times 10^{-14}$ m/s for HDPE geomembranes far exceeds the requirements for any containment and conveyance project. The main controlling factor is always the loss of water through seams and punctures resulting from damage during or after installation. In fact, leakage, and not permeability, is the primary concern when designing geosynthetic containment structures. Leakage can occur through poor seams, pinholes from manufacture, and puncture holes from handling, placement, or in-service loads. Leakage from the geosynthetic liner/barrier systems is minimized by giving attention to design, specification, testing, quality control and quality assurance.

The geosynthetic barrier should be designed as per its role as a primary or secondary liner keeping in view in-service conditions, installation damage and durability. Note that the installation of geomembrane or geosynthetic clay liners is a primary design consideration. Placement, handling and soil covering operations can also affect the geosynthetic design. Liquid depths also govern the design of the liner system. If a designer assumes that the surface liner system will not even leak, then the following provisions must be made (Richardson, 2002):

1   The use of such details as battens and conventional pipe penetration details that cannot be leak tested must be avoided. All components of the containment system must be pressure or vacuum tested.
2   The liner must be protected from harm during its surface life. Thus, if one can see the geomembrane, it must be assumed that a defect and resultant leakage will appear during the liner's service life. It is more reasonable to assume that the surface impoundment liner has a very minor rate of leakage, and design to accommodate that leakage as follows:

  •   If the contained liquid may harm the environment, then a secondary liner/collection system should be used to monitor the performance of the primary liner.
  •   If the contained liquid does not harm the environment, then the ability of leakage to drain away from the bottom of the liner must be ensured. This may require a designed underdrain where the natural subgrade soil has a low permeability.

The following points are some general design indications concerning the use of geosynthetics in ponds, reservoirs and canals (Duquennoi, 2002):

1   The bottom of the structure should form a slight slope, between 1% and 2% lengthways, and between 2% and 3% sideways.
2   The embankment slopes should be designed according to the state-of-the-art soil mechanics; it has to be underlined so that the geomembrane lining systems cannot be used to reinforce slopes. For many applications, a 1V:2H (1 vertical by 2 horizontal) slope should be advised, and 2V:3H has to be considered as a maximum.
3   The embankment top should be wide enough to enable anchoring of the geosynthetic; a minimum anchoring length is generally 2 m for ponds and reservoirs and 1 m for canals, but specific designs are to be taken into account. It is generally not recommended to lay a geomembrane directly on the subgrade, except in particular cases when the risks of geomembrane puncturing and underliner pore water or gas pressure have been catered for. A better way to prevent these risks is to specifically design the underliner systems. The underliner water drainage can be performed either by gravel layers, gravel-filled drainage trenches, or geosynthetic draining strips. Depending on the volume of water to be drained, the perforated geopipes may supplement the gravel-based drainage systems. The drain pipes are always connected to a main collecting pipe or manhole and then to pumped or gravity outlet.
4   A protective layer may be interposed between the geomembrane and the subgrade when the latter is not smooth enough to guarantee the geomembrane safety, especially below the high water head. Geotextiles are now generally preferred because of their possible combined functions of gas drainage and mechanical protection of underliner.
5   The core of a lining system is, of course, the impermeable material, that is, either a geomembrane or a GCL. The design and choice of a lining system should be decided considering economic, hydraulic, mechanical and durability aspects in addition to ease of installation and seam performance aspects as per the site-specific requirements.
6   Besides single geomembrane or geosynthetic clay lining systems, it is possible to install the double lining systems using two geomembranes with a drainage layer in between them. This solution is still rare in liquid containment and conveyance applications, and is only applied where the risk of leaks must be extremely reduced.
7   One of the best ways to prevent anticipated aging of geosynthetics in general and geomembranes in particular is to limit their exposition to weather action by covering them. The purpose of overliner layers is also to prevent liner damage by floating or transported solids (e.g. ice and wood), by operating vehicles or machines (e.g. mobile pumping equipment), by burrowing animals and plant roots, and by vandalism or accidental human intervention. One usual design is to protect the geomembrane by a geotextile and then to cover it by a layer of granular material. Granular layers may be composed of several sub-layers differing in particle-size distribution, from the finest-grained (e.g. fine sand) directly over the geotextile up to the coarsest-grained material (e.g. rip-rap) on top of the granular layer. Other usual designs may consist of concrete covers, using precast blocks or

slabs, in situ poured reinforced concrete layers or even shotcrete. Another purpose of the overliner covers is to prevent the geomembrane lining system uplift under wind action. Some installation procedures may include temporary ballast over the geomembrane in order to prevent uplift during installation, before installing permanent overliner protection layers.

8  Preventing hydraulic actions such as fluvial erosion in the case of canals requires the use of specifically designed systems, which are generally geosynthetic systems: geocells, geomattresses, geoarmours or geomats. These systems may also be used alone to prevent bank erosion, without covering any geosynthetic lining system.

9  All geosynthetic systems are to be anchored on top of the embankment slopes or on the slope itself, depending on the overall design. The most common anchoring design is the anchor trench, which is generally a square section trench in which the geomembrane is laid on one side and at the bottom; the trench is then backfilled with non-puncturing soil. It is generally recommended that the anchor trenches should be deeper and wider than 0.5 m; and they should be situated at least 0.7 m from the edge of the slope. For canals, especially, the excess geomembrane width related to anchoring design may generate excess cost; anchoring characteristics must then be precisely derived from calculation or alternative techniques may be applied such as tying the geomembrane to stakes. In some applications (e.g. deep reservoirs) intermediary anchoring may be required alongside the slope (Fig. 4.43).

10  Access roads and tracks are sometimes required, especially in large containment ponds where the vehicles have to access the bottom of the pond for maintenance or exploitation purposes. Special attention has to be given to the protection of geomembranes under the road and to the stability of the road over geosynthetics. The subgrade has to be shaped to take the access road into account; extra protection of the geosynthetic lining system should be designed.

11  Connection to concrete structures usually poses the problem of waterproofing continuity. A lot of technical solutions are available, depending mainly on the geomembrane type. Metallic fixations are generally used in association with metallic and elastomeric plates and/or geomembrane overlaps.

Figure 4.44 shows the design details of some typical lining systems for ponds, reservoirs and canals based on the case studies presented by Duquennoi (2002). It is important to underline that all the above-mentioned points are closely inter-related in terms of design. For example, it is impossible to select a geomembrane without taking the characteristics of the overliner protection layer into account, and conversely to design a geomembrane protection layer without considering the type

(a)                                              (b)

*Figure 4.43* Typical examples of anchor trenches on slopes.

of geomembrane. A geosynthetic lining system has thus to be designed as a whole, including subgrade preparation, underliner and overliner layers and specific features such as the ones described above. Moreover, different geosynthetic lining systems may be equivalent in terms of hydraulic, chemical or mechanical criteria and the difference may be finally only related to installation needs, economic criteria or availability. As we can see, the basics of geosynthetic systems for liquid conveyance and containment are fairly simple, however applying them to specific works may be complex and requires more information and experience than has been briefly presented here.

### 4.4.2.3   Earth dams

Safety is the main concern with earth dams. Although the design of an earth dam is a complex art, with each situation different from the other, the basic steps involved in the design, as mentioned below, are quite easy to follow:

1   A thorough exploration of the foundation and abutments, and an evaluation of the quantities and characteristics of all construction materials available within a reasonable distance of the site.
2   Selection of possible trial design.
3   An analysis of safety of the trial design.
4   The modification of the design in order to meet stability requirements.
5   The preparation of the detailed cost estimation.
6   The final selection of the design which seems to offer the best combination of economy, safety and convenience in construction.

Although a conventional design incorporates these steps to a great extent, some recent developments in embankment and dam construction have imposed several challenges in order to achieve perfection and an economical cross-section both in terms of time and money. In current construction practice, the use of geosynthetics, in conjunction with conventional earth dam construction materials, has become common. This imposes a challenging task to civil engineering practices. Further, use of geosynthetics in earth dams affects their construction procedure and stability. In fact, an efficient use of geosynthetics requires special attention. The properties of geosynthetics must be evaluated (see Chapter 2) based on specific criteria and functional requirements, such as acting as a water barrier, filter, drainage medium, protective layer or reinforcement.

In general, the design procedure is guided by the International Commission on Large Dams (ICOLD). A desktop analysis may be undertaken based on the available guidelines. Once a desktop analysis is completed and suitable geosynthetics are identified, these are subjected to soil-geosynthetic compatibility testing, before making a final selection, which includes consideration of minimum strength and deformation requirements of the geosynthetics. These parameters need to be taken into account of both the short-term loading expected during installation and construction as well as post-construction loads and deformations. While the overall embankment settlement may be low, local stresses and strains may be high due to differential settlements or shrinkage of the soil. It is for this reason that a geosynthetic needs to maintain its restraining characteristics, even after local concentration of stresses and strains takes place. There may be a substantial change in pore size of the geotextiles due to elongation (Legge, 1986). However, the main concern is the extent to which woven tape and staple fibre product pores elongate when the fabric is placed in tension. Note that all dams shall be designed on the understanding that there is a significant risk that the core will crack and that the possibility of internal erosion of the core has to be allowed for in the filter design (McKenna, 1989).

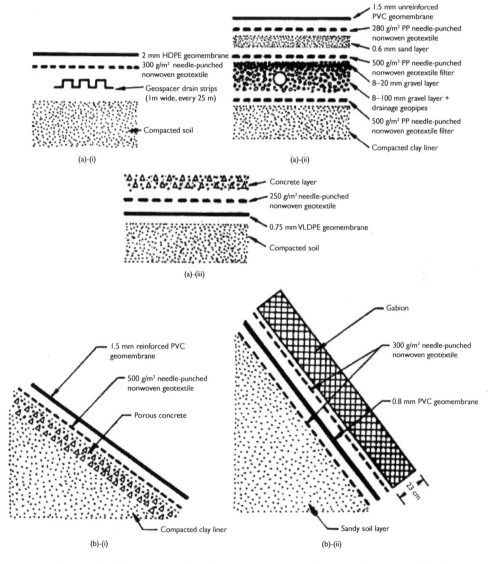

*Figure 4.44* Design details of some typical lining systems for ponds, reservoirs and canals: (a) bottom lining system; (b) slope lining system (after Duquennoi, 2002).

The geotextile filters used in dams play a critical role and, therefore, must be carefully selected. The designers of dams must not use the simplistic filter criteria that are sometimes used for non-critical applications. For example, if the soil in contact with the filter has a high coefficient of uniformity – a common situation for earth dams – some simplistic filter criteria may lead to the selection of a geotextile filter that does not prevent the soil piping – a typical cause of dam failure. In cases where some particular soil characteristics make it difficult for the geotextile filters to strictly meet the filter criteria, the filtration tests simulating the conditions expected in the field can be conducted to evaluate the candidate geotextile filters.

When a transmissive geosynthetic is used to provide the drainage in a dam, it is important that the transmissivity of the geosynthetic be measured in a laboratory test that simulates the conditions in which the geosynthetic will be used in the field. The design engineer must perform the calculations to determine the maximum stresses that are expected on the geosynthetic in the field, and the laboratory team must plan a test where the field boundary conditions, including the maximum applied stresses, are accurately reproduced.

The durability of geosynthetics should always be an essential consideration when they are used in dams. It is also a key requirement in waste disposal applications. Based on the knowledge accumulated in designing and constructing lining systems for waste containment, the geosynthetics available today clearly have adequate durability for safe use in dams.

## 4.4.3   Application guidelines

In containment facilities (landfills, canals, ponds, reservoirs, dams, etc.), the installation of geomembrane basal liner is critical and relatively complicated, compared with installations of other geosynthetic products. The construction work involves the operation of heavy earth moving equipment as well as the minute handling of sensitive geosynthetic products. The construction personnel must be highly quality conscious. For effective installation of the geomembrane and other geosynthetic materials, special care is required; given below are general guidelines that should be followed during installation:

1   The subgrade/clay soil liner surface should be firm and unyielding with no abrupt elevation changes, voids and cracks, vegetation, roots, sharp-edged stones, construction debris, ice, standing water, and any other deleterious material that may cause damage to the geomembrane. If stones that could puncture the geomembrane exist in the clay soil liner/subgrade, they must be removed prior to the installation of geomembrane. Deviations from the theoretical plane surface should not exceed 20 mm over a distance of 4 m. The ruts caused by compaction equipment may not be deeper than 5 mm. The subgrade should be protected from desiccation, flooding and freezing. If required, the protection may consist of a thin plastic protective cover installed over the complete subgrade until the placement of the geomembrane liner begins. If the subgrade surface is too rough for the direct placement of the geomembrane, a nonwoven needle-punched geotextile can be placed on the subgrade prior to the placement of the geomembrane.

2   The method and equipment used to place the panels must not damage the geomembrane or the supporting subgrade surface. The personnel working on the

geomembrane must not wear shoes that can damage the geomembrane or engage in actions that could result in damage to the geomembrane. Adequate temporary loading and/or anchoring (i.e. sandbags, tyres, etc.) must be done to prevent uplift of the geomembrane by wind.

3   There should be an intimate hydraulic contact between the geomembrane liner and the underlying soil subgrade/clay soil liner. To achieve the intimate contact, the surface of the soil subgrade/clay liner on which the geomembrane is placed should be smooth-rolled with a steel-drum roller.

4   Since the field seaming of the geomembrane panels is a critical aspect of their successful functioning as a barrier to liquid flow, it must be handled very carefully. In general, the seams should be oriented parallel to the slope, that is, along, not across, the slope. All geomembrane sheets, regardless of type, should be seamed by thermal methods: fusion (hot wedge) welding and extrusion welding.

5   The geomembrane liner must be attached properly to structures as shown in Fig. 4.45. Welding or bonding the geomembrane to the structure may provide an attachment that has a lower possibility of leakage. If the geomembrane is attached directly to the structure, sealants are usually not required. The surface of the structure for which the geomembrane is to be attached should be constructed or formed without irregularities to limit the damage to the geomembrane. Edges or corners of the structures should be rounded. If a structure cannot be constructed or formed without irregularities, then a cushion/sealant as a protective layer should be placed between the geomembrane and the structure.

6   The geomembrane should be placed and backfilled in a way that minimizes wrinkles. A field-deployed and seamed geomembrane must be backfilled with soil (generally drainage material with or without a geotextile protection layer) or covered with a subsequent layer of geosynthetics (generally a geonet or geocomposite drain) in a timely manner after its acceptance by the construction quality

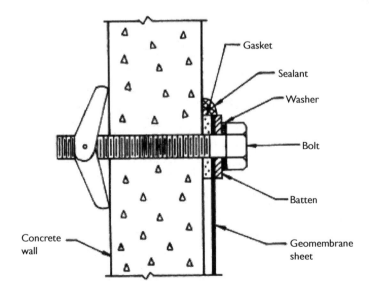

*Figure 4.45* A typical attachment of a geomembrane sheet to a concrete wall.

assurance (CQA) personnel. Large voids under the geomembrane should be filled to stop the geomembrane from becoming overly stressed. Note that geonets are always covered with a geotextile, that is, they are never directly soil covered, since the soil particles would fill the apertures of the geonet rendering it useless.

7 The GCL should lie flat on the underlying surface, with no wrinkles or folds, especially at the exposed edges of the panels. Only as much geosynthetic clay liner should be deployed per working day as can be covered by suitable cover soils. The sealing between installed rolls of GCLs should be made by overlapping. The lengthwise seams should typically be overlapped a minimum 150 mm, and the widthwise seams a 500 mm. The GCL should be placed so that the seams are parallel to the direction of the slope. If a trench is used for anchoring the end of the GCL, the soil backfill should be placed in the trench to provide resistance against pullout. The GCL should be sealed around structures embedded in the subgrade and the pipe penetrations as explained in ASTM D 6102-12 (ASTM, 2012b).

8 Clay soil liners and covers should be compacted wet of optimum water content. Cover soils should be free of sharp edged stones or other foreign matter that could damage the geomembrane or the geosynthetic clay liner. The lift thickness should not be more than 6 in. (150 mm) after compaction. This differs from the lift thickness of 9–18 in. (230–300 mm) for embankments and other geotechnical applications. The cover soil should be prevented from entering the geosynthetic overlap zones. Note that the construction equipment should never be allowed to move directly on any deployed geomembrane.

9 Geomembranes and other involved geosynthetics should usually be terminated by a horizontal runout, an anchor trench or a combination thereof (see Chapter 7). The runout and the anchor trench are covered with soil and suitably compacted to hold installed geosynthetics in place against the applied loads. The holding capacity comes mainly from the frictional resistance between the geosynthetic and the soil, and it depends on several factors such as runout length, cover soil depth, shape and depth of anchor trench, types of soil underlying and overlying the geosynthetic, and the type of geosynthetic used.

10 All construction operations at a landfill site are sensitive to weather conditions. Obviously, the placement of a clay liner is impossible during heavy rain, snowfall or frost, and partly finished clay blankets must be protected against water and against desiccation due to dry wind and sunshine when the construction work is interrupted at weekends, due to bad weather or for any other reasons. For such a temporary protection, thin plastic membranes are used. The installation of GMBs requires favourable weather as well. It cannot be done in the rain. The minimum temperature for seaming polyethylene sheets is 5°C. Sufficient time has to be allocated to the placement of geomembranes to cope with unavoidable delays due to unfavourable weather.

### 4.4.4 Case studies

#### Case study 1 (landfill)

Designing a constructible composite liner system, for the side slopes of a landfill, is a challenging task. To meet this challenge at the Lopez Canyon Sanitary Landfill, Los Angeles, USA, an entirely geosynthetic composite liner (GCL) and a leachate

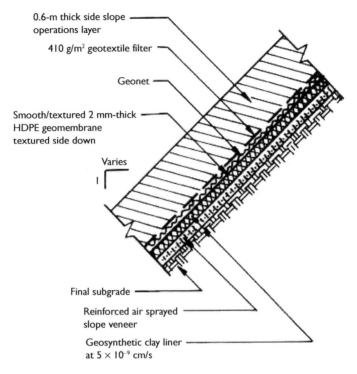

0.6-m thick side slope
operations layer

410 g/m² geotextile filter

Geonet

Smooth/textured 2 mm-thick
HDPE geomembrane
textured side down

Varies

Final subgrade

Reinforced air sprayed
slope veneer

Geosynthetic clay liner
at 5 × 10⁻⁹ cm/s

*Figure 4.46* A geocomposite liner system for steep Canyon landfill side slopes in Los Angeles, USA (after Snow *et al.*, 1994).

collection and removal system (LCRS) were developed in the year 1991 (Snow *et al.*, 1994). A schematic cross-section of the geosynthetic alternative developed for the side slopes of disposal area C is shown in Fig. 4.46. The veneer of concrete was specified to have a compressive strength of 170–205 kPa and was sprayed on to the graded canyon side slopes to provide support and a smooth surface for the composite liner. A PE geonet was used in lieu of granular soil to provide an LCRS on the side slopes. The primary advantages of the geonet are simple installation and a high drainage capacity resulting in a low liquid head on the composite liner. Construction of the geosynthetic side-slope liner system was subjected to large temperature variations, high winds and the steep slopes at the site. The familiarity of the geosynthetics installer with these conditions on other landfills in the area was a significant benefit to the project. A total of about 15500 m² and 77000 m² of geosynthetic composite side-slope liner system was placed during Phase I and II of the liner system construction, respectively. In phase I, the GCL joints were simply overlapped with no additional preparation, while in phase II, the GCL joints were prepared by the addition of powdered bentonite at the rate of 1.5 kg/m² in the overlap areas. Performance of the liner system under dynamic loading was observed during the Northridge earthquake, of Richter magnitude 6.6, which struck Los Angeles on 17 January 1994. The Lopez Canyon site is located less than 15 km from the earthquake epicentre. The recording stations nearby measured horizontal peak ground accelerations of up to 0.44g. The observations made that same day indicated that the geosynthetic side-slope liner system performed very well.

### Case study 2 (reservoir)

The Gennevilliers (France) recreation reservoir was constructed in 1986 over an old landfill and gravel quarry and designed for recreational use such as sailing. Figure 4.47 shows the typical section of the reservoir. Many technical difficulties arose from the fact that it was built on a landfill, which in itself represents a reference case. The results of preliminary geotechnical studies led to the design of a geomembrane lining system. The subgrade preparation consisted of bank consolidation, followed by compacting and grading of a 30–40 cm thick sand layer over the entire surface of the reservoir. The eventuality of a water table rise, together with possible remaining waste gases, led to the design of a water and gas drainage system consisting

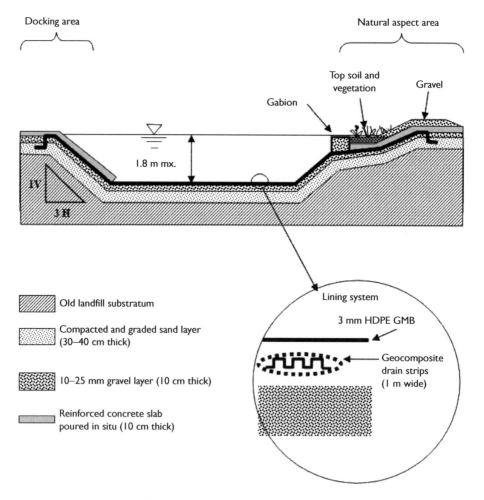

*Figure 4.47* The Gennevilliers (France) recreation reservoir: typical cross-section and specific features (not to scale) (after Duquennoi, 2002).

of a 10-cm thick 10–25 mm gravel layer, enhanced by the geocomposite drain strips. The gravel layer was connected to a central drain pipe and to peripheral gas vents. A 3-mm HDPE geomembrane was selected with the following criteria: tensile properties to resist the differential settlement, static and dynamic puncture resistance, and roll width and length to minimize the overall seam length. Note that the geomembrane was not associated with any geotextile in this project, whether as an underliner nor as a protection overliner. The geomembrane protection systems have been designed on the banks only. They generally consisted of 10-cm thick reinforced concrete slabs which were poured in situ. Wherever the landscaping purposes required specific bank works, such as gabions and vegetation, the concrete slabs were topped with gravel and topsoil. The geomembrane was unprotected at the bottom of the reservoir.

### Case study 3 (canal)

The Mulhouse (France) canal was constructed in 1997 to convey water from the city of Mulhouse water treatment plant to the Hardt irrigation canal (Potié, 1999; Duquennoi, 2002). The canal is 9 km long with a trapezoidal section and a total output of 7 $m^3$/s (Fig. 4.48). The preparation of the subgrade consisted of excavating and grading the alluvial gravelly soil. A 3.5-mm bituminous geomembrane, factory-surfaced with an overliner geotextile, was installed directly on the subgrade (Fig. 4.49). The geotextile was designed to drain the eventual infiltration water under the extruded concrete cover. Since the concrete cover was to be poured with a sliding formwork machine, the eventuality of geotextile sliding caused by the machine had to be prevented by bonding the geotextile and the geomembrane together. The bituminous geomembrane was selected because of its higher puncture resistance and

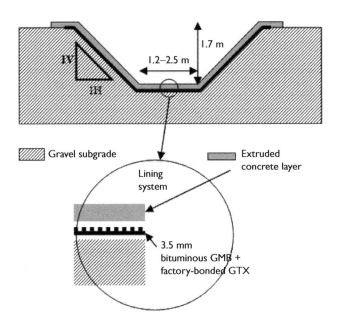

*Figure 4.48* The Mulhouse (France) canal: typical cross-section and specific features (not to scale) (after Duquennoi, 2002).

*Figure 4.49* The Mulhouse (France) canal: geomembrane-geotextile composite installation (after Duquennoi, 2002).

its ease of installation under the harsh climatic conditions. Welding had to be done using 40-cm wide bituminous geomembrane strips, because of the surface geotextile. The anchoring of the geomembrane on top of the embankment was not necessary because of the ballasting effect of the concrete layer. A protecting layer of extruded concrete was laid on the geosynthetic lining system using a sliding formwork machine, specifically designed for this work. The rate of installation, including the lining system installation and extruded concrete pouring, was 300 m/day.

## 4.5 TUNNELS

### 4.5.1 General description

Tunnels are used for various purposes in civil engineering, including traffic movement and the fluid flow. Waterproof tunnels are required at some sections of the highway and railway alignments. A crack-free concrete lining is needed for a waterproof tunnel. Geotextiles and geomembranes are commonly used in modern-day tunnel technology to construct waterproof tunnels.

Figure 4.50 shows the cross-section of a tunnel vault with the general arrangement of the lining system. The shotcrete lining placed over the excavated surface provides a smooth surface for the geosynthetics. In addition, the rock surface is supported by the shotcrete immediately after excavation so that the radially acting forces can be accepted adhesively (Wagner and Hinkel, 1987). Nonwoven geotextile, generally needle-punched geotextile, acts as a drainage layer and as protection for a waterproofing geomembrane. It also acts as a cushion (stress-relieving layer) to significantly reduce the formation of cracks in the inner concrete lining by allowing free shrinkage deformation of the concrete during the setting process.

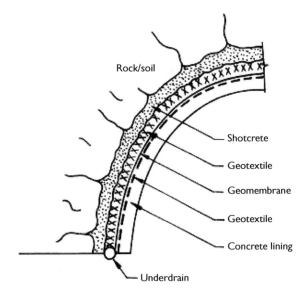

Rock/soil

Shotcrete

Geotextile

Geomembrane

Geotextile

Concrete lining

Underdrain

*Figure 4.50* Cross-section of a tunnel vault showing the general arrangement of the lining system.

Note that the geomembrane sheet sealing with a protective nonwoven geotextile drainage layer has predominated over the conventional sealing methods such as asphalt membranes or spray applied glass fibre-reinforced plastic or bitumen-latex based products. The geosynthetic system not only meets the demands of rapid tunneling rates but also the demands for rough construction treatment.

Note that the tunnels are also used to convey water, under pressure or in free-flow, for power and/or water supply works. Such tunnels, called *hydraulic tunnels*, are usually lined either with steel, cast *in situ* concrete or sprayed concrete. The most common concrete linings are rarely free from defects such as voids and cracks. During operation, when the concrete linings are exposed to the dynamic action of water, they suffer from deterioration resulting in the formation of fissures/cracks, increased permeability through body and joints, and increased roughness. The use of geomembranes allows the combination of continuous watertightness and improved friction properties with the strength of the concrete lining at a much lower cost. Geomembrane systems have reached a high degree of refinement and reliability and when adequately designed and installed, they are durable and reliable under high water heads and demanding environments. More details about the engineering aspects of geosynthetic-based lining system for new and old hydraulic tunnels can be found in Cazzuffi *et al.* (2012).

## 4.5.2 Analysis and design concepts

Geosynthetic design in tunnel applications requires that the geosynthetic system must provide watertight integrity for the life of the tunnel. It must withstand different kinds of stress and strain both during installation and after construction. It also must withstand variable and aggressive chemical environments. As both installation and

*Table 4.5* Geosynthetic properties for the tunnel applications, measured as per the relevant ASTM standards (after Gnilsen and Rhodes, 1986).

| Property | Minimum Specifications |
|---|---|
| A Geotextile (nonwoven polypropylene) | |
| Thickness (mm) | 4.0 |
| Mass per unit area (g/m²) | 500 |
| Grab strength (N) | 1150 |
| Elongation (%) | 80 |
| Trapezoid tear strength (N) | 440 |
| Burst strength (kPa) | 2760 |
| Chemical resistance (pH value) | 2–13 |
| Flammability | Self-extinguishing |
| B Geomembrane (PVC-soft) | |
| Thickness (mm) | 1.5 |
| Ultimate tensile strength (kPa) | 7600 |
| Ultimate elongation (%) | 300 |
| Brittleness temperature | ±7 °C |
| Flammability | Self-extinguishing |
| Dimensional stability 6 h @ 80 °C (%) | 2 |

service conditions are severe, it is considered essential that the geomembrane should be exceptionally resistant to tearing, puncturing and abrasion. The geotextile should fulfill the following criteria (Posch and Werner, 1993):

- *Mechanical resistance*: The geotextile must have certain minimum values for mechanical strength and elasticity. These are needed to absorb the stresses resulting from the installation and concreting pressures, the deformation of the inner lining of the tunnel due to load shifting and temperature variations, and the joint water pressure increasing locally over time.
- *Chemical resistance*: The geotextile should resist all kinds of rock water, calcium hydroxide, and other constituents such as the binding agents in concrete and grout.
- *Water permeability in the plane*: Residual water, consisting of seepage and leakage, must be reliably drained in the geotextile plane to the bottom drain.

The typical geosynthetic properties required in tunnel applications are given in Table 4.5.

### 4.5.3 Application guidelines

Waterproofing of tunnels can be successfully carried out using the geosynthetics. Even a completely submerged tunnel can be waterproofed. Geomembrane sheet sealing with a protective nonwoven geotextile drainage layer has become the predominant sealing system worldwide. The staging of construction activities should be designed such that the installation of geosynthetics becomes a separate and continuous operation. The geosynthetic installation process should minimize interruption of other tunnel construction activities. Based on the reported case studies (Benneton *et al.*, 1993; Davies, 1993; Posch and Werner, 1993), the major construction steps can be summarized as follows:

- excavation of rock and/or soil;
- grouting to stop/minimize inflowing water, if present;
- supporting the exposed surface by shotcrete (gunite);
- fastening the thick needle-punched nonwoven geotextile, as a protective screen as well as a drainage medium, to the shotcrete by means of PVC plastic discs (plates) and fasteners (nails);
- fixing the geotextile to underdrains on each side of tunnel base;
- placement of a geomembrane (usually PVC) to PVC plastic discs by means of hot air welding;
- spot-bonding of a protective shield (3-mm thick PVC) to the geomembrane;
- placement of the concrete liner against the geomembrane;
- providing additional seals consisting of an expansion product (e.g. butyl bentonite) at the concrete restart points.

### 4.5.4  Case studies

The Arlberg Tunnel, Austria's outstanding and most important road tunnel, was constructed in 1974–1978 (Posch and Werner, 1993). The design methodology was the New Austrian Tunneling Method (NATM), using the following three different types of lining systems for tunnel waterproofing over an approximately 14-km long section (Fig. 4.51):

1  a 300–400-mm thick concrete lining without any sealing measures or separation layers (34.3% of the tunnel length)
2  a 300–400-mm thick concrete lining and PP needle-punched nonwoven geotextile of 700 g/m$^2$ with 3.6 kN puncture resistance (Polyfelt TS 008), functioning as a separation and drainage layer (12.4% of the tunnel length)

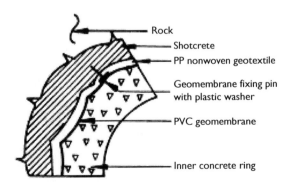

*Figure 4.51* Typical cross-section of the tunnel lining in Arlberg Tunnel, Austria (after Posch and Werner, 1993).

3   a 300–400-mm thick concrete lining and waterproofing system, which is a combination of a 1.5-mm thick PVC geomembrane and a 500 g/m² PP needle-punched nonwoven geotextile protection and drainage layer (53.3% of the tunnel length)

Needle-punched PP nonwoven geotextiles were selected based on mechanical resistance, chemical resistance, and in-plane water permeability criteria. The concrete lining was constructed in 12-m long ring sections of waterproof pumped concrete B30. The concrete consisted of 240–250 kg of Portland cement PZ 275, 40–60 kg of fly ash and aggregate with a maximum size of 32–60 mm.

Ten years after completion, this almost classic example of the NATM revealed remarkable results in terms of reduced crack formation in the concrete lining. A 40% reduction in crack formation was found in the geotextile and concrete lined tunnel sections. In this section, the geotextile has been the separation and drainage layer between the shotcrete and inner concrete lining. Geotextile samples were removed 12 years after construction, and the following results were obtained:

- The mass per unit area has increased by 9% due to sedimentation of fine particles.
- The coefficient of normal permeability was reduced from 5.72 to 3.25 mm/s.
- There was no indication of clogging caused by the mineral precipitation.

Note that the use of a nonwoven geotextile increased the cost of the concrete lining by only 9%. The insignificant difference in the number of cracks in the geotextile lined section, compared with the geotextile-geomembrane lined section, made this design an alternative to lined waterproofing. Particularly, the use of a geomembrane increases the liner costs by 70%. In the limited area where the cracks occurred in the geotextile lined section, the effective drainage of moisture prevented the seepage of water into the tunnel through the cracks. This was true even when the drains at the floor of the tunnel showed seepage occurring behind the lining. It has been recommended that a geotextile should be used even in those sections which do not show any immediate ingress of rock water during excavation of the tunnel, because the water leakage path in the transition zone may change over time from the sealed to the unsealed area.

## Chapter summary

1   A filter consists of any porous material that has openings small enough to prevent movement of soil into the drain, and that is sufficiently pervious to offer little resistance to seepage. When a geosynthetic (nonwoven geotextile or lightweight woven geotextile) is used as a filter in the drainage applications, it prevents upstream soils from entering adjacent granular layers or subsurface drains.

2   The geosynthetic filters are designed to avoid blocking, blinding and clogging phenomena. The clogging, in general, takes place very slowly, and the blinding of the filter is far more detrimental than clogging. To achieve the satisfactory filter performance by geosynthetics, especially geotextiles, the following criteria must be fulfilled during the design life of application under consideration: permeability criterion, retention criterion, anti-clogging criterion, and survivability and durability criterion.

3 Geotextiles with high in-plane drainage ability and several geocomposites are used as a drain in the place of traditional granular drains. The primary function of the geosynthetic in subsurface drainage applications is filtration. Prefabricated vertical drains are nowadays often installed within the saturated soil mass to provide a vertical drainage.

4 The geosynthetic, particularly a woven geotextile, should be placed with the machine direction following the direction of water flow.

5 The problem of soil movement due to erosive forces by moving water and/or wind as well as by seeping water is called soil erosion.

6 Fundamentally, the function of armouring systems is to protect and hold in place the filtering system, which is nowadays almost always a geotextile.

7 Geosynthetic nets and meshes available in various forms have proven to be successful in both temporary and long-term erosion control projects. Three-dimensional erosion control geosynthetic mats and geocells that are nowadays commercially available with various dimensions can be used in permanent erosion control systems.

8 The fluid barrier is generally the required function to be performed by the geosynthetic in almost all the containment facilities. Liquid containment and conveyance facilities, such as ponds, reservoirs and canals are required in several areas including hydraulic, irrigation and environmental engineering.

9 Geosynthetic design in tunnel applications requires that the geosynthetic system must provide watertight integrity for the life of the tunnel.

## Questions for practice

(Select the most appropriate answer to the multiple-choice questions from Q 4.1 to Q 4.17.)

4.1 Which one of the following dictates the filtering efficiency of the filter system at the onset of equilibrium conditions?
(a) Structure of the geotextile filter
(b) Structure of the soil immediately adjacent to the geotextile filter
(c) Quality of water being filtered
(d) All of the above

4.2 A drainage geocomposite behind a concrete retaining wall is beneficial mainly because it
(a) reduces the plasticity of the backfill
(b) increases the durability of the concrete
(c) reduces the lateral pressures on the wall.
(d) improves the compaction requirements for the backfill

4.3 If $k_n$ be the coefficient of cross-plane permeability of the geotextile and $k_s$ the coefficient of permeability of the protected soil, then for a dam clay core, the geotextile filter criterion can be
(a) $k_n > k_s$
(b) $k_n > 10k_s$
(c) $k_n > 20k_s$
(d) $k_n > 100k_s$.

4.4 If the cross-section of a strip drain is 100 mm × 5 mm, then the equivalent drain diameter will be approximately
(a) 5 mm
(b) 33 mm
(c) 67 mm
(d) 100 mm.

4.5 In the drainage system of retaining walls, the geosynthetic drain should be located away from the wall in
(a) an inclined orientation
(b) a parallel orientation
(c) a perpendicular orientation
(d) both (b) and (c).

4.6 Geotextiles are used mainly for filtration function in
(a) canals
(b) dams
(c) tunnels
(d) ponds.

4.7 Which one of the following will have the highest efficiency in terms of soil loss under rain splash?
(a) Synthetic product
(b) Geocomposite based on jute and synthetic products
(c) Jute product
(d) None of the above

4.8 Which one of the following is an advantage of the mulching mat when used for slope erosion control?
(a) It holds the seeds in their place.
(b) It prevents evaporation from the soil slope and helps the plants to grow more successfully.
(c) It maintains the ground temperature and reduces damage.
(d) All of the above.

4.9 Which one of the following is the most important component of a landfill?
(a) Cover system
(b) Liner system
(c) Leachate collection and removal system
(d) A gas collection and control system.

4.10 The barrier in a liner or cover system of the landfill may consist of
(a) compacted clay liner (CCL)
(b) geomembrane (GMB) sheet
(c) geosynthetic clay liner (GCL)
(d) all of the above.

4.11 In the sanitary landfills, the thickness of daily soil cover is generally
(a) 5 cm
(b) 15 cm

(c) 30 cm

(d) 60 cm.

4.12 The main purpose of using the geotextiles in canals and rivers is to
(a) replace or improve the traditional filters
(b) increase the load-bearing capacity
(c) distribute the load
(d) relieve the pore water pressures.

4.13 In the construction and rehabilitation of earth dams, a geotextile filter is not used in
(a) downstream drainage blanket
(b) downstream chimney drain
(c) upstream side of the core
(d) all of the above.

4.14 In order to warrant a sufficient robustness of the geomembrane in handling, the specified minimum thickness of approved geomembranes in landfill liner systems is generally
(a) 2.5 mm
(b) 5.0 mm
(c) 7.5 mm
(d) 10 mm.

4.15 Which one of the following is the primary concern when designing the geosynthetic containment structures?
(a) Permeability
(b) Strength
(c) Leakage
(d) Stiffness

4.16 In the case of liquid containment ponds, to shield the geomembrane liners from ozone, UV light, temperature extremes, ice damage, wind stresses, accidental damage and vandalism, the minimum thickness of soil cover should be
(a) 15 cm
(b) 30 cm
(c) 75 cm
(d) 100 cm.

4.17 In the tunnel lining projects, which of the following geosynthetics is generally not used?
(a) GCL
(b) GTX
(c) GMB
(d) all of the above.

4.18 List the advantages of a geosynthetic filter over the graded granular filter.

4.19 Define the 'filtering efficiency'. Explain its significance.

4.20 It is a misconception that the geosynthetic filter can replace the function of a granular filter completely. Do you agree with this statement? Justify your answer.

4.21 What are the essential properties of soil to be determined for the successful use of a geotextile in the filtration application?

4.22 Define the following phenomena: 'blocking', 'blinding' and 'clogging'.

4.23 Blinding of the geosynthetic filter is far more detrimental than its clogging. Why?

4.24 Why would there be a need for strength criteria for geosynthetics that are used in hydraulic applications?

4.25 What are the essential features that must be considered in the design of filters?

4.26 The permeability and retention criteria for filters are, in principle, contradictory if they have to be fulfilled simultaneously. Do you agree with this statement? Give a proper justification in support of your answer.

4.27 What is the soil-tightness number? Explain its significance.

4.28 What are the main conditions known where soils are likely to cause excessive clogging of geotextiles? In such instances, what is the logical recommendation?

4.29 Describe the mechanism of filter failures resulting from washout of fines.

4.30 A geotextile-wrapped trench drain is to be constructed to drain a soil mass. Determine the appropriate hydraulic properties of the geotextile to function as a filter in a non-critical application with the following soil properties:
$C_u = 2.5$
$D_{15} = 0.15$ mm
$D_{85} = 0.7$ mm
$k_s = 1 \times 10^{-5}$ m/s
Percentage passing 75 μm = 7%

4.31 Describe the structure of a drainage geocomposite.

4.32 What is the effect of stiffness of the geosynthetic on the discharge capacity of the geosynthetic drain?

4.33 A geosynthetic has to be selected to provide drainage behind a retaining wall with a vertical backface. The estimated vertical flow into the drain is 0.0018 m³/s. Determine the required transmissivity of the geosynthetic. Would an ordinary single layer of nonwoven geotextile be adequate?

4.34 What do you mean by a geometrically tight filter?

4.35 Intimate contact between the geosynthetic filter and the base soil is necessary. Is there any reason behind this practice? Justify your answer.

4.36 Describe the major construction steps for geotextile-wrapped underdrains.

4.37 Describe the typical arrangement of a geocomposite buried drain.

4.38 What is the application of a turf reinforcement mat (TRM)?

4.39 Compare the roles of a geojute and a geotextile in an erosion control application.

4.40 What is a geocontainer? List its applications.

4.41 Present a comprehensive review of the various existing short-term and long-term erosion control systems.

4.42 Why is it recommended to place a layer of aggregates between the geotextile and the riprap?

4.43 What are the special considerations for designing a geotextile filter to be used in erosion control systems?

4.44   Consider a waterway revetment system with the following parameters:
Wave height, $H = 1.0$ m
Unit weight of protective covering material (free concrete blocks), $\gamma_c$ = 24 kN/m³
Assume that the permeability of the geotextile is greater than the permeability of soil to be protected. Determine the minimum depth of the protective covering required to protect the soil. Take the unit weight of water as $\gamma_w = 10$ kN/m³.

4.45   Determine the average size of the armour stone in a revetment system with geotextile filter required to protect a 22° slope from waves up to 3.0 m high assuming that no overtopping of the revetment occurs. Take the specific gravity of stone, $G_s = 2.71$.

4.46   "Although a geotextile functions as a filter in silt fence applications, consideration of its strength is also important in selection of geotextiles." Do you agree with this statement? Justify your answer.

4.47   What is the role of a cushion layer (a granular layer) between the overlying stones and the underlying geotextile?

4.48   How will you install a turf reinforcement mat (TRM) for erosion control of slopes?

4.49   What is the typical size of an anchoring trench?

4.50   How does an engineered landfill differ from an open dump of wastes?

4.51   Regarding the siting of a lined landfill, what are the major features to be considered?

4.52   List the main elements or systems comprising a modern municipal solid waste landfill. Describe their functions.

4.53   What is the purpose and function of a landfill liner system?

4.54   List the advantages and disadvantages of a geosynthetic clay liner over a compacted clay liner.

4.55   In your opinion, what are the advantages and disadvantages of a geosynthetic clay liner over a geomembrane liner?

4.56   Which type of geomembrane is the most chemically resistant of all geomembranes?

4.57   How does the seepage pattern through a geomembrane liner differ from the patterns through clay soil and composite soil liners?

4.58   What is the primary engineering assignment in designing, constructing and operating landfills?

4.59   Explain the importance of 'advection' and 'diffusion' in landfill design and construction?

4.60   How will you make the stability checks for geomembrane lined slopes?

4.61   What special precautions are required for the construction of composite liner systems for the side slopes of a landfill?

4.62   What factors will you consider for the selection of a geomembrane liner for ponds, reservoirs and canals?

4.63   How will you attach the geomembrane liner to structures?

4.64   Describe the clay liner sealing methods around a structure and a pipe penetration.

4.65 What precautions are required while installing a geomembrane liner?
4.66 Describe some general design indications concerning the use of geosynthetics in ponds, reservoirs and canals.
4.67 Write the basic steps for the design of earth dams with geosynthetics.
4.68 Why are geosynthetics used extensively in earth dam construction and rehabilitation?
4.69 Draw a cross-section of a typical tunnel vault with the general arrangement of the lining system.
4.70 What are the functions served by a nonwoven geotextile layer installed adjacent to the geomembrane layer in a waterproof tunnel?
4.71 Write down the typical geosynthetic properties required in tunnel applications.
4.72 Describe the major steps for construction of waterproofed tunnels with geosynthetics.

## References

AASHTO (American Association of State Highway and Transportation Officials) (2013). *Geotextile Specification for Highway Applications*, AASHTO Designation: M 288-06 (2011), *Standard Specifications for Transportation Materials and Methods of Sampling and Testing*, Thirty-third Edition, Part 1B: Specifications M 280-R 63, American Association of State Highway and Transportation Officials, Washington, DC, USA.

ACZ (1990). *Application and Execution Aspects of Sand Tube Systems*. Internal Report of ACZ Marine Contractors B.V., Gorinchem, The Netherlands.

Agnew, W. (1991). Erosion control product selection. *Geotechnical Fabrics Report*, April 1991, pp. 24–27.

Ahn, T.B., Cho, S.D. and Yang, S.C. (2002). Stabilization of soil slope using geosynthetic mulching mat. *Geotextiles and Geomembranes*, 20, 2, pp. 135–46.

ASTM (American Society for Testing and Materials) (2011). ASTM D5567-94, *Standard Test Method for Hydraulic Conductivity Ratio (HCR) Testing of Soil/Geotextile Systems*. ASTM International, West Conshohocken, PA, USA.

ASTM (2012a). ASTM D5101-12. *Standard Test Method for Measuring the Filtration Compatibility of Soil-Geotextile Systems*. ASTM International, West Conshohocken, PA, USA.

ASTM (2012b). ASTM D6102-12, *Standard Guide for Installation of Geosynthetic Clay Liners*. ASTM International, West Conshohocken, PA, USA.

Bagchi, A. (1994). *Design, Construction, and Monitoring of Landfills*. John Wiley & Son, Inc., New York.

Benneton, J.P., Mahuet, J.L. and Gourc, J.P. (1993). Geomembrane waterproofing of dry tunnel under two rivers, Lyon metro tunnel, France. In: *Geosynthetics Case Histories* edited by G.P. Raymond and J.P. Giroud on behalf of ISSMFE Technical Committee TC9, Geotextiles and Geosynthetics, pp. 118–119.

Broms, B.B., Chu, J. and Chora, V. (1994). Measuring the discharge capacity of band drains by a new drain tester. *Proceedings of the 5th International Conference on Geotextiles, Geomembranes and Related Products*. Singapore, pp. 803–806.

Cancelli, A., Monti, R. and Rimoldi, P. (1990). Comparative study of geosynthetics for erosion control. *Proceedings of the 4th International Conference on Geotextiles and Geomembranes*. The Hague, The Netherlands.

Cazzuffi, D., Scuero, A. and Vaschetti, G. (2012). Hydraulic tunnels. Chapter 15, *Handbook of Geosynthetic Engineering*, Second Edition, Shukla, S.K., Editor, ICE Publishing, London, pp. 303–315.

Cedergren, H.R. (1989). *Seepage, Drainage, and Flownets*. Third Edition, John Wiley and Sons, New York.

Chen, Y.H. and Cotton, G.K. (1988). Design *of roadside channels with flexible linings*. Federal Highway Administration Report, HEC-15/FHWA-1P087-7, McLean, Va., USA.

Christopher, B.R. (1998). The first step in geotextile filter deign. *Geotechnical Fabrics Report*, March 1998, pp. 36–38.

Christopher, B.R. and Holtz, R.D. (1985). *Geotextile Engineering Manual*. Report No. FHWA-TS-86/203, Federal Highway Administration, Washington, D.C.

Corbet, S.P. (1992). The design and specification of geotextiles and geocomposites for filtration and drainage. *Proceedings of the Conference Geofad' 92: Geotextiles in filtration and drainage*, Cambridge, Thomas Telford, UK, pp. 29–40.

Daniel, D.E. (1993). 'Landfill and Impoundments', in *Geotechnical Practice for Waste Disposal*, Daniel, D.E., Editor, Chapman & Hall, pp. 97–112.

Daniel, D.E., Koerner, R.M., Bonaparte, R., Landreth, R.E., Carson, D.A. and Scranton, H.B. (1998). Slope stability of geosynthetic clay liner test plots. *Journal of Geotechnical and Geoenvironmental Engineering*, ASCE, **124**, 7, pp. 628–637.

Davies. P.L. (1993). Geomembrane waterproofing of wet hydroelectric tunnel, Drakensberg hydroelectric tunnel, South Africa. In: *Geosynthetics Case Histories* edited by G.P. Raymond and J.P. Giroud on behalf of ISSMFE Technical Committee TC9, Geotextiles and Geosynthetics, pp. 114–115.

Duquennoi, C. (2002). Containment ponds, reservoirs and canals. Chapter 13, *Geosynthetics and Their Applications*, Shukla, S.K., Editor, Thomas Telford, London, pp. 299–325.

Faure, Y. and Mlynarek, J. (1998). Geotextile filter hydraulic requirements. *Geotechnical Fabrics Report*, May 1998, pp. 30–33.

Fourie, A. (1998). The determination of appropriate filtration design parameters. *Geotechnical Fabrics Report*, April 1998, pp. 44–47.

Fox, P.J., Rowland, M.G. and Scheithe, J.R. (1998). Internal shear strength of three geosynthetic clay liners. *Journal of Geotechnical and Geoenvironmental Engineering*, ASCE, **124**, 10, pp. 933–944.

Frobel, R.K. (1996). Geosynthetic clay liners, Part four: interface and internal shear strength determination. *Geotechnical Fabrics Report*, IFAI, Vol. 14, No. 8, pp. 20–23.

Fuller, C. (1992). Concrete blocks gain acceptance as erosion control systems. *Geotechnical Fabrics Report*, pp. 24–37.

Giroud, J.P. (1982). Filter criteria for geotextiles. *Proceedings of the 2nd International Conference on Geotextiles*. Las Vegas, pp. 103–108.

Giroud, J.P. (1984). Geotextiles and geomembranes definitions, properties and design, IFAI, St Paul, Minnesota.

Giroud, J.P. (1992). Geosynthetics in dams: Two decades of experience. *Geotechnical Fabrics Report*, pp. 22–28.

Giroud, J.P. (1994). Quantification of geosynthetic behaviour. *Proceedings of the 5th International Conference on Geotextiles, Geomembranes and Related Products*. Singapore, pp. 1249–1273.

Giroud, J.P. (1996). Granular filters and geotextile filters. *Proceedings of GeoFilters '96*, Montreal, pp. 565–680.

Giroud, J.P. and Bonaparte, R. (1989). Leakage through liners constructed with geomembranes – Part I. Geomembrane liners. *Geotextiles and Geomembranes*, **8**, 1, pp. 27–67.

Gnilsen, R and Rhodes, G.W. (1986). Innovative use of geosynthetics to construct watertight D.C. subway tunnels. *Geotechnical Fabrics Report*, Vol. 4, No. 4.

Heibaum, M.H. (1998). Protecting the geotextile filter from installation harm. *Geotechnical Fabrics Report*, June-July 1998, pp. 26–29.

Hewlett, H.W.M., Boorman, L.A. and Bramley, M.E. (1987). *Design of Reinforced Grass Waterways*, Report 116, CIRIA, London.

Hoare, D.J. (1982). Synthetic fabrics as soil filters. *Journal of the Geotechnical Engineering Division, ASCE*, **108**, GT10, pp. 1230–1245.

Holtz, R.D., Christopher, B.R. and Berg, R.R. (1997). *Geosynthetic Engineering*. BiTech Publishers Ltd., Canada.

Hudson, R.Y. (1959). Laboratory investigation of rubble mound breakwaters. *Journal of the Waterways and Harbors Division, ASCE*, **85**, 3, pp. 93–121.

Hunt, J.R. (1982). The development of fin drains for structure drainage. *Proceedings of the Second International Conference on Geotextiles*, Las Vegas, Nevada, pp. 25–36

ICOLD (International Commission on Large Dams) (1991). *Watertight geomembranes for dams*. State of the art. International Commission on Large Dams Bulletin 78, Paris, p.140.

ICOLD (1993). *Embankment Dams Upstream Slope Protection*. International Commission on Large Dams Bulletin 91, Paris, p. 122.

Ingold, T.S. (2002). Slopes – erosion control. Chapter 9, *Geosynthetics and Their Applications*, Shukla, S.K., Editor, Thomas Telford, London, pp. 223–235.

IRC (Indian Roads Congress) (2011). IRC:56, *Recommended Practices for Treatment of Embankment and Roadside Slopes for Erosion Control*. Indian Roads Congress, New Delhi, India.

BIS (Bureau of Indian Standards) (2001). IS:14986, *Guidelines for Application of Jute Geotextile for Rain Water Erosion Control in Road and Railway Embankments and Hill Slopes*. Bureau of Indian Standards, New Delhi, India.

Kenney, T.C. and Lau, D. (1985). Internal stability of granular filter. *Canadian Geotechnical Journal*, 22, 215–225

Kjellman, W. (1948). Accelerating consolidation of fine-grained soils by means of cardboard wicks. *Proceedings of the Second International Conference on Soil Mechanics and Foundation Engineering*, Rotterdam, pp. 302–305.

Koerner, R.M. and Hwu, B.-L. (1991). Stability and tension considerations regarding cover soils on geomembrane lined slopes. *Geotextiles and Geomembranes*, 10, 4, pp. 335–355.

Lafleur, J., Mlynarek, J. and Rollin, A.L. (1993). Filter criteria for well graded cohesionless soils. *Proceedings of the Geofilters '92*, Karlsruhe, Germany, pp. 97–106.

Landreth, R.E. and Carson, D.A. (1991). *Inspection Techniques for the Fabrication of Geomembrane Field Seams*. USEPA Technical Guidance Document, EPA/530/SW-91/051, Cincinnati, OH, 174 p.

Lawson, C.R. (1982). Filter criteria for geotextiles: relevance and use. *Journal of the Geotechnical Engineering Division, ASCE*, **108**, GT10, pp. 1300–1317.

Lawson, C.R. (1986). Geotextile filter criteria for tropical residual soils. *Proceedings of the 3rd International Conference on Geotextiles*. Vienna, pp. 557–562.

Lawson, C. (1998). Retention criteria and geotextile filter performance. *Geotechnical Fabrics Report*, August 1998, pp. 26–29.

Legge, K.R. (1986). Testing of geotextiles. *Proceedings of SAICE Filters Symposium*. Johannesburg.

Luettich, S.M., Giroud, J.P. and Bachus, R.C. (1992). Geotextile filter design guide. *Geotextiles and Geomembranes*, **11**, 4–6, pp. 19–34.

McKenna, J.M. (1989). Properties of core materials, the downstream filter and design. *Clay Barriers for Embankment Dams*, Thomas Telford, London.

Mlynarek, J. and Fannin, J. (1998). Introduction to Designing with Geotextiles for Filtration. *Geotechnical Fabrics Report*, January-February 1998, pp. 22–24.

Palmeira, E.M. and Fannin, R.J. (2002). Soil-geotextile compatibility in filtration. *Proceedings of the 7th International Conference on Geosynthetics*, France, pp. 853–974.

Pilarczyk, K.W. (1984a). Discussion on revetment design. *Proceedings of the International Conference on Flexible Armoured Revetments Incorporating Geotextiles*, pp. 209–215.

Pilarczyk, K.W. (1984b). Prototype tests of slope protection systems. Discussion on revetment design. *Proceedings of the International Conference on Flexible Armoured Revetments Incorporating Geotextiles*, pp. 239–254.

Pilarczyk, K.W. (2000). *Geosynthetics and Geosystems in Hydraulic and Coastal Engineering.* A.A. Balkema, Rotterdam, Netherlands.

Posch, H. and Werner, G. (1993). Geosynthetics used in dry tunnel lining systems, Arlberg Tunnel, Austria. In: *Geosynthetics Case Histories* edited by G.P. Raymond and J.P. Giroud on behalf of ISSMFE Technical Committee TC9, Geotextiles and Geosynthetics, pp. 112–113.

Qian, X, Koerner, R.M. and Gray, D.H. (2002). *Geotechnical Aspects of Landfill Design and Construction.* Prentice Hall, New Jersey.

Repetto, P.C. (1995). 'Geo-environment', in *The Civil Engineering Handbook*, Chen, W.F., Editor, CRC Press, Boca Raton, FL, pp. 883–902.

Richardson, G.N. (1997). GCL internal shear strength requirements. *Geotechnical Fabrics Report*, pp. 20–25.

Richardson, G.N. and Christopher, B. (1997). Geotextiles in drainage systems. *Geotechnical Fabrics Report*, April, pp. 17–28.

Richardson, G.N. (2002). Surface impoundment design goals. *Geotechnical Fabrics Report*, pp. 15–18.

Sherard, J.L., Decker, R.S. and Ryker, N.L. (1972). Piping in earth dams of dispersive clay. *Proceedings of the ASCE Specialty Conference on Performance of Earth and Earth-Supported Structures.* New York: ASCE, Pt. 1, pp. 589–626.

Shukla, S.K. (2014). *Core Principles of Soil Mechanics*, ICE Publishing, London.

Shukla, S.K. (2015). *Core Concepts of Geotechnical Engineering*, ICE Publishing, London.

Singh, D.N. and Shukla, S.K. (2002). Earth dams. Chapter 12, *Geosynthetics and Their Applications*, Shukla, S.K., Editor, Thomas Telford, London, pp. 281–298.

Sivaramakrishnan, R. (1994). Jute geotextiles as revetment filter for river bank protection. *Proceedings of the 5th International Conference on Geotextiles, Geomembranes and Related Products.* Singapore, pp. 899–902.

Snow, M.S., Kavazanjian Jr., E. and Sanglerat, T.R. (1994). Geosynthetic composite liner system for steep canyon landfill side slopes. *Proceedings of the 5th International Conference on Geotextiles, Geomembranes and Related Products.* Singapore, 1994.

Tchobanoglous, T., Theisen, H. and Vigil, S. (1993). *Integrated Solid Waste Management, Engineering Principles and Management Issues.* McGraw-Hill, Inc.

Terzaghi, K. and Peck, R.B. (1948). *Soil Mechanics in Engineering Practice.* John Wiley & Sons, New York.

Theisen, M.S. (1992). Geosynthetics in erosion and sediment control. *Geotechnical Fabrics Report*, pp. 26–35.

Theisen, M.S. and Richardson, G.N. (1998). Geosynthetic erosion control for landfill final covers. *Geotechnical Fabrics Report*, pp. 22–27.

Tutuarima, W.H. and Van Wijk, W. (1984). Profix mattresses – an alternative erosion control system. *Proceedings of the International Conference on Flexible Armoured Revetments Incorporating Geotextiles*, pp. 335–348.

USEPA (US Environmental Protection Agency) (1993). Report of Workshop on *Geosynthetic Clay Liners.* EPA/600/R-93/171, August 1993. Office of Research and Development, US Environmental Protection Agency, Washington, DC.

Wagner, H. and Hinkel, W. (1987). The New Austrian Tunneling Method. *Proceedings of the Tunnel/Underground Seminar*, ASCE, New York.

Wang, D.W. (1994). Filter criteria of woven geotextiles for protective works. *Proceedings of the 5th International Conference on Geotextiles, Geomembranes and Related Products.* Singapore, pp. 763–766.

Werner, G., Puhringer, G. and Frobel, R.K. (1990). Multiaxial stress rupture and puncture testing of geotextiles. *In the Proceedings of the Fourth International Conference on Geotextiles, Geomembranes, and Related Products.* The Hague, Vol. 2, pp. 765–770.

Wilson, C.B. (1992). Repair to Reeves lake dam, Cobb County, Georgia. *Proceedings of the 1992 Annual Conference of the Association of State Dam Safety Officials*, Baltimore, pp. 77–81.

Wooten, R.L., Powledge, G.R. and Whiteside, S.L. (1990). CCM overtopping protection on three Park way dams. *Proceedings of Hydraulic Engineering National Conference, ASCE.* San Diego, pp. 1152–1157.

Wu, K.J. and Austin, D.N. (1992). Three dimensional polyethylene geocells for erosion control and channel linings. *Geotextiles and Geomembranes*, pp. 611–620.

Zanzinger, H. and Gartung, E. (2002). Landfills. Chapter 11, *Geosynthetics and Their Applications*, Shukla, S.K., Editor, Thomas Telford, London, pp. 259–279.

## Answers to selected questions

4.1    (b)
4.3    (d)
4.5    (a)
4.7    (c)
4.9    (b)
4.11   (b)
4.13   (c)
4.15   (c)
4.17   (a)

# Chapter 5

# Transportation applications of geosynthetics

## 5.1 INTRODUCTION

The transportation structures, such as roads, runways and railway tracks, often have to be constructed across weak and compressible soil subgrades. It is therefore common practice to distribute the traffic loads in order to decrease the stresses on the soil subgrade. This is generally done by means of placing a granular layer over the soil subgrade. The granular layer should have good mechanical properties and enough thickness. The long-term interaction between the fine soil subgrade and the granular layer, under the dynamic loads, is likely to cause the pumping erosion of soil subgrade and penetration of the granular particles into the soil subgrade, giving rise to permanent deflections and eventually to failure of pavement structures. Geosynthetics are being used nowadays to solve such problems.

Based on the type of pavement surfacing provided, roads can be classified as *unpaved roads* and *paved roads*. If roads are not provided with permanent hard surfacing (i.e., asphaltic/bituminous or cement concrete pavements), they are called unpaved roads. Such roads have stone aggregate layers, placed directly above the soil subgrades, and they are at most surfaced with sandy gravels for reasonable ridability; thus the granular layer serves as a base course and a wearing course at the same time. If permanent hard pavement layers are added to unpaved roads, then called *paved roads*, their behaviour under traffic loading changes significantly. Note that unpaved roads can be utilized as temporary roads or permanent roads, whereas the paved roads are, in most cases, utilized as permanent roads which usually remain in use for 10 years or more.

This chapter presents the details of applications of geosynthetics in unpaved roads, paved roads and railway tracks. Each of these applications is described, focusing on the basic details, the analysis and design concepts, the application guidelines, and some case studies.

## 5.2 UNPAVED ROADS

### 5.2.1 Basic description

Geosynthetics, especially geotextiles and geogrids, have been used extensively in unpaved roads to make their construction economical by reducing the thickness of the granular layer as well as to improve their engineering performance and extend their life. A geosynthetic layer is generally placed at the interface of the granular layer and the soil subgrade (Fig. 5.1). Reinforcement and separation are two major functions to be served by the geosynthetic layer (see Table 5.1). As discussed in Section 1.6.2., if the soil subgrade is soft, that is, the California bearing ratio (CBR) of the soil subgrade is low, say its unsoaked value is less than 3 (or soaked value is less than 1), then the reinforcement will be the primary function because of adequate tensile strength mobilization in the geosynthetic through a large deformation, that is, deep ruts (say, greater than 75 mm) in the soil subgrade. Geosynthetics, used with soil subgrades with an unsoaked CBR higher than 8 (or soaked CBR higher than 3), will have negligible amount of reinforcement occurring, and in such cases the primary function will uniquely be separation. For soils with intermediate unsoaked CBR values between 3 and 8 (or soaked CBR values between 1 and 3), there will be an interrelated group of separation, filtration and reinforcement functions, may be called the *stabilization* function of the geosynthetic. Geosynthet-

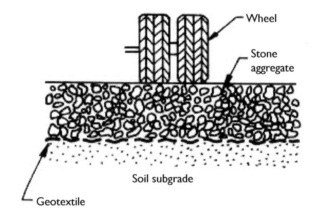

*Figure 5.1* A typical cross-section of a geosynthetic-reinforced unpaved road.

*Table 5.1* Primary function of the geosynthetic layer in unpaved road construction based on the field CBR value.

| Soil subgrade description | California bearing ratio (CBR) | | Primary function of the geosynthetic | Cost justification for use of the geosynthetic |
|---|---|---|---|---|
| | Unsoaked | Soaked | | |
| Soft | Less than 3 | Less than 1 | Reinforcement | Significantly less granular material utilization |
| Medium | 3 to 8 | 1 to 3 | Stabilization (An interrelated group of separation, filtration and reinforcement functions) | Less granular material utilization and longer lifetime |
| Firm | Greater than 8 | Greater than 3 | Separation | Much longer lifetime |

ics, especially geotextiles and some geocomposites, may also provide performance benefits from their filtration and drainage functions by allowing the excess pore water pressure, caused by traffic loads in the soil subgrade, to dissipate into the granular base course, and in case of poor-quality granular materials, through the geosynthetic plane itself.

By providing a geosynthetic layer, improvement in the performance of an unpaved road is generally observed in either of the following:

1   for a given thickness of the granular layer, the traffic can be increased;
2   for the same traffic, the thickness of the granular layer can be reduced, in comparison with the required thickness when no geosynthetic is used.

The introduction of a geotextile layer can typically save one-third of the granular layer thickness of the roadway over moderate to weak soils. Giroud *et al.* (1984) reported reductions of about 30–50% of the thickness of the aggregate layer with the inclusion of geogrids. Improvement in the performance of unpaved roads can also be observed in the form of reduction in permanent (i.e. non-elastic) deformations to the order of 25–50% with the use of geosynthetics, as reported by several researchers in the past (De Garidel and Javor, 1986; Milligan *et al.*, 1986; Chaddock, 1988; Chan *et al.*, 1989; Hirano *et al.*, 1990).

## 5.2.2   Analysis and design concepts

Several design methods are available for unpaved road constructions with geosynthetics. Research work is still continuing for the development of new design methods and also for improvement of the existing ones. Some of the manufacturers have developed their own unpaved road design charts for use with their particular geosynthetics. All these design charts recommend greater savings of granular material required in construction as the soil subgrade becomes softer, thus showing logical results. A design method based on a specific, well-defined geosynthetic property, such as the geosynthetic modulus, is generally acceptable by all. Such a design method is described as a *reinforcement function design method*.

### 5.2.2.1   Reinforcement function design method (RFDM)

Giroud and Noiray (1981) presented a design method for the geotextile-reinforced unpaved roads by combining the quasi-static analysis and the empirical formula. This method evaluates the risk of failure of the foundation soil and of the geotextile. The geotextile is considered to function as only reinforcement. The failure of the granular layer is not considered, thus it is assumed that

1   the friction coefficient of the granular layer is large enough to ensure the mechanical stability of the layer.
2   the friction angle of the geotextile in contact with the granular layer under the wheels is large enough to prevent the sliding of the granular layer on the geotextile.

It is also assumed that

1   the thickness of the granular layer is not significantly affected by the subgrade soil deflection.
2   the granular layer provides a pyramidal distribution with depth of the *equivalent tyre contact pressure*, $p_{ec}$, applied on its surface (Fig. 5.2(a)).

Therefore,

$$p_{ec}LB = (B + 2h_0 \tan \alpha_0)(L + 2h_0 \tan \alpha_0)(p_0 - \gamma h_0) \tag{5.1}$$

in the absence of geotextile, and

$$p_{ec}LB = (B + 2h \tan \alpha)(L + 2h \tan \alpha)(p - \gamma h) \tag{5.2}$$

in the presence of geotextile.

   In Equations (5.1) and (5.2), $L$ and $B$ are the length dimensions of the equivalent rectangular tyre contact area; $h_0$ is the thickness of granular layer in the absence of geotextile; $h$ is the thickness of granular layer in the presence of geotextile; $\alpha_0$ is the load diffusion angle in the absence of geotextile; $\alpha$ is the load diffusion angle in the presence of geotextile; $p_0$ is the pressure at the base of granular layer in the absence of geotextile; $p$ is the pressure at the base of the granular layer in the presence of geotextile; and $\gamma$ is the unit weight of granular fill material.

   The equivalent tyre contact pressure is given as

$$p_{ec} = \frac{P}{2LB} \tag{5.3}$$

where $p$ is the axle load.

   From Equations (5.1), (5.2) and (5.3), the following equations are obtained:

$$p_0 = \frac{P}{2(B + 2h_0 \tan \alpha_0)(L + 2h_0 \tan \alpha_0)} + \gamma h_0 \tag{5.4}$$

in the absence of geotextile, and

$$p = \frac{P}{2(B + 2h \tan \alpha)(L + 2h \tan \alpha)} + \gamma h \tag{5.5}$$

in the presence of geotextile.

   The load diffusion angles, $\alpha$ and $\alpha_0$, may vary in their values, however they are assumed both equal to $\tan^{-1}(0.6)$ in the present design method. This assumption implies that the presence of the geotextile layer does not modify significantly the load transmission mechanism through the granular layer.

On the application of the wheel load, the geotextile exhibits a wavy shape; consequently, it is stretched. This happens if the soil subgrade, having a low permeability, is saturated, and behaves in an undrained manner under the traffic loading. This incompressible nature of the soil subgrade results in settlement under the wheels and heave between and beyond the wheels (Fig. 5.2(b)). Under such a situation, the volume of soil subgrade displaced downwards by settlement must be equal to the volume of soil displaced upwards by heave, which may be called the *volume conservation* of the undrained soil subgrade. In the stretched position of the geotextile, the pressure against its concave face is higher than the pressure against its convex face. This reinforcing mechanism is known as the *membrane effect* of the geotextile, which provides the following two beneficial effects:

1   confinement of the soil subgrade between and beyond the wheels;
2   reduction of the pressure applied by the wheels on the soil subgrade.

The pressure applied on the soil subgrade by the portion $AB$ of the geotextile is:

$$p^* = p - p_g \tag{5.6}$$

where $p_g$ is the reduction of pressure resulting from the use of a geotextile. The pressure reduction, $p_g$, is a function of the mobilized tension in the geotextile, which depends on its elongation, thus its deflected shape is significant.

Since the soil subgrade confinement provided by the geotextile helps in keeping the deflection small for all applied pressures less than the ultimate load-bearing capacity, $q_u$, of the soil subgrade, as given by Equation (5.7) below, the pressure $p^*$ can be as large as $q_u$.

$$q_u = (\pi + 2)c_u + \gamma h \tag{5.7}$$

where $c_u$ is the undrained cohesion or shear strength of the soil subgrade.

$$p^* = q_u = (\pi + 2)c_u + \gamma h \tag{5.8}$$

From Equations (5.6) and (5.8), one gets

$$p - p_g = (\pi + 2)c_u + \gamma h \tag{5.9}$$

In the absence of a geotextile, an equation similar to Equation (5.9) can be obtained by equating $p_0$ to the elastic bearing capacity of the soil subgrade given as

$$q_e = \pi c_u + \gamma h \tag{5.10}$$

in order to avoid the large deflection under the wheel. Thus,

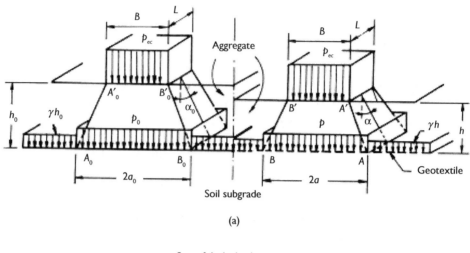

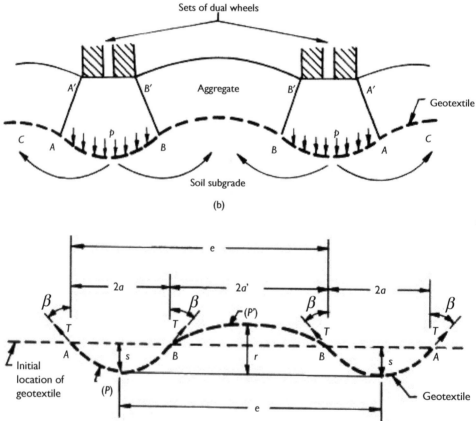

*Figure 5.2* (a) Load diffusion model; (b) kinematics of subgrade deformation; (c) shape of the deformed geotextile (after Giroud and Noiray, 1981).

$$p_0 = \pi c_u + \gamma b \tag{5.11}$$

in the absence of the geotextile.

Equations (5.4) and (5.11) lead to

$$c_u = \frac{P}{2\pi(B + 2b_0 \tan \alpha_0)(L + 2b_0 \tan \alpha_0)} \tag{5.12}$$

which is applicable in the absence of geotextile.

The shape of the deformed geotextile is assumed to consist of portions of parabolas connected at $A$ and $B$, points located on the initial plane of the geotextile (Fig. 5.2(c)). The reduction of pressure, $p_g$, is due to tension of the geotextile in parabola $(P)$. In fact, $p_g$ is a uniform pressure applied on $AB$ and is equivalent to the vertical projection of tension $T$ of the geotextile at points $A$ and $B$:

$$ap_g = T \cos \beta \tag{5.13}$$

According to the property of parabolas:

$$\tan \beta = \frac{a}{2s} \tag{5.14}$$

From the definition of secant modulus, $E$ (in N/m), of the geotextile,

$$T = E\varepsilon \tag{5.15}$$

where $\varepsilon$ is the per cent elongation.

Combining Equations (5.13), (5.14) and (5.15),

$$p_g = \frac{E\varepsilon}{a\sqrt{1 + \left(\dfrac{a}{2s}\right)^2}} \tag{5.16}$$

Equations. (5.5), (5.9) and (5.16) lead to

$$(\pi + 2)c_u = \frac{P}{2(B + 2b \tan \alpha)(L + 2b \tan \alpha)} + \frac{E\varepsilon}{a\sqrt{1 + \left(\dfrac{a}{2s}\right)^2}} \tag{5.17}$$

which is applicable in the presence of geotextile.

In Equations (5.12) and (5.17), the following expressions can be used for $L$ and $B$:

$$L = \frac{B}{\sqrt{2}}, \text{ and} \tag{5.18a}$$

$$B = \sqrt{\frac{P}{p_c}} \tag{5.18b}$$

for on-highway trucks

$$L = \frac{B}{2}, \text{ and} \tag{5.19a}$$

$$B = \sqrt{\frac{P\sqrt{2}}{p_c}} \tag{5.19b}$$

for off-highway trucks, where $p_c$ is the tyre inflation pressure

Solving Equation (5.12) for $h_0$, and Eq. (5.17) for $h$ allows us to determine the reduction of granular layer thickness, $\Delta h$, due to reinforcement function of the geotextile as per the quasi-static analysis. Thus,

$$\Delta h = h_0 - h \tag{5.20}$$

A further assumption is that the value of $\Delta h$ remains unchanged under the repeated traffic loading, thus allowing it to uncouple the reinforcement effect and its analysis from the cyclic nature of loading. Therefore,

$$h' = h'_0 - \Delta h \tag{5.21}$$

where $h'$ is the required granular layer thickness of the unpaved road in the presence of geotextile and under traffic loading, and $h'_0$ is the required granular layer thickness of the unpaved road in the absence of geotextile and under traffic loading.

Under traffic loading, the required granular layer thickness, $h'_0$, of the unpaved road without geotextile is determined using an empirical method originally developed by Webster and Alford (1978) for a rut depth of $r = 0.075$ m, and simplified by Giroud and Noiray as

$$h'_0 = \frac{0.19 \log_{10} N_s}{(CBR)^{0.63}} \tag{5.22}$$

where $N_s$ is the number of passes of standard axle with a load $P_s = 80 kN$; and CBR is the California bearing ratio of soil subgrade.

Giroud and Noiray extended Equation (5.22) to other values of axle load and rut depth using the following relationships:

$$\frac{N_s}{N_p} = \left(\frac{P}{P_s}\right)^{3.95}$$ (5.23)

$$\log_{10} N_s \rightarrow \left[\log_{10} N_s - 2.34(r - 0.075)\right]$$ (5.24)

where $\rightarrow$ indicates 'replaced by'.

They also introduced the undrained cohesion of the soil subgrade using the following empirical relationship:

$$c_u(\text{in Pa}) = 30000 \times \text{CBR}$$ (5.25)

Combining Equations (5.22), (5.23) and (5.25), and replacing $\log_{10} N_s$ as per Equation (5.24), the following expression is obtained:

$$h_0' = \frac{119.24 \log_{10} N + 470.98 \log_{10} P - 279.01r - 2283.34}{c_u^{0.63}}$$ (5.26)

This formula is based on extrapolation and therefore, it should not be used when the number of passages exceeds 10000.

A design chart for a particular set of parameters, based on the analysis presented here, is shown in Fig. 5.3. The following two features of this chart are noteworthy:

1   $\Delta h$ can never be higher than $h_0$.
2   No granular layer is needed on top of the geotextile when the $\Delta h$ versus $c_u$ curve is above the $h_0'$ versus $c_u$ curve.

The design chart provides values of $\Delta h$ and $h_0'$. The subtraction of $\Delta h$ from $h_0'$, according to Equation (5.21), results in the value of granular layer thickness, $h'$. A set of curves, giving the geotextile elongation, $\varepsilon$, versus subgrade soil cohesion, $c_u$, in the design chart allows the user of the chart to check that, in the considered case, the geotextile is not subjected to excessive elongation.

The reduction in granular layer thickness can be on the order of 20–60%, as in some typical cases considered by Giroud and Noiray. To be safe, it is recommended not to use the design chart in Fig. 5.3 for a number of passes larger than 10000.

Note that among the assumptions made in the RFDM, the adoption of different limit bearing pressures for the soil subgrade in unreinforced and reinforced cases leads to results that may seem theoretically inconsistent. According to the RFDM, the computed performance of an unreinforced road should be similar to that of the same road reinforced with a zero-modulus geotextile, which is not a fact. It has been recognized in practice that even very low modulus geotextiles are beneficial in reducing the granular layer thickness because of their *separation function*. The RFDM does not consider this reinforcing mechanism in analysis for the granular layer thickness. In addition to the determination of the granular layer thickness by the RFDM, computations should be completed with verifications of the tensile resistance and lateral anchorage of the geotextile (Bourdeau and Ashmawy, 2002). However, mainly because of simplicity,

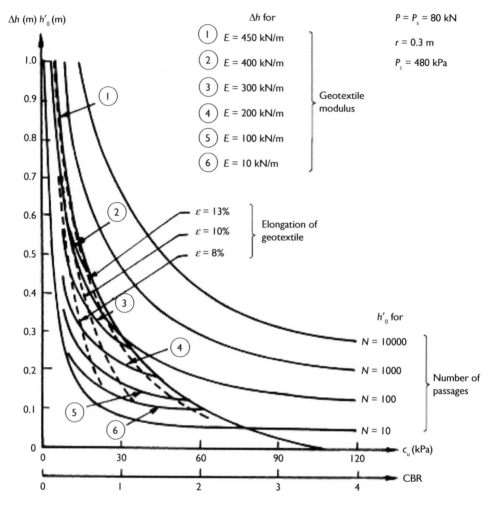

*Figure 5.3* Design chart for the geotextile-reinforced unpaved road related to on-highway truck with standard axle load of 80 kN (after Giroud and Noiray, 1981).

**EXAMPLE 5.1**

Consider:

Number of passes, $N = N_s = 340$
Single axle load, $P = P_s = 80$ kN
Tyre inflation pressure, $p_c = 480$ kPa
Subgrade soil CBR = 1.0
Modulus of geotextile, $E = 90$ kN/m
Allowable rut depth, $r = 0.3$ m

What is the required thickness of the granular layer for the unpaved road in the presence of geotextile?

**SOLUTION**

The design chart, presented in Fig. 5.3, provides the following:

$h'_0 = 0.35$ for CBR $= 1.0$ and $N = 340$

$h' = 0.15$ for CBR $= 1.0$ and $E = 90\,kN/m$

The required thickness of granular layer for the unpaved road in the presence of geotextile is calculated using Equation (5.21) as

$h' = h'_0 - \Delta h = 0.35 - 0.15 = 0.20\,m$

From the design chart, elongation of the geotextile, $\varepsilon = 10\%$. It should be checked that the elongation at failure of the geotextile, as obtained from the practical test, is larger than this value.

Note: This example was explained by Giroud and Noiray (1981).

the RFDM is widely used for designing geosynthetic-reinforced unpaved roads for a common range of parameters.

Based on field observations on unpaved roads with geosynthetics, Fannin and Sigurdsson (1996) reported that Giroud and Noiray's RFDM is found to be appropriate to unpaved roads that do not experience compaction of the granular base layer during traffic loading. Compaction will lead to an overprediction of performance at small ruts. It was also reported that the separation appears to be very important on the thinnest granular base course, where the geotextiles outperform the geogrid. The geogrid outperforms the geotextiles on the thicker granular base layers, for which reinforcement rather than separation dominates in a system that is less deformed by vehicle loading.

Since for roads, the geosynthetic reinforcement needs to support repeated loads, it is the response of the geosynthetic to rapid and cyclic loads that should be considered for design purposes. This aspect has been considered in the design method developed by Giroud and Han (2004a) for determining the thickness of the base course of geogrid-reinforced unpaved roads. This method considers distribution of stress, strength of base course material, interlock between geogrid and base course material, and geogrid stiffness in addition to the conditions considered in earlier method by Giroud and Noiry (1981): traffic volume, wheel loads, tire pressure, subgrade strength, rut depth, and influence of the presence of a reinforcing effect of geosynthetic on the failure mode of the unpaved road or area. The calculation of base course thickness is made using a unique equation, whereas more than one equation is required in earlier method by Giroud and

Noiry (1981). The calibration of this design method using data from field wheel load tests and laboratory cyclic plate loading tests on unreinforced and reinforced base courses has also been presented by Giroud and Han (2004b). Note that the design method presented by Giroud and Han (2004a) can be used for geotextile-reinforced unpaved roads and for unreinforced roads with appropriate values of relevant parameters.

### 5.2.2.2   Separation function design method (SFDM)

Steward *et al.* (1977) presented a design method for geosynthetic-reinforced unpaved roads, considering mainly the separation function of the geosynthetic, which is more important for thin roadway sections with relatively small live loads where ruts, approximating less than 75 mm, are anticipated. This design method is based on a theoretical analysis and empirical (laboratory and full-scale field) tests, and it allows the designer to consider vehicle passes, equivalent axle loads, axle configurations, tyre pressures, soil subgrade strengths and rut depths, along with the following limitations:

1   The granular layer must be (a) cohesionless (non-plastic), and (b) compacted to CBR = 80.
2   Vehicle passes less than 10000.
3   Geotextile survivability criteria must be considered (see Chapter 7).
4   Soil subgrade undrained shear strength is less than about 90 kPa (CBR < 3).

Steward *et al* presented design charts to determine the required thickness of the granular layer (Fig. 5.4). The main concept involved in developing these design charts is the presentation of the stress level acting on the soil subgrade in terms of the bearing capacity factor, similar to those commonly used for the design of shallow foundations (continuous footings) on cohesive soils using the following expression for ultimate bearing capacity, $q_u$:

$$q_u = c_u N_c + \gamma D \qquad\qquad (5.27)$$

where $c_u$ is the undrained cohesion of the soil subgrade; $N_c$ is the bearing capacity factor; $\gamma$ is the unit weight of the granular material above the geosynthetic layer; and $D$ is the depth of granular layer.

The bearing capacity factor is adjusted when a geosynthetic, especially geotextile, is introduced between the soft soil subgrade and the granular base course, as per the values given in Table 5.2.

The following points are worth mentioning:

1   The design method like SFDM, which assumes no reinforcing effect, is generally conservative.
2   The geotextile recommended for use in unpaved roads should meet the minimum hydraulic requirements in addition to minimum installation survivability requirements, as discussed in Chapter 7.

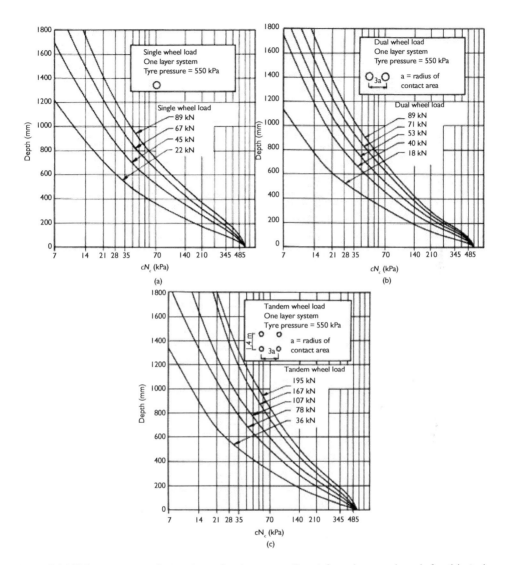

*Figure 5.4* US Forest service design charts for the geotextile-reinforced unpaved roads for: (a) single wheel load; (b) dual wheel load, and (c) tandem wheel load (after Steward *et al.*, 1977).

*Table 5.2* Bearing capacity factors for different ruts and traffic conditions both with and without geotextile separators (after Steward *et al.*, 1977).

| Field site situation | Ruts (mm) | Traffic (passes of 80 kN axle equivalents) | Bearing capacity factor, $N_c$ |
|---|---|---|---|
| Without geotextile | Less than 50 | Greater than 1000 | 2.8 |
| | Greater than 100 | Less than 100 | 3.3 |
| With geotextile | Less than 50 | Greater than 1000 | 5.0 |
| | Greater than 100 | Less than 100 | 6.0 |

**EXAMPLE 5.2**

Consider:
  Number of passes, $N$ = 6000
  Single axle load, $P$ = 90 kN
  Tyre inflation pressure, $p_c$ = 550 kPa
  Cohesive subgrade soil, CBR = 1.0
  Modulus of geotextile, $E$ = 90 kN/m
  Allowable rut depth, $r$ = 40 mm

What is the required thickness of granular layer for the unpaved road without geotextile and with geotextile?

**SOLUTION**

  Single wheel load = (90 kN)/2 = 45 kN
  From Table 5.2, for 6000 passes and 40 mm rut,
  $N_c$ = 2.8 without the geotextile layer, and
  $N_c$ = 5.0 with the geotextile layer.
  Using Equation (5.25), for CBR = 1.0,

  $c_u$ = 30 kPa

Without a geotextile layer,

  $cN_c$ = (30)(2.8) = 84 kPa

With a geotextile layer

  $cN_c$ = (30)(5.0) = 150 kPa

The design chart, presented in Fig. 5.4(a), provides the following:
Without the geotextile layer, the thickness of granular layer is

  $h_0$ ≈ 500 mm

With the geotextile layer, the thickness of granular layer is

  $h$ ≈ 350 mm

Note that about 150 mm granular layer thickness can be saved by placing a geotextile layer as a separator at the interface of soil subgrade and the granular base layer in unpaved roads.

3 Richardson (1997a) presented a simple separation function design method (SFDM) for the geosynthetic-reinforced unpaved roads, as described in the following steps:

*Step 1:* Use a granular layer thickness that produces a subgrade pressure $p = 4c_u$. This results in sufficient granular material being placed initially to fill the ruts that will develop as the geotextile/granular layer deforms.

*Step 2:* Determine the minimum survivability requirements for the geotextile.

*Step 3:* Determine the minimum hydraulic requirements for the geotextile.

*Step 4:* Select a suitable geotextile that meets the criteria in Steps 2 and 3. Almost any woven and nonwoven geotextile can be used if it meets the requirements in Steps 2 and 3.

In the first step of design, one can use a simple 60° angle for estimating the distribution of the applied surface load through the granular layer.

4 The design method, suggested by Richardson, is based on the field observations made by Fannin and Sigurdsson (1996) on stabilization of unpaved roads with geosynthetics. This method can be used in routine applications; however, it is suggested that the design values should be compared with those obtained from other methods, especially for a few initial problems as this would act as a confidence-building measure.

### 5.2.2.3 Modified California bearing ratio (CBR) design method

This method uses a multiplier to the in situ CBR of the soil subgrade, in order to get an equivalent CBR when using a geosynthetic layer at the interface of soft soil subgrade and the granular fill layer. The multiplier is assumed to be equal to the *reinforcement ratio*, which is the ratio of loads at the specified deflection, as determined from the load versus deflection test in the CBR mould both with and without geosynthetic at the interface of soft soil subgrade and granular fill layer. It has been observed that the reinforcement ratio increases as both the deflection and the water content in soil subgrade increase. The thickness of the granular fill layer is calculated using the following equation for both the cases without and with geosynthetic, taking in situ CBR, and modified CBR, respectively at the specified deflection (US Army Corps of Engineers Modified CBR Design Method, WES TR 3–692 as reported by Koerner, 1994).

$$h = \left(0.1275 \log_{10} N + 0.087\right) \left(\frac{P}{8.1 \times CBR} - \frac{A}{\pi}\right)^{1/2} \tag{5.28}$$

where $h$ is the design thickness of the granular layer in inches; $N$ is the anticipated number of vehicle passes; $P$ is the single or equivalent single wheel loads in pounds; and $A$ is the tyre contact area in square inches.

**EXAMPLE 5.3**

Consider:
  Number of passes, $N = 10000$
  Single wheel load, $P = 20000$ lb
  Tyre contact area = 12 in. × 18 in.
  Subgrade soil CBR = 1.0
  Reinforcement ratio, as determined from the modified CBR test, $R = 2$

What is the required thickness of granular layer for the unpaved road without geotextile and with geotextile?

**SOLUTION**

Using Equation (5.28), in the absence of geotextile, the required thickness is

$$h_0 = \left(0.1275\log_{10}10000 + 0.087\right)\left(\frac{20000}{8.1\times1.0} - \frac{12\times18}{\pi}\right)^{1/2} \approx 29.2 \text{ in.}$$

Modified CBR = (subgrade soil CBR) × $R$ = 1.0 × 2.0 = 2.0

In the presence of geotextile, the required thickness is

$$h = \left(0.1275\log_{10}10000 + 0.087\right)\left(\frac{20000}{8.1\times2.0} - \frac{12\times18}{\pi}\right)^{1/2} \approx 20.4 \text{ in.}$$

Note that the savings in granular layer thickness is 8.8 in. (≈ 30%) in the presence of a geotextile layer at the interface of the soil subgrade and the granular base layer in an unpaved road.

### 5.2.3  Application guidelines

A geosynthetic layer, generally a geotextile layer, is typically placed directly on the soil subgrade followed by placement and compaction of an adequate depth of granular layer. The construction practices being adopted must ensure that the geosynthetic will survive installation, and the construction sequencing will not lead to failure of the existing soil subgrade. Some specific guidelines for the field installation of geosynthetic layers are given below.

1  The designated site should be prepared by clearing topsoil and vegetation. The area for placement of geosynthetic layer must be as smooth and clean as possible to prevent the geosynthetic damage and permit the uniform granular layer thickness.
2  Prior to placement of the geosynthetic layer, the prepared subgrade surface, that is, the formation level must be provided an appropriate cross slope.
3  During the geosynthetic placement, care must be taken not to damage the material or disturb the prepared subgrade. Care must also be taken to minimize wrinkles and folds in the geosynthetic.

*Table 5.3* Overlap requirement of geotextile for different CBR values (after AASHTO, 2011).

| Soil CBR | Minimum Overlap |
|---|---|
| Greater than 3 | 300–450 mm |
| 1–3 | 0.6–1 m |
| 0.5–1 | 1 m or sewn joint |
| Less than 0.5 | Overlap is not recommended. Sewn joint should be provided. |
| All roll ends | 1 m or sewn joint |

4   Parallel rolls of geosynthetic should be overlapped or sewn as required. Recommended overlaps are given in Table 5.3. Overlaps of parallel rolls should occur at the centerline and at the shoulder. Overlaps should not be made along the anticipated main wheel path locations. Overlaps at the end of the rolls should be in the direction of the granular fill placement with the previously placed roll on top. Continuous visual inspection of all field seams and overlaps should be done throughout the installation to ensure that there are no voids in the seam or overlap area. Repairs that may be required during installation can be accomplished by patching by taking a piece of the primary geosynthetic that extends approximately 300 mm beyond each edge of the area to be repaired.

5   For very weak soil subgrades, the granular fill thickness should be limited to prevent construction-induced failure. The first lift of granular fill should be graded down to a thickness of 300 mm or the maximum design thickness. All remaining lifts of granular fill should be placed in lifts not exceeding 220 mm loose thickness. The maximum granular particle size in the initial lift should be limited to less than 1/4 of the lift thickness.

6   At no time equipment should be allowed on the geosynthetic with less than 150 mm of granular fill between the wheels and the geosynthetic. Construction vehicles should be limited in size and weight such that the rutting in the initial lift is less than 75 mm. The turning of construction vehicles should not be permitted on the first granular lift over the geosynthetic.

7   Care should always be taken to push the granular cover material over the top of the geosynthetic, rather than into the overlaps. Also, the material is pushed in such a way that the geosynthetic is not pulled, and wrinkles are not caused in front of the cover material. The compaction of granular fill by vibratory roller must be closely monitored where the subgrade is susceptible to liquefaction.

8   The ruts that develop during construction must be filled in with additional granular material; otherwise the geosynthetic may eventually become exposed at the crown between the ruts. In no case the ruts should be bladed down.

9   It must be ensured that there is no contamination of the granular layer from the fine-grained subgrade soils during construction and service life. It has been observed that the addition of 10–20% clayey fines to clean gravel reduces the bearing capacity of the gravel to that of the clay. It also must be ensured that the

subgrade must be free to drain as it consolidates under the traffic and roadway induced stresses. Thus, the geosynthetic must act as a filter to prevent the movement of soil fines into the granular subbase/base while allowing water to drain from the subgrade.

10  Edges of the geosynthetic layer must be attached to the side drains for allowing the water at the subgrade level to join the drain while moving along the plane of the geosynthetic layer.

## 5.2.4  Case studies

The use of geosynthetics in unpaved roads on soft soils makes it possible to increase the load-bearing capacity of the soft soil and the granular fill. A geosynthetic layer in unpaved road allows the passage of heavily loaded vehicles over the granular fills of reduced thicknesses, placed on the soft soil subgrades. This, in turn, allows decreased consumption of materials, transport expenses, and duration of construction.

One of the first roads in the USSR, where a domestic geotextile was first used, was a temporary road in Smolenskaya region (Kazarnovsky and Brantman, 1993). Construction of the road had to be accelerated to evacuate populated localities from areas to be flooded, when the reservoir of Vazuzskaya hydrosystem was being filled with water, and also to allow for the movement of construction vehicles. A temporary road, about 20 km long, was to be constructed within the shortest period of time and with a minimum thickness of fill. The site was characterized as soft plastic loam soils with a high groundwater level and a prolonged stagnation of water above the ground. To accelerate construction of the road, which at the same time provided a passage for construction vehicles delivering sand for the fill, the geotextile 'Dornit' $\phi$-1 was used on several sections of the road where the soil consisted of a light silty loam and had water content, at the time of construction, equal to 27–30%. The used geotextile, manufactured from waste synthetic fibres by needle-punching, had the following characteristics: mass per unit area = 600 g/m²; roll width = 1.7 m; nominal thickness = 4.5 mm; tensile strength = 12 kN/m (roll direction) and 6 kN/m (cross-roll direction) and strain at failure = 70% (roll direction) and 130% (cross-roll direction).

The construction procedures of the road sections on which the geotextile was used included the following:

- a rough grading of the soft subgrade by a bulldozer going back and forth with a lowered blade;
- unrolling the geotextile across the fill axis with 300 mm overlaps between adjacent rolls;
- filling and grading of 450–500-mm thick medium-grained sand layer (containing gravel and 2% of silty and clayey particles), followed by compaction with a lightweight roller.

The difference in driving conditions on sections with the geotextile and without it could be observed immediately after installation of the geotextile. It was actually impossible to perform work after 8 to 10 passes of dump trucks along the same ruts in the section where the geotextile was not used and the sand fill thickness was limited

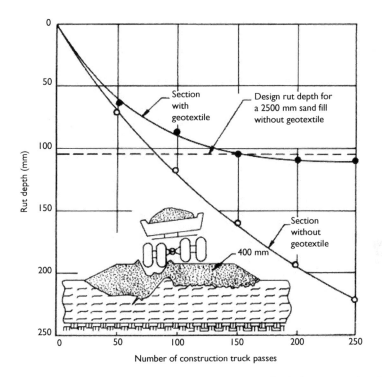

*Figure 5.5* Change in the rut depths depending on the number of vehicles passing (sand fill thickness = 400 mm) (after Kazarnovsky and Brantman, 1993).

to 400 mm for the comparison. On the road section where the geotextile was placed under the sand fill, the rut depth did not exceed 100–120 mm and intermixing of the fill sand with the subgrade did not occur. On each working day, 300 vehicles, mostly dump trucks, travelled in both directions. After the road was used for several months, the ruts on the road section without the geotextile had to be graded continually. Every morning, before the main traffic drove on the road, both the sections with and without the geotextile were graded with a bulldozer. Measurement of the ruts showed that, on the section without a geotextile, ruts from 200–250 mm deep were formed, and the traffic speed slowed to 5 km/h. On the road section with a geotextile, the rut depths were only 100–120 mm in spite of the fact that the dump trucks traveled along the same rut (Fig. 5.5). The traffic was able to maintain a speed of 25–30 km/h and the vehicles could pass each other using the whole width of the fill.

## 5.3   PAVED ROADS

### 5.3.1   Basic description

Pavements are civil engineering structures used for the purpose of operating wheeled vehicles safely and economically. Paved roadways that include carriageways and shoulders have been constructed for more than a century. Their basic design methods and construction techniques have undergone some changes, but the development of geosynthetics in the past 35–40 years has provided strategies for enhancing the overall performance of paved roadways. Governments in most countries devote unprecedented time and resources to roadway construction, maintenance and repair. Efforts are also being made to apply the newfound technology to the old pavement problems.

### Geosynthetic layer at the soil subgrade level

Geosynthetic layers are used in paved roads usually at the interface of the granular base course and the soft soil subgrade during the initial stage of their construction, may be called the *unpaved age*, as a stabilizer lift, to allow construction equipment access to weak soil subgrade sites, and to make proper compaction of the first few granular soil lifts. In the case of thicker granular bases, the geosynthetic layer may be placed within the granular layer, preferably near the mid-level to achieve the optimum effect. The presence of a geosynthetic layer at the interface of the granular base course and the soft soil subgrade improves the overall performance of paved roads with their long operating life because of its *separation, filtration, drainage* and *reinforcement* functions (Holtz *et al.*, 1997; Shukla, 2006).

During construction as well as during the operating life of paved roads, contamination of the granular base course by the fines from the underlying soft soil subgrade leads to promoting pavement distress in the form of structural deficiencies (loss of vehicular load-carrying capacity) or functional deficiencies (development of conditions such as rough riding surface, cracked riding surface, excessive rutting, potholes, etc., causing discomfort) that result in early failure of the roadway (Perkins *et al.*, 2002). This is mainly because of the reduction of the effective granular base thickness, by contamination, to a value less than the design value already adopted in practice. This problem may not exist in the presence of a geosynthetic layer at the interface of the granular base course and the soft soil subgrade because of its role as a separator and/or a filter (Fig. 5.6).

Geosynthetics, especially the bitumen-impregnated geotextiles, are used to improve paved roads as a separator and/or a fluid barrier by providing capillary breaks to reduce the frost action in frost-susceptible soils (fine-grained soils – silts, clays and related mixed soils). The paved roads can also be improved by providing membrane-encapsulated soil layers (MESL) as a moisture-tight barrier beneath the wearing course with an aim to reduce the effects of seasonal water content changes in soils (Fig. 5.7). If good-quality granular materials are not available for base/subbase courses, then the concept of MESL can be used to construct base/subbase courses of paved roadways even using locally available poor-quality soils. Commercially available thin-film geotextile composites (Fig. 5.8) are also used as moisture barriers in roadway construction to prevent or minimize the moisture changes in the pavement subgrades.

Pavement distress can also be caused by inadequate lateral drainage through the granular base course. It has been observed that adequate drainage of a pavement extends its life by up to 2–3 times that of a similar pavement having inadequate drainage (Cedergren, 1987). A geosynthetic layer, especially a thick geotextile or a drainage

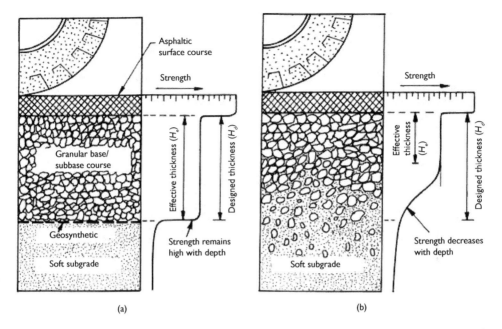

*Figure 5.6* Concept of geosynthetic separation in paved roadways: (a) paved roadway with geosynthetic separator; (b) paved roadway without geosynthetic separator (modified from Rankilor, 1981).

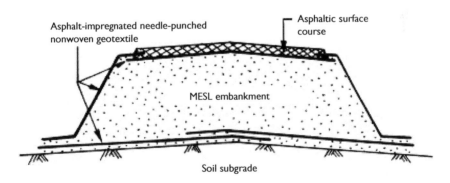

*Figure 5.7* Concept of the membrane-encapsulated soil layer (MESL) as a base/subbase course in paved roadways.

geocomposite, can act as a drainage medium to intercept and carry water in its plane to side drains on either side of the pavement.

The use of a geosynthetic layer also helps in enhancing the structural characteristics and in controlling the rutting of the paved roadway through its reinforcement function. Note that the principal reinforcing mechanism of the geosynthetic in

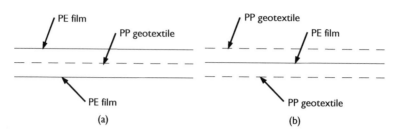

*Figure 5.8* Thin-film geotextile composites: (a) PP geotextile sandwiched between two PE films; (b) PE film sandwiched between two PP geotextiles.

paved roads is its *confinement effect*, not its *membrane effect*, which is applicable for unpaved roads allowing extensive rutting. The lateral confinement provided by the geosynthetic layer resists the tendency of the granular base courses to move out under the traffic loads imposed on the asphaltic or cement concrete wearing surface. In the case of paved roads on the firm subgrade soils, prestressing the geosynthetic by external means can significantly increase the lateral confinement to the granular base course. It also significantly reduces the total and differential settlements of the reinforced soil system under applied loads (Shukla and Chandra, 1994). Prestressing the geosynthetic can be an effective technique to adequately improve the behaviour of geosynthetic-reinforced paved roads in general situations, if it is made possible to adopt the prestressing process in the field in an economical manner.

### Geosynthetic layer at the overlay base level

Commonly a paved road becomes a candidate for maintenance when its surface shows significant cracks and potholes. The cracks in the pavement surface cause numerous problems, including the following:

- riding discomfort for the users;
- reduction of safety;
- infiltration of water and subsequent reduction of the load-bearing capacity of the subgrade;
- pumping of soil particles through the crack;
- progressive degradation of the road structure in the vicinity of the cracks due to stress concentrations.

The construction of bituminous/asphalt overlays is the most common way to renovate both flexible and rigid pavements. Most overlays are done predominantly to provide a waterproofing and pavement crack retarding treatment. A minimum thickness of the asphalt concrete overlay may be required to provide an additional support to a structurally deficient pavement. An asphalt overlay is at least 25 mm thick and it is placed on top of the distressed pavement. Overlays are economically practical, convenient, and effective. The cracks under the overlay rapidly propagate through to the new surface. This phenomenon is called the *reflective cracking*, which is the major drawback of asphalt overlays. Because the asphalt overlays are otherwise an excellent option, research and development has focused on preventing the reflective cracking.

The reflective cracks in an asphalt overlay are basically a continuation of the discontinuities in the underlying damaged pavement. When an overlay is placed over a crack, the crack grows up to the new surface. The causes of crack formation and enlargement in asphalt overlays are numerous but the mechanisms involved may be categorized as (Fig. 5.9): *traffic induced*, *thermally induced* and *surface initiated*. Surface cracking in overlays can occur from the traffic induced fatigue as a result of the repeated bending condition in the pavement structure or shear effect causing the pavement on one side of a crack (in the old layer) to move vertically relative to the other side of the crack during traffic movement. High axle loads or increased traffic can further increase the stresses and strains in the pavement that lead to the surface cracking. In the case of an asphalt overlay on top of a concrete pavement, the cracks may be reflected to the overlay as the concrete slabs expand and contract under varying temperatures. The expansion and contraction of the overlays and upper asphalt layers can lead to tension within the surfacing which can also lead to the surface cracking. The stresses are at their maximum at the pavement surface where the temperature variation is the greatest. In this case, the cracks are initiated at the surface and propagate downwards. Note that the term 'reflective cracking' is often used to describe all these types of cracking.

Methods for controlling the reflective cracking and extending the life of overlays consider the importance and effectiveness of overlay thickness and proper asphalt mixture specification. Asphalt mixes have been improved and even modified by adding a variety of materials. In the past, a number of potential solutions have also been

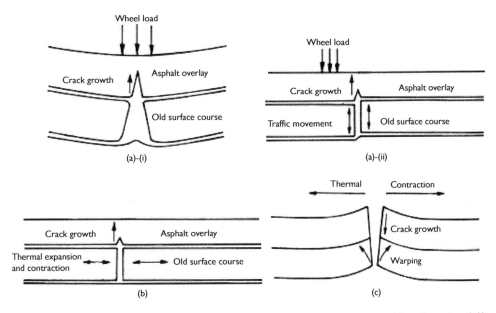

*Figure 5.9* Mechanisms of crack formation and propagation in the asphalt overlay: (a) traffic induced, (i) repeated bending, (ii) shear effect; (b) thermally induced; (c) surface initiated.

evaluated including the unbound granular base 'cushion courses' and wire mesh reinforcement. All have been found either marginally effective or extremely costly.

The most basic way to slow down the reflective cracking is to increase the overlay thickness. In general, as the overlay thickness increases, its resistance to reflective cracks increases. However, the upper limit of overlay thickness is highly governed by the expense of asphalt and the increase in height of the road structure.

Asphalt additives do not stop reflective cracking, but do tend to slow down the development of cracks and convert a large crack in the old pavement into multiple small cracks in the overlay. Mixing glass fibres, metal fibres or polymers in asphalt prior to paving creates *modified or optimized asphalt*, which is not always specified because it is much more expensive than unimproved asphalt and the relationship between the investment and the improvement has not been established.

The crack resistance of the overlay can also be enhanced via the interlayer systems. An interlayer is a layer between the old pavement and the new overlay, or within the overlay, to create an overlay system. The benefits of a geosynthetic interlayer include the following:

- waterproofing the pavement;
- delaying the appearance of reflective cracks;
- lengthening the useful life of the overlay;
- added resistance to fatigue cracking;
- saving up to 50 mm of overlay thickness.

A geosynthetic layer, especially a geotextile layer, is used beneath the asphalt overlays, ranging in thickness from 25 to 100 mm, of asphalt concrete (AC) and Portland cement concrete (PCC) paved roads. The geotextile layer is generally combined with asphalt sealant or tack coat to form a membrane interlayer system known as the *paving fabric interlayer*. Fig. 5.10 shows the layer arrangement in paved roads with paving fabric interlayer. When properly installed, a geotextile layer beneath the asphalt overlay mainly performs the following (Holtz *et al.*, 1997; Shukla and Yin, 2004):

- fluid barrier (if impregnated with bitumen, that is asphalt cement), protecting the underlying layers from degradation due to infiltration of road-surface moisture;
- cushion, that is, the stress-relieving layer, for the overlays, retarding and controlling some common types of cracking, including reflective cracking.

A paving fabric, in general, is not used to replace any structural deficiencies in the existing pavement. However, the above functions combine to extend the service life of overlays and the roadways with reduced maintenance cost and increased pavement serviceability.

Pavements typically allow 30–60% of precipitation to infiltrate and weaken the road structure. The fluid barrier function of the bitumen-impregnated geotextile may be of considerable benefit if the subgrade strength is highly moisture-sensitive. In fact, the excess moisture in the subgrade is the primary cause of premature road failures. Heavy vehicles can cause extensive damage to roads, especially when the soil subgrade is wet and weak. Pore water pressure can also force soil fines into the voids within

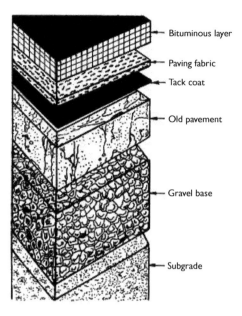

*Figure 5.10* Typical cross-section of a paved roadway with a paving fabric interlayer.

the subbase/base layer, thus weakening them if a geotextile is not used as a separator/
filter. Therefore, efforts should be made to keep the soil subgrade at fairly constant
and low water content by stopping the moisture infiltration into the pavement and
providing proper pavement drainage.

A stress-relieving interlayer retards the development of reflective cracks in the
overlay by absorbing the stresses induced by underlying cracking in the old pavement. The stress is absorbed by allowing slight movements within the paving fabric
interlayer inside the pavement without distressing the asphalt concrete overlay significantly. In fact, the addition of a stress-relieving interlayer reduces the shear stiffness
between the old pavement and the new overlay, thus creating a buffer zone (or break
layer) that gives the overlay a degree of independence from movements in the old
pavement. Pavements with paving fabric interlayers also experience much less internal
crack developing stress than those without. This is why the fatigue life of a pavement
with a paving fabric interlayer is many times that of a pavement without, as shown in
Fig. 5.11. A stress-relieving interlayer also waterproofs the pavement, so when cracking does occur in the overlay, water cannot worsen the situation.

Geotextiles generally have performed best when used for load-related fatigue distress, for example, closely spaced alligator cracks. Fatigue cracks, mainly caused by
too many flexures of the pavement system, should be less than 3 mm wide for the best
results. Geotextiles used as a paving fabric interlayer to retard the thermally induced
fatigue cracking, caused by actual expansion and contraction of underlying layers,
mostly within the overlay, have, in general, been found to be ineffective. For obtaining the best results on existing cracked pavements, the geotextile layer is laid over the

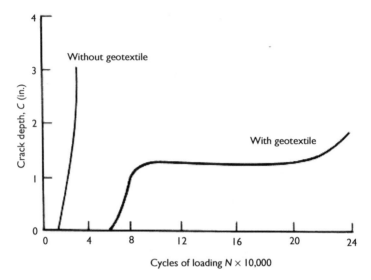

*Figure 5.11* Fatigue response of asphalt overlay (after IFAI, 1992).

entire pavement surface or over the crack, spanning it by 15–60 cm on each side, after placement of an asphalt leveling course followed by an application of asphalt tack coat, and then asphalt overlay is placed above (Fig. 5.10). This construction technique is adopted keeping in view that much of the deterioration that occurs in overlays is the result of unrepaired distress in the existing pavement prior to the overlay.

The selection of a geosynthetic for use in asphalt overlays is complicated by the variable deterioration conditions of the existing roadway systems. The deterioration may range from simple alligator cracking of the pavement surface to significant pot-holes caused by failure of the underlying subgrade. It is important to note that an overlay system as well as a paving fabric interlayer may fail if the deficiencies already present in the existing pavements are not corrected prior to the placement of overlay and/or paving fabric.

The selected paving grade geosynthetic must have the ability to absorb and retain the bituminous tack coat sprayed on the surface of the old pavement and to effectively form a permanent fluid barrier and cushion layer. The most common paving grade geosynthetics are lightweight needle-punched nonwoven geotextiles, with a mass per unit area of 120–200 g/m$^2$. Woven geotextiles are ineffective paving fabrics, because they have no interior plane to hold asphalt tack coat and so do not form an impermeable membrane. They also do not perform well as a stress-relieving layer to help reduce cracking.

Tests should be performed to determine the bitumen (asphalt cement) retention of paving fabrics for their effective application. In the most commonly used test procedure, after taking weights individually, the test specimens are submerged in the bitumen at a specified temperature, generally 135 °C, for 30 min. Specimens are then hung to drain in the oven at 135 °C for 30 min from one end and also 30 min from the other end to obtain a uniform saturation of the fabric. Upon completion of specimen submersion in bitumen and draining, the individual specimens are weighed and bitumen retention, $R_B$, is calculated as follows (ASTM D 6140-00 (ASTM, 2014)):

$$R_B = \frac{W_{sat} - W_f}{\gamma_B A_f} \tag{5.29}$$

where $W_{sat}$ is the weight of saturated test specimen (kg); $W_f$ is the weight of paving fabric (kg); $A_f$ is the area of test specimen (m²); and $\gamma_B$ is the unit weight of bitumen/asphalt cement at 21°C (kg/l). The average bitumen retention of specimens are calculated and reported in l/m².

Paving fabrics precoated with modified bitumen are also available commercially in the form of strips. These products perform the same functions of waterproofing and stress relief as the field impregnated paving fabrics; however, they are more expensive. Their applications are economical if only limited areas of the pavement need a paving fabric interlayer system. For waterproofing and covering potholes, precoated paving fabrics may be good.

Heavy-duty geocomposites and bituminous membranes are used, especially over cracks and joints of cement concrete pavements that are overlaid with asphalt concrete. Geogrids and geogrid-geotextile composites are also commercially available for overlay applications to function as *reinforcement interlayer* for holding the crack, if any, together, and dissipating the crack propagation stress along its length. It has been reported that a reinforcement geogrid, as shown in Fig. 5.12, if used beneath the overlay, can reduce the crack propagation by a factor of up to 10 when traffic-induced fatigue is the failure mechanism (Terram Ltd., UK). The study conducted by Ling

*Figure 5.12* Asphalt reinforcement geogrid.

and Liu (2001) shows that the geogrid reinforcement increases the stiffness and the load-bearing capacity of the asphalt concrete pavement. Under the dynamic loading, the life of the asphalt concrete layer is prolonged in the presence of geosynthetic reinforcement. The stiffness of the geogrid and its interlocking with the asphalt concrete contributes to the restraining effect.

Note that choosing the proper application sites for paving geosynthetic is a function of the structural integrity of existing pavement and the crack types – not its surface condition. For successful performance, proper installation must occur on a pavement without significant vertical or horizontal differential movement between the cracks or the joints and without the local deflection under the design loading (Marienfeld and Smiley, 1994).

## 5.3.2   Analysis and design concepts

### Geosynthetic layer at the subgrade level

The ruts with a depth in excess of approximately 25 mm are generally not acceptable in paved roadways, which are utilized as the permanent roadways for safe, efficient and economical transport of passengers and goods. If a geosynthetic layer is used only for the construction lift (or stabilization lift), then the thickness of the granular subbase/base layer, required to adequately carry the design traffic loads for the design life of the paved roadway, is generally not reduced. Paved roadways with geosynthetic layers are usually designed for structural support using the normal pavement design methods, as described by various agencies (AASHTO, 1993; AASHTO, 2008; IRC, 2011; IRC, 2012), without providing any allowance for the geosynthetic layers.

If the soil subgrade is susceptible to pumping and granular base course intrusion, an additional granular layer thickness above that required for the structural support is needed. In the presence of a geosynthetic layer, especially a nonwoven geotextile, at the interface of the granular subbase/base layer and the soil subgrade, the required additional granular layer thickness can be reduced by approximately 50% keeping the project cost-effective (Holtz *et al.*, 1997). Savings of the granular material can also be made by placing a geosynthetic layer in the granular stabilizer lift that can tolerate even 75 mm of rutting under the construction equipment. The stabilizer lift with a geosynthetic layer is generally designed, taking it to be a geosynthetic-reinforced unpaved road that has been described in Section 5.2.2 in detail.

As a final design step, the recommended geosynthetic should be checked to meet both the minimum hydraulic requirements and the minimum survivability requirements, as discussed in Section 7.18 (see Chapter 7).

### Geosynthetic layer at the asphalt overlay base level

The fluid barrier function of the geosynthetic should be achieved in the field applications, keeping in view the fact that the water (coming from rain, surface drainage or irrigation near pavements), if allowed to infiltrate into the base and subgrade, can cause pavement deterioration by one or more of the following processes:

- softening the soil subgrade
- mobilizing the soil subgrade into the road base stone, especially if a separation/ filtration geosynthetic is not used at the road base and subgrade interface
- hydraulically breaking down the base structures, including stripping bitumen-treated bases and breaking down chemically stabilized bases
- freeze/thaw cycles.

The selected paving grade geosynthetic should meet the physical requirements described in Section 5.3.1. Prior to laying the paving fabric, the tack coat should be applied uniformly to the prepared dry pavement surface at the rate governed by the following equation (IRC, 2002):

$$Q_d = 0.36 + Q_s + Q_c \qquad (5.30)$$

where $Q_d$ is the design tack coat quantity (kg/m²); $Q_s$ is the saturation content of the geotextile being used (kg/m²) to be provided by the manufacturer; and $Q_c$ is the correction based on the tack coat demand of the existing pavement surface (kg/m²).

The quantity of tack coat is critical to the final membrane system. Too much tack coat will leave an excess between the fabric and the new overlay, resulting in a potential sliding failure surface and the potential bleeding problems, while too little will fail to complete the bond and create the impermeable membrane. In fact, the misapplication of the tack coat can make the difference between paving fabric installation success and failure. The asphalt tack coat forms a low-permeability layer in the fabric and bonds the system to the existing pavement and overlay. The fabric allows a slight movement of the system, while holding the tack-coat layer in place and maintaining its integrity.

The actual quantity of tack coat depends on the relative porosity of the old pavement and the amount of bitumen sealant required to saturate the paving fabric being used. The quantity of sealant required by the existing pavement is a critical consideration. The saturation content of the fabric depends primarily on its thickness and porosity; that is, its mass per unit area. Note that the more the mass per unit area of the geotextile, the greater tack coat is required to saturate the fabric. For typical paving fabrics in the 120–135 g/m² mass per unit area range, most manufacturers recommend a fabric-bitumen absorption of about 900 g/m², or application rates of about 1125 g/m². For the full waterproofing and stress-relieving benefits, the paving fabric must absorb at least 725 g/m² of bitumen. The remaining part of the applied bitumen helps in bonding the system with the existing pavement and the overlay. An additional tack coat may be required in the overlap to satisfy the saturation requirements of the fabric.

A review of projects with unsatisfactory paving fabric system performance shows the importance of the tack coat to the whole system. From a study of the records of 65 projects, which were completed over a 16-year period, it is clear that the tack coat applications were too light (less than 725 g/m²) in an overwhelmingly high percentage of failure cases, as shown in Fig. 5.13. In the laboratory tests, it has been observed that the waterproofing benefit of a paving fabric is negligible until the fabric absorbs at least 725 g/m² of tack coat (Fig. 5.14). Inadequate tack coat may result in rutting, shoving, or occasionally complete delamination of the

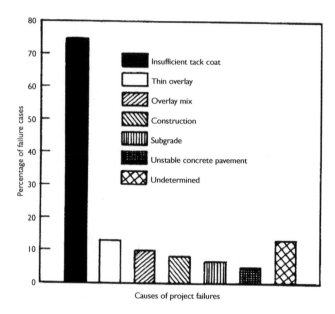

Figure 5.13 Causes in 65 project failures investigated in the United States between 1982–1997 (after Baker, 1998).

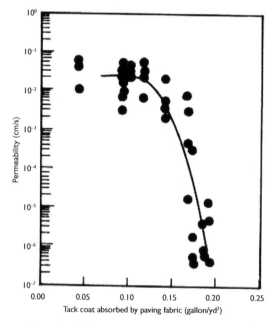

Figure 5.14 Laboratory-prepared paving-fabric tests, demonstrating the permeability's sensitivity to the amount of tack coat on the paving fabric (after Marienfeld and Baker, 1998).

overlay. It has been found that the structural problems such as overlay slippage and delamination begin to occur where the tack coat quantity absorbed by the fabric is less than about 450 g/m².

In addition to low application amounts of tack coat, there can be another set of conditions that may result in a low tack amount in paving fabric. Inadequate rolling or low overlay temperatures may create conditions in which the tack may not be taken up by the fabric. In fact, overlays less than 40 mm thick are seldom recommended with paving fabrics, in part, because of their rapid heat loss.

Controlled studies have shown that an overlay thickness designed to retard the reflective cracking can be reduced by up to 30 mm for equal performance, with additional waterproofing advantage if a paving fabric interlayer is included in the system (Marienfeld and Smiley, 1994). Equations are available that enable the designer of a geosynthetic-reinforced overlay to design an appropriate overlay thickness and the corresponding geosynthetic. The major drawback to currently available design techniques is that they allow us to address the potential failure modes (traffic load induced, thermally induced, and surface initiated) separately, but not together. In reality, all three modes occur together – a condition that can only be evaluated using the sophisticated finite element modeling (Sprague and Carver, 2000). For the routine applications, the thickness of the overlay may not be reduced with the use of a geosynthetic interlayer, and one can design the overlay as per the guidelines in the design standards on overlays without geosynthetic interlayer. This is mainly because the major purpose of introducing a paving fabric is to enhance the performance of the pavement, not to reduce the thickness of the overlay.

### 5.3.3  Application guidelines

A paving fabric interlayer system is looked upon as an economical tool, which effectively solves the general pavement distress problems. It is easy to install and readily complements any paving operation. The ideal time to place a paving fabric interlayer system is in the very early stages of hairline cracking in a pavement. It is also appropriate in the new pavement construction to provide a waterproof pavement from day one.

The installation of a paving fabric generally follows the same pattern wherever it is used. There are four basic steps in the proper installation of an overlay system with a geosynthetic interlayer. The surface preparation is followed by the application of tack coat, installation of the geosynthetic and finally the placement of the overlay. These steps along with the general guidelines are described below.

*Step 1 – Surface preparation:* The site surface is prepared by removing loose materials and sharp/angular protrusions, and sealing the cracks, as necessary. The prepared surface should be leveled, dry, and free of dirt, oil and loose materials. The cracks, 3 mm wide or greater, should be cleaned with pressurized air or brooms and filled with a liquid asphalt crack sealant. This will prevent the tack coat from entering the cracks and reducing available tack coat for saturation of the fabric. Very large cracks should be filled with a hot or cold asphalt mix. A commercial crack filler can also be used. Cracks should be level with the pavement surface and not overfilled. If the quality of the existing pavement is poor, a leveling course of asphalt concrete is placed over it prior to the placement of the paving fabric interlayer system. On existing cement concrete pavement, a layer of asphalt concrete should be provided before laying the fabric. The surface on which a moisture barrier interlayer is placed must have a grade which will drain the water off the pavement.

*Step 2 – Tack coat application:* Proper application of tack coat is crucial; any mistake can lead to early failure of the overlay. Straight paving-grade bitumen is the best and the most economical choice for paving fabric tack coat. Cutbacks and emulsions which contain solvents should not be used for the tack coat; if they are used, they must be applied at a higher rate and allowed to cure completely. Minimum air and pavement temperature should be at least 10°C or more for placement of the tack coat (IRC: SP: 59–2002). The temperature of tack coat should be sufficiently high to permit a uniform spray pattern. It should be spread at between 140°C and 160°C, to permit uniform spray and to prevent damage to the paving fabric. The target width of tack coat application should be equal to the paving fabric width plus 150 mm. Tack coat should be restricted to the area of immediate fabric lay-down.

Besides proper quantity, uniformity of the sprayed bitumen tack coat is of great importance. Application of hot bitumen should be done preferably by means of a calibrated distributor spray bar for better uniformity. Hand spraying and brush application may be used in locations of fabric overlap. When hand spraying, close attention must be paid to spraying a uniform tack coat.

*Step 3 – Geosynthetic placement:* The paving fabric is placed prior to the tack coat cooling and losing tackiness. The paving fabric is placed onto the tack coat with its fuzzy side down leaving the smooth side up using mechanical or manual lay-down equipment capable of providing a smooth installation without wrinkling or folding. Today most paving fabrics are applied using the tractor-mounted rigs. Slight tension can be applied during the paving fabric installation to minimize the wrinkling. However, the stretching is not recommended, because it will reduce the thickness, thus changing the bitumen retention properties of the fabric. Too little elongation may result in wrinkles. Too much elongation produces excessive stretch, thinning the geosynthetic so that it may not be thick enough to absorb the tack coat, and leaving excess tack coat that may later bleed through the bituminous concrete on a hot day. Wrinkles and overlaps can cause cracks in the new overlay if not properly handled during the construction process.

Overlaps and all overlapped wrinkles for fabric and grid composites should have an additional tack coat placed. The tack coat must be sufficient to saturate the two layers and make a bond. If not done correctly, a slip plane may exist at each overlapped joint, resulting in a possible crack of the asphalt from the fabric. Overlaps should be no more than 150 mm on longitudinal and transverse joints. This is different for grids, and each manufacturer has its own recommendations for the overlaps. Paving multiple lanes has inherent installation problems. It is best to install in one lane and pave it for traffic prior to installing in another lane. 150 mm of fabric should be left unpaved for overlap on the adjacent panel of fabric to be installed.

A paving reinforcement geogrid is installed into a light asphalt binder or it may be attached to the existing pavement by mechanical means (nailing) or by adhesives, thus preventing the geogrid from being lifted by paving equipment passing over it. When a composite of geogrid and geotextile is installed, the tack coat is applied in the same way as in paving a geotextile alone.

Installing the geosynthetic around curves without producing excessive wrinkles is the most difficult task for installers of paving synthetics. However, with the proper procedures, it can be accomplished with ease. Attempts should not be made to roll the geosynthetic around a curve by hand. It may wrinkle too much. Placing fabric around a limited curve with machinery is preferable. Some minor wrinkles may occur. Grids have low elongation and thus do not stretch in curves. In most cases, the grid will need to be installed by hand or in short sections by machine to avoid wrinkles (Barazone, 2000).

Excess tack coat, which bleeds through the paving fabric, is removed by spreading the hot mix, or sand should be spread over it. Any traffic on the geosynthetic should be carefully controlled. Sharp turning and braking may damage the fabric. For safety reasons, only construction traffic should be allowed on the installed paving fabric.

*Step 4 – Overlay placement:* All areas with paving geosynthetic placed are paved on the same day. In fact, the asphalt concrete overlay construction should be done immediately after the placement of paving geosynthetic. Asphalt can be placed by any conventional means. Compaction should take place immediately after dumping in order to ensure that the different layers are bonded together.

The temperature of asphalt concrete overlay should not exceed about 160 °C to avoid damage to the paving fabric. Overlays should not be attempted with its temperature less than 120 °C and air temperature less than 10°C. Adequate overlay thickness, if used, generates enough heat to draw the tack coat up, into and through the paving fabric, thus making a bond. In fact, the heat of the overlay and the pressure applied by its compaction force the tack coat into the paving fabric and complete the process. If sufficient residual heat after compaction is not present, the bonding process is disrupted, resulting in slippage and eventual overlay failure. Thickness of the asphalt overlay should not be less than 40 mm. Compacting the asphalt concrete immediately after placement helps to concentrate the heat and supply the pressure to start the process of the bitumen moving up into and through the fabric. This is very important when using a thinner overlay as it cools more rapidly. In cold weather, a thicker overlay may be necessary to achieve the same objective.

A paving fabric interlayer can also be used beneath the seal coat or other thin surface applications. In such applications, sufficient heat is not there to reactivate the asphalt sealant. Therefore, the installed paving fabric must be trafficked or rolled with a pneumatic roller to push the fabric completely into the asphalt sealant. Sand can be applied lightly to avoid bitumen tackiness during trafficking. Once the paving fabric has absorbed the asphalt sealant, the seal surface treatment is applied exactly as it would be over any road surface.

It is suggested that the first-time users of paving fabric interlayer should obtain help from the paving fabric manufacturers and installers, keeping in view the site and material variables.

### 5.3.4 Case studies

Shukla *et al.* (2004) described the use of a geotextile layer between the soil subgrade and the granular subbase course in the paved roadway project, about 76 km long, undertaken in India in the Varanasi zone. In the whole length of the roadway, the bituminous pavement construction was completed with nonwoven geotextile layer at the subgrade level. This project formed a part of the Golden Quadrilateral connecting four metropolitan cities of India, namely, Delhi, Mumbai, Chennai and Kolkatta with a total length of about 6000 km. The National Highway, NH-2, in Varanasi zone, lies broadly in flat to rolling terrain. The entire section traverses the flat flood plains of the Ganga and the Sone rivers. During the normal monsoons, the drainage does not appear to pose any serious problem. In urbanized areas, where the road level is generally the same as the habitation on both sides and is sometimes slightly lower, water flows on the pavement/shoulders. The area faces severe flood situation once in 10 to 15 years, resulting in blockage of the traffic. Such floods were experienced during the years 1971, 1978, 1987 and 1996. The project area has a tropical climate. The mean annual rainfall in the area is 1500 mm, of which 80 percent falls during the monsoons (mid-June to end of September). The project area is mainly covered by the Indo-Gangetic alluvium. The soil subgrade consists mainly of the fine-grained clayey/silty materials with soil class CL (silty clays with low compressibility) according to the Indian soil classification system (A-4 to A-6 according to AASHTO classification system). The plasticity index of the soils is in the range of 0–22%. However, the major soil samples tested have a plasticity index in the range of 8–14%. Soils have negligible swelling characteristics.

The carriageway, in general, is two-lane with 8.75 m (including 1.5 m shoulder) width except when it passes through the urban centres where the width is less than 8.75 m. The shoulders are unpaved for most of the length of the roadway and are of varying widths in the range 1.0–1.5 m. The height of embankment is generally less than 1 m. A level ground is mostly available in the urban sections.

PP continuous filament needle-punched nonwoven geotextile (Polyfelt TS50) was selected to be installed at the subgrade level (see Fig. 5.15) to function as a separator and/or drainage medium. This geotextile has the following properties: mass per unit area = 212 g/m²; thickness under pressure 2 kPa = 1.99 mm; tensile strength (MD/CMD) = 19.5/12.3 kN/m; tensile elongation (MD/CMD) = 35/108%; CBR puncture resistance = 2465 N; apparent opening size < 0.075 mm; and permittivity = 0.0198s⁻¹. The aim of geotextile application was to prevent intrusion/pumping of the subgrade soil particles into the subbase/base course. It would also intercept and carry water in its plane to side drains on either side of the pavement. These two functions were intended to improve the overall performance of the paved roadways and increase their operating life.

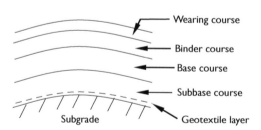

Figure 5.15 Typical cross-section of the asphaltic/bituminous pavement with a geotextile layer at the subgrade level.

The pavement design was carried out for flexible pavements (bituminous pavements) without considering any reduction in the granular layers due to presence of the geotextile layer. The design was based on the design traffic calculated on the basis of the traffic and axle load surveys conducted in the feasibility phase. The values of vehicle damage factor (VDF) for single axles adopted for the design were 6.83 in the direction Varanasi-Aurangabad and 9.0 in the opposite direction.

The roadway was designed with the following typical main characteristics:

- design speed = 100 km/h
- number of two-lane carriageways = 2
- width of median between carriageways = 1.5–5 m
- lane width = 3.5 m
- width of paved shoulder = 1.5 m with the same pavement as the carriageway
- width of earthen shoulder outside the paved shoulder = 1 m
- cross-slope of paved areas = 2.5%
- height above the flood level = 1.5 to 2.0 m
- minimum radius of curve = 360 m (Almost all curves had a radius of more than 500 m.)
- California bearing ratio (CBR) of the subgrade = 5.0%
- design load = 150 million equivalent standard axles (ESALs)

The construction started with excavation work at the areas where the levels were higher than the subgrade levels. Unsuitable materials (debris, loose material, boulders, etc.), if found, were disposed of. Embankment works were carried out using crawler dozer, motor grader, vibration roller and water trucks. Prior to the filling works, batter peg markers were installed to indicate the limit of embankment/subgrade at the regular intervals of 50 m as a guide. The suitable material obtained from excavation and approved borrow area was used at the embankment areas by spreading the material in layers (not exceeding 200 mm), with crawler dozer/motor grader, and compacting it to 95% of the maximum dry density (MDD) using a vibratory roller up to 500 mm below the subgrade level. For upper portion, it was compacted up to 97% of MDD. Upon completion of the embankment filling up to the subgrade level, the side slopes were shaped and trimmed. The subgrade was provided with a cross-slope of 2.5% prior to placement of the geotextile layer. As per the design calculations, four different layers were provided on the geotextile layer. The order of these layers from the bottom layer was as follows: 210-mm thick granular subbase course (GSBC), 250-mm thick wet mix macadam (WMM), 190-mm thick dense bituminous macadam (DBM), and 50-mm bituminous concrete (BC).

Figure 5.16(a) shows the installed geotextile layer along with the spreading operation of the granular subbase material over it by the motor grader. The compacted subgrade surface as well as the sewn joints in geotextile layer can also be observed in this figure. Figure 5.16(b) shows the compaction of the granular subbase layer by the vibratory roller. The granular subbase course after the final compaction by rubber-tyred roller is shown in Fig. 5.16(c).

A portion of the roadway was completed at the end of March 2003 and opened for traffic (see Fig. 5.16(d)). By using a nonwoven geotextile layer as separation and drainage layer, the pavement layers were constructed conveniently and economically

*Figure 5.16* National Highway, NH-2, Varanasi zone, India: (a) a nonwoven geotextile layer with the granular subbase material being spread over it; (b) compaction of the granular subbase material by a vibratory roller; (c) granular subbase course after final compaction by a rubber-tyred roller; (d) the completed roadway opened to traffic (after Shukla *et al.*, 2004).

even over poor soil subgrades. Due to the permanent separation, the settlements and the balance of earthworks quantity were predicted more accurately. The geotextile layer contributed to the construction speed of the granular lifts without any local shear failure of the soil subgrade. The author has observed that the roadway has been functioning well structurally as well as functionally without any major problem.

## 5.4 AIRFIELD PAVEMENTS AND PARKING LOTS

Geosynthetics are used in airfield pavements and parking lots for their improvement in the same way as described in the previous sections for road pavements. One of the basic differences lies in the fact that airfield pavements and parking lots are wider than road pavements. In the wide pavements with free-draining bases, the bases must be tied into an effective edge-drain or underdrain system to help drain the water quickly. Another simple solution is to keep the water from entering the pavement base from the start using a paving fabric interlayer. When properly installed, the paving fabric interlayer keeps the water out of the pavement base for the maximum pavement life. It is very popular for wide pavements, especially airfields and parking lots, where the path to under-drains or edge drains can be distant. The use of paving fabric interlayer is based on the fact that it is much easier to handle the surface water than the water in the pavement base.

## 5.5   RAILWAY TRACKS

### 5.5.1   Basic description

Railway tracks serve as a stable guideway to trains with appropriate vertical and horizontal alignment. To achieve this role, each component of the track system (see Fig. 5.17) must perform its specific functions satisfactorily in response to the traffic loads and environmental factors imposed on the system.

Geosynthetics play an important role in achieving higher efficiency and better performance of modern-day railway track structures. They are nowadays used to correct some track support problems. Acceptance and use of geotextiles for track stabilization is now common practice in the USA, Canada and Europe. Geotextiles are also being used in high maintenance locations such as turnouts, rail crossings, switches, and highway crossings. One of the most important areas served by geotextiles is beneath the mainline track for stabilization of marginal or poor subgrade, which can suffer from severe mud-pumping and subsidence.

In normal static ground conditions, such as in the standard drainage, the cohesive nature of clay and silty clay soils allows the soil particles to bridge over the fine-sand size pores, while allowing water to filter through. Under the dynamic conditions caused by pulsating train loading, the action known to the engineers as pumping occurs beneath the ballast. This is a phenomenon where clay and silt soil particles are washed upwards into the ballast under the pressure from train loading. The subgrade mud-pumping and the load-bearing capacity failure beneath the railway tracks are problems that can be handled by the use of geotextiles, geogrids, and/or geomembranes at the ballast-subgrade interface (see Fig. 5.17). The design difficulty lies in the choice of the most suitable geosynthetic.

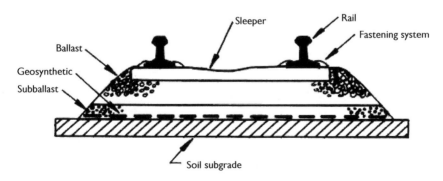

*Figure 5.17* Components of a railway track structure.

Note that not all fines originate from the ground below. As the ballast ages, the stone deteriorates through abrasive movement and weathering, producing silty fines which reduce the performance of the ballast until it needs cleaning or replacing.

Four principal functions are provided when a properly designed geosynthetic is installed within the track structure. These are

- separation, in new railway tracks, between the soil subgrade and the new ballast;
- separation, in rehabilitated railway tracks, between the old contaminated ballast and the new clean ballast;
- filtration of soil pore water rising from the soil subgrade beneath the geosynthetic, due to rising water conditions or the dynamic pumping action of the wheel loadings, across the plane of the geosynthetic;
- lateral confinement-type reinforcement in order to contain the overlying ballast stone;
- lateral drainage of water entering from above or below the geosynthetic within its plane leading to side drainage ditches.

The separation function of the geotextile in railway tracks is to prevent the ballast, which is both expensive and difficult to replace, from being pushed down into the soil subgrade and effectively lost. Similarly, the geotextile needs to prevent the soil subgrade working its way up into the ballast, contaminating it, and causing loss of ballast effectiveness.

The drainage has been found to be the most critical aspect for achieving the long-term stability in the railway track structure. Excess moisture in the track is found to reduce the subgrade strength and provide easy access for soil fines to foul the open ballast. The drainage function of the geotextile is to allow any ground water within the subgrade to escape upwards and through the geotextile towards the side drains (Fig. 5.18). If water is trapped beneath a geotextile, it may weaken the soil foundation to a very significant extent – which is why a geotextile is made permeable. A well-engineered geotextile allows groundwater to escape upwards easily, involving a reduc-

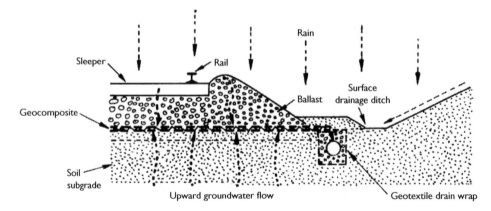

*Figure 5.18* A typical drainage system in railway tracks (after Jay, 2002).

tion in excess porewater pressures generated from the repeated applied axle loads of a passing train. At the same time, during rainfall, the downward flow of water is encouraged by the geotextile to be shed into trackside drains. If the water table is not available at perhaps 600 mm or more below the formation, then the upward release of water pressure offered by a permeable geotextile is not required. Instead, an impermeable membrane can be used, because it keeps the rain water from reaching the soil subgrade, as well as separating the soil and ballast.

Railway track specifications seem to favour relatively heavy nonwoven needled-punched geotextiles because of their high flexibility and in-plane permeability (transmissivity) characteristics. The logic behind the high flexibility is apparent, since the geotextiles must deform around relatively large ballast stone and not fail or form a potential slip plane. In-plane drainage itself is not a dominant function, because any geotextile that acts as an effective separator and filter would preserve the integrity of the drainage of the ballast. Note that the geotextiles, generally installed at 300-mm depth below the base of the sleepers/ties, are likely to be damaged during any subsequent rehabilitation so that the geotextile life only needs to extend the full rehabilitation cycle (Raymond, 1999).

The extension of existing railway routes or construction of new rail routes, to take higher axle loads as well as higher volume and speed of traffic, requires that the load-carrying system should be strengthened. In particular, the load-carrying capacity of the railway subballast must be increased. A subgrade protective layer between the subgrade and the subballast can bring about an increase in load-carrying capacity and reduce settlement. A high-strength geogrid works effectively as a subgrade protective layer, even at small deformations, because they can absorb high tensile forces at the low strain. The geogrid layer has a stiffening effect on the track structure, reducing the rate of track settlement to that approaching a firm foundation. The elastic deflections are also reduced, smoothing out the variable track quality. If the construction of a railway track project has to be completed in a very limited time-span, then in the present-day track construction technology, use of geogrid layers is the best option.

Excess water may create a saturated state in ballast and subballast and cause significant increases in track maintenance costs. Because each source of water requires different drainage methods, the sources must be identified in order to determine the effective drainage solutions. The use of geotextile-wrapped trench drains (i.e. French drains) and fin drains can provide rapid and cost-effective solutions for the requirements of subsurface and side drainage in railway tracks. The details of the drainage system are already described in Section 4.2 of Chapter 4.

Note that the function of a geosynthetic beneath a railway track is fundamentally different from that within highway and railway pavements as described in the previous sections. The following essential differences must be kept in mind while designing the railway track structure (Tan and Shukla, 2012):

- The ballast used to support the sleeper is very coarse, uniform and angular.
- The regular repeated loading from the axles can set up resonant oscillations in the subgrade, making the wet subgrade with fine soils very susceptible to mud-pumping.
- The rail track system produces long-distance waves of both positive and negative pressure into the ground ahead of the train itself.

## 5.5.2 Analysis and design concepts

Geosynthetics in the railway tracks are designed to perform several functions: separation, filtration, drainage and reinforcement/confinement. Keeping these functions in view, the following design procedure can be followed for the design of a geotextile layer (Tan and Shukla, 2012).

- Design the geotextile as a separator – this function is always required. Burst strength, grab strength, puncture resistance and impact resistance should be considered.
- Design the geotextile as a filter – this function is also usually required. The general requirements of adequate permeability, soil retention and long-term soil-to-geotextile flow equilibrium are needed as in all filtration designs. Note, however, that the railway loads are dynamic; thus the pore water pressures must be rapidly dissipated. For this reason, a high permittivity is required.
- Consider the geotextile flexibility if the cross-section is raised above the adjacent subgrade. Here a very flexible geotextile is an advantage in laterally confining the ballast stone in its proper location. Quantification of this type of lateral confinement is, however, very subjective.
- Consider the depth of the geotextile beneath the bottom of the sleeper/tie. The very high dynamic load of railway, acting on the ballast, imparts accelerations to the stone that gradually diminish with depth. If the geotextile location is not deep enough, it will suffer from abrasion at the points of contact with the ballast. The studies conducted at the Canadian rail sites indicate that the major damage occurs within 250 mm of the tie, and at depths greater than 350 mm, damage is not noticeable (see Fig. 5.19). From this information, it can be safely concluded

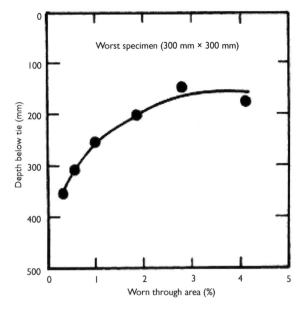

*Figure 5.19* Depth below the tie/sleeper base of the exhumed geotextile versus the damage assessment in terms of complete worn-through area from Canadian rail sites (after Raymond, 1999).

that the minimum depth for geotextile placement is 350 mm for abrasion protection. If this depth is excessive, a highly abrasion-resistant geotextile must be used. An example of abrasion damage to geotextile due to inadequate ballast thickness is shown in an exhumed geotextile in Fig. 5.20.

- The last step is to consider the survivability of geotextile during installation. To compact the ballast under ties, the railroad industry uses a series of vibrating steel prongs forced into the ballast. Considering both the forces exerted and the vibratory action, high geotextile puncture resistance is required. Hence it is necessary to keep the geotextile deep or to use special high puncture resistant geotextile.

Note that the selected geotextile must meet the following four durability criteria:

- Tough enough to withstand the stresses of installation process. The properties required are tensile strength, burst strength, grab strength, tear strength and resistance to UV light degradation for two weeks exposure, with negligible strength loss.
- Strong enough to withstand static and dynamic loads, high pore water pressures and severe abrasive action to which it is subjected during its useful life. The properties required are puncture resistance, abrasion resistance and elongation at failure.
- Resistant to excessive clogging or blinding, allowing water to pass freely across and within the plane of the geotextile, while at the same time filtering out and retaining fines in the subgrade. The properties required are cross-plane permeability (permittivity), in-plane permeability (transmissivity) and apparent opening size (AOS).
- Resistant to rot, and attacks by insects and rodents. It must be resistant to chemicals such as acids and alkalis, and spillage of diesel fuel.

*Figure 5.20* Abrasion failures of geotextiles placed too close to the track structures (after Raymond, 1982).

*Table 5.4* Properties of geotextiles recommended by AREA (American Railway Engineering Association).

| Test Methods | Regular | Heavy | Extra Heavy |
|---|---|---|---|
| Puncture resistance, N | 500 | 675 | 900 |
| Abrasion resistance, N | 675 | 810 | 1080 |
| Grab strength, N | 900 | 1080 | 1440 |
| Elongation, % | 50 | 50 | 50 |
| Trapezoidal tear strength, N | 450 | 540 | 720 |
| Cross-plane permeability, cm/s | 0.2 | 0.2 | 0.2 |
| Permittivity, l/s | 0.5 | 0.4 | 0.3 |
| In-plane transmissivity, m²/min × 10⁻⁴ | 2 | 4 | 6 |
| AOS (Apparent Opening Size), US standard sieve, microns | 70 | 70 | 70 |

Note: The properties are measured as per the relevant ASTM standards.

A standard specification for the use of geotextiles in railway track stabilisation has been developed and published by the American Railway Engineering Association (AREA, 1985). The specification recommends the minimum physical property values for three categories of nonwoven geotextiles: regular, heavy and extra heavy. The selection of one of these, based on the subgrade conditions, is somewhat subjective. Therefore, many designers recommend the heavy and extra heavy geotextiles, as the cost of geotextiles is small compared to the overall cost of track rehabilitation work being done at the time of installation.

Table 5.4 provides the index properties recommended by AREA for average roll values that should be considered when specifying geotextiles for railway tracks.

Raymond (1982, 1983a,b, 1986a), and Raymond and Bathurst (1990) evaluated a number of exhumed geotextiles beneath Canadian and US railroads and found that many were pockmarked with abrasion holes. Their studies examined the soil fouling content, change in permeability ratio, change in geotextile strength and the like geotextile properties. The primary functions/properties of railway rehabilitation geotextile have been established as separation, drainage, filtration, abrasion resistance and elongation. Based on extensive laboratory tests on both unused and exhumed geotextiles from the railway track installations in Canada, and the USA, the following manufacturing specifications for the geotextiles were recommended for railway rehabilitation works (Raymond, 1999):

- *Mass* – 1050 g/m² or greater for track rehabilitation without the use of capping sand;
- *Type* – needle-punched nonwoven with 80 penetrations per square centimetre (80 p/cm²) or greater;
- *Fiber size* – 0.7 tex or less;
- *Fiber strength* – 0.4 newton per tex (N/tex) or greater;
- *Fiber polymer* – polyester (PET);
- *Yarn length* – 100 mm or greater;
- *Filtration opening size* – 75 microns or less;

- In-plane coefficient of permeability – 0.005 cm/s or greater;
- *Fibre bonding by resin treatment or similar* – not less than 5% nor more than 20% by weight of low modulus acrylic resin or other suitable non-water soluble resin that leaves the geotextile pliable;
- *Elongation* – 60% or more to ASTM D4632/4632M-08 (ASTM, 2013a);
- *Colour* – must not cause 'snow blindness' during installation;
- *Abrasion resistance* – 1050 g/m² geotextile must withstand 200 kPa on 102 mm burst sample after 5000 revolutions of H-18 stones, each loaded with 1000 grams of rotary platform double head abraser (ASTM D3884–09 (ASTM, 2013b));
- Width and length without seaming—to be specified by client;
- *Seams* – no longitudinal seams permitted;
- *Wrapping* – 0.2 mm thick black polyethylene or similar;
- *Packaging* – must be weatherproofed and clearly identified at both ends stating manufacturer, width, length, type of geotextile and date of manufacture.

Note that the surficial inspection of Canadian tracks in 1980–81 showed that all the geotextiles installed in the previous five years with a mass per unit area less than 500 g/m² had failed. Nonwoven geotextiles with a mass per unit area greater than 500 g/m² showed considerably less distress. These assessments were confirmed by excavations of failed geotextiles at several locations. Finally, a mass per unit area of 1050 g/m² was found to be desirable for a rehabilitation geotextile, placed on the undercut surface, to remain durable so as to continue to function as separation and filtration layer. The geotextiles installed without a capping of sand and meeting the specifications given above have shown excellent durability after 18 years of service in the very physically harsh environment of the North American tracks (Raymond, 1999).

Jay (2002) has reported that the use of geotextiles directly on clay and silt soils beneath the ballast may slow the pumping process, but will not prevent it altogether. In this situation, it has been recommended to use a blanket of well graded, fine-to-medium sand subballast overlain by a geotextile (Fig. 5.21). The fine sand filters the

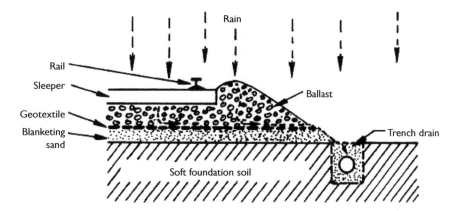

*Figure 5.21* A typical cross-section of railway track showing the use of a geotextile and sand blanket below the ballast.

silty clay subgrade, and the geotextile filters the sand, preventing intermixing with the ballast. A sand blanket layer of 15 mm will form an effective filter, but for practical construction reasons, a thickness of not less 50 mm is usually specified. This solution is safe even over the wet ground, over a high water table, or even over the artesian groundwater conditions which can occur, for example in railway cuttings. The benefit of the geotextile is to reduce the need for a deep layer of blanketing sand, thus reducing the cost.

### 5.5.3 Application guidelines

For the optimum performance of a geotextile layer, it must be properly installed. The geotextile can be installed under existing tracks in a number of ways, but is usually placed in conjunction with undercutting, plowing, or sledding operations as described by Walls and Newby (1993). In some instances, the track sections are removed by the crane during rehabilitation of the trackbed, with geotextiles being installed at the same time.

The following points must be considered during the installation (Raymond 1986b, 1999; Tan and Shukla, 2012):

1   The surface, over which the geotextile is being placed, should be prepared and contoured to remove debris and trackbed irregularities, with cross-fall gradients to facilitate the drainage of water from the track centreline to adjacent ditches and drains.
2   When joining the geotextiles, an overlap of at least 0.5 m is recommended.
3   The geotextile should be placed so that water entering the geotextile can drain away from the track.
4   Geotextiles should be installed at a depth of not less than 200 mm, and preferably 300 mm, below the tie/sleeper base and that tamping should not be permitted until that depth of ballast is in place. This is to prevent damage from normal tamping operations as shown in Fig. 5.22. Ensuring sufficient ballast depth can be obtained through the use of sand bags filled with ballast.

*Figure 5.22* Installation of geotextile and sledding of subballast (after Tan and Shukla, 2012).

5   Geotextile installations should be provided with day-lighted French drains on both sides of the track with inverts 15 mm below the load-bearing surface and the geotextile edges should be turned down into the French drains. If an outlet is not provided for the internal drainage, the geotextile will drain water into the load-bearing area, resulting in a worse condition than that obtained when the geotextile has not been used at all.

6   In order to allow a smooth transition from the rehabilitated track with geotextile to the unrehabilitated areas, it is highly recommended to provide a transition zone of about 6 m where the track is undercut and not provided with a geotextile. This will decrease the sudden change of the track modulus and reduce the associated dynamic traffic loadings.

## 5.5.4   Case studies

### Case study 1

Walls and Newby (1993) reported a case study on the use of geosynthetics for railway track rehabilitation in Alabama, USA. In May 1983, the existing track was inspected by the railway personnel and consultants who identified several problems that needed to be corrected, namely: increase the load-bearing capacity of the soil subgrade, prevent contamination of the ballast with subgrade fines and dissipate the high pore water pressures built up by cyclic train loading. Since the geosynthetics are used separately to solve each of these problems, a combination of geotextiles and geogrids was considered to maximize the benefits of the two products in the following ways:

- geogrid to provide tensile reinforcement and shear resistance to increase the effective load-bearing capacity of the subgrade;
- geogrid to interlock with and confine the ballast, increasing its resistance to both vertical and lateral movement;
- geotextile for separation and filtration to prevent contamination of the ballast and providing quick relief of porewater pressures.

The design called for the removal of the fouled ballast to a depth of approximately 0.3 m below the base of the tie/sleeper. Rather than disposing of the old ballast, it was to be placed along the edges of the embankment to widen the existing shoulders. Next a 380 g/m² nonwoven geotextile was to be placed on top of the remaining subballast followed by a geogrid placed directly on top of the geotextile, and 0.3 m of new ballast. The specifications for the geotextile required that it should be a nonwoven needle-punched engineering grade fabric comprising PET fibres with a linear density of 0.8 to 1 tex per filament. The minimum fibre tenacity was to be 4 mN/tex and the minimum fibre length was to be 150 mm. The geotextile selected for this project was Quline 160 manufactured by Wellman Quline. The geogrid specified was Tensar SS2 geogrid manufactured by the Tensar Corporation. The important properties of the Tensar geogrids made it suitable for ballast reinforcement. These geogrids have open grid geometry for interlock capability, high junction strength to resist lateral deformation of the grid cross members and high tensile modulus to reinforce at low strain levels.

Although the design thickness of ballast was selected arbitrarily at the time, recent developments in geogrid design technology for railway track bed indicate that about 600 mm of reinforced ballast and subballast would be required for a weak subgrade having a CBR value of approximately 1% compared to 1 m of unreinforced ballast and subballast. However, since the actual depth of existing ballast and subballast was considerably greater than 300 mm, it was assumed that the remaining ballast and subballast provided enough support for the new reinforced ballast while fulfilling the total reinforced thickness requirement to prevent overstressing of the weak subgrade. Although other stabilization alternatives were proposed, it was decided to use the geotextile-geogrid combination because of its relatively low cost (just one sixth of the estimated total cost of the relocation alternative) and the proven performance of geogrids in other reinforcement applications. Construction of the reinforced track was carried out in December, 1983. Following an initial observation period of approximately 3 months in which it was determined that the reinforced track structure was performing satisfactorily, it was decided to gradually increase the allowable speed up to 56 km/h and ultimately to 80 km/h. The rehabilitated track was reported to be in service for four years without any reoccurrence of stability problems and only routine track maintenance had been required. This case study has supported the results of large-scale laboratory tests on tie-ballast loading that the geogrids can increase the life of ballasted track between the maintenance cycles, particularly over weak soil subgrades.

### Case study 2

Fernandes *et al.* (2008) presented a study on the use of geosynthetic reinforcement to reduce the consumption of good-quality sub-ballast material in a railway track in the state of Minas Gerais, Brazil. Six test sections on the railway were instrumented and monitored for a period of 2 years. Nonwoven geotextile and a geogrid were used as reinforcement in different positions in the subballast. Six experimental railway test sections, each 25 m long, were constructed as part of the research programme. These sections were localised in Line 2 of the Variante Capitao Eduardo, Vitoria–Minas railway, between kilometres 40 and 42. This railway is intensely trafficked, with approximately 16 compositions (2 locomotives of 160 t plus 100 wagons with approximately 100 t each: 300 kN axle load) daily. The steel sleepers were of type UIC865, with dimensions 2.2 m × 0.3 m × 0.02 m. The rail type was TR 68.

The results of in situ tests and strain measurements showed that the presence of the geosynthetic reinforcement reduced the compressibility of the system. Reduced breakage of the ballast material and greater abrasion resistance were observed in the test sections with geosynthetics. This can be attributed to the greater lateral confinement provided by the reinforcement layer. Similar track performance was observed for the two types of geosynthetic tested. However, significant mechanical damage was observed in the geotextile used after 600 days in service. This suggests that the long-term performance of the geotextile might be severely compromised by increasing the mechanical damage, and that the use of a thicker, more durable geotextile would have been more appropriate. The results obtained validated the use of an alternative, low-cost, sub-ballast material in combination with geosynthetic reinforcement. This solution can be attractive for the mining industry in situations where reasonably good quality mining wastes are plentiful, and conventional track construction materials are scarce or expensive.

## Chapter summary

1   In road constructions, when a geosynthetic layer is placed at the interface of the granular layer and the soil subgrade, the primary function of the geosynthetic is reinforcement if the soaked CBR of the soil subgrade is lower than 1, separation if the soaked CBR of the soil subgrade is higher than 3, and stabilization if the soaked CBR of the soil subgrade lies between 1 and 3. The stabilization function of the geosynthetics is an interrelated group of separation, filtration and reinforcement functions.

2   The commonly used design methods for unpaved roads are reinforcement function design method (RFDM), separation function deign method (SFDM) and modified CBR design method. These methods are based on several assumptions. The design method like SFDM, which assumes no reinforcing effect, is generally conservative.

3   At no time equipment should be allowed on the geosynthetic with less than 150 mm of granular fill between the wheels and the geosynthetic.

4   The edges of the geosynthetic layer must be attached to the side drains for allowing the water at the subgrade level to join the drain while moving along the plane of the geosynthetic layer

5   The principal reinforcing mechanism of the geosynthetic in paved roads is confinement effect, not the membrane effect, which is applicable for unpaved roads allowing a large rutting.

6   A geosynthetic layer, especially a geotextile layer, is used beneath the asphalt overlays, ranging in thickness from 25 to 100 mm, of the paved roads. The geotextile layer is generally combined with asphalt sealant or tack coat to form a membrane interlayer system known as the paving fabric interlayer, which mainly performs the fluid barrier and cushion functions; the former protects the underlying layers from degradation caused by moisture, and the later retards and control the reflective and other cracks.

7   The most common paving grade geosynthetics are lightweight needle-punched nonwoven geotextiles, with a mass per unit area of 120–200 g/m$^2$. For the full waterproofing and stress-relieving benefits, the paving fabric must absorb at least 725 g/m$^2$ of bitumen.

8   Geosynthetics in railway tracks are designed to perform several functions: separation, filtration, drainage and reinforcement/confinement. The separation function of the geotextile installed at the interface of ballast layer and the soil subgrade in railway tracks prevents the ballast from being pushed down into the soil subgrade and effectively lost. The geogrid within the railway track interlock with and confine the ballast, increasing its resistance to both vertical and lateral movement.

9   The specification for the use of geotextiles in railway track stabilization developed by the American Railway Engineering Association recommends the minimum physical property values for three categories of nonwoven geotextiles: regular, heavy and extra heavy.

10  The geotextile layer should be installed at a depth of not less than 200 mm, and preferably 300 mm, below the tie/sleeper base. It should be placed in a way so that water entering the geotextile can drain away from the track.

## Questions for practice

(Select the most appropriate answer to the multiple-choice questions from Q 5.1 to Q 5.15.)

5.1 The major functions served by the geotextile in unpaved roads are
   (a) separation and filtration
   (b) separation and reinforcement
   (c) reinforcement and filtration
   (d) filtration and drainage.

5.2 If the soil subgrade is soft, that is, the California bearing ratio (CBR) of the soil subgrade is low, say its unsoaked value is less than 3.0, then the geotextile layer at the subgrade level in unpaved roads will primarily function as a
   (a) separator
   (b) filter
   (c) drainage medium
   (d) reinforcement.

5.3 Which one of the following is an incorrect assumption of RFDM for unpaved roads, suggested by Giroud and Noiray (1981)?
   (a) The friction coefficient of the granular layer is large enough to ensure the mechanical stability of the granular layer.
   (b) The friction angle of the geotextile in contact with the granular layer under the wheels is large enough to prevent sliding of the granular layer on the geotextile.
   (c) Thickness of the granular layer is significantly affected by the subgrade soil deflection.
   (d) The granular layer provides a pyramidal distribution with depth of the equivalent tyre contact pressure, applied on its surface.

5.4 When a geotextile, within the unpaved road, gets deformed in a curved shape under loading, the pressure against its concave face is
   (a) equal to the pressure against its convex face
   (b) higher than the pressure against its convex face
   (c) lower than the pressure against its convex face
   (d) equal to or higher than the pressure against its convex face.

5.5 In unpaved roads, the overlaps of parallel geosynthetic rolls should not occur at
   (a) the anticipated main wheel path locations
   (b) the centerline of the roadway
   (c) the shoulder of the roadway
   (d) all of the above.

5.6 In the case of paved roads, the principal mechanism responsible for the reinforcement function of the geotextile is
   (a) shear stress reduction effect
   (b) membrane effect
   (c) confinement effect
   (d) interlocking effect.

5.7  When properly installed, a geotextile layer beneath the asphalt overlay mainly function as
(a)  reinforcement
(b)  fluid barrier
(c)  cushion
(d)  both (b) and (c).

5.8  The most common paving grade geosynthetics are lightweight needle-punched nonwoven geotextiles with a mass per unit area of
(a)  60–120 g/m$^2$
(b)  60–120 kg/m$^2$
(c)  120–200 g/m$^2$
(d)  120–200 kg/m$^2$

5.9  The minimum thickness of bituminous overlays recommended with paving fabrics is
(a)  20 mm
(b)  40 mm
(c)  75 mm
(d)  100 mm.

5.10  Which one of the following is the best and the most economical choice for paving fabric tack coat?
(a)  Paving-grade bitumen
(b)  Cutback
(c)  Emulsion
(d)  None of the above

5.11  To avoid damage to the paving fabric, the temperature of asphalt concrete overlay can be kept a maximum of
(a)  50 °C
(b)  100 °C
(c)  160 °C
(d)  260 °C.

5.12  Most railway track specifications call for
(a)  thin needle-punched nonwoven geotextiles
(b)  thick needle-punched nonwoven geotextiles
(c)  thin thermally bonded nonwoven geotextiles
(d)  thick thermally bonded nonwoven geotextiles.

5.13  For abrasion protection, the minimum depth of ballast below the tie for geotextile placement in a railway track is
(a)  100 mm
(b)  150 mm
(c)  300 mm
(d)  500 mm.

5.14  In railway tracks, when joining the geotextiles, the minimum overlap generally recommended is

(a) 0.3 m
(b) 0.5 m
(c) 1.0 m
(d) 1.5 m.

5.15 Select the incorrect statement related to the application of geotextile-geogrid composites in railway tracks.
(a) Geotextile performs its fluid barrier function to prevent contamination of the ballast.
(b) Geogrid provides tensile reinforcement and shear resistance to increase the effective bearing capacity of the subgrade.
(c) Geogrid confines the ballast through interlocking, thus increasing its resistance to both vertical and lateral movement.
(d) None of the above.

5.16 What are the benefits of using a geotextile layer or layers in the construction of unpaved roads?

5.17 Make a list of the mechanical properties of a geotextile that are of greatest importance when using it as a separator in an unpaved road at the interface of a stone base and relatively soft foundation soil.

5.18 Does the reduction of granular fill thickness in the unpaved road, due to the presence of a geosynthetic layer at the interface of granular-fill and the soft subgrade soil, depend on the traffic loading?

5.19 Describe the basic principles of the following design methods for geosynthetic-reinforced unpaved roads:
(b) Reinforcement function design method
(c) Separation function design method
(d) CBR design method

5.20 Consider:
Number of passes, $N = N_s = 1000$
Single axle load, $P = P_s = 80$ kN
Tyre inflation pressure, $p_c = 480$ kPa
Subgrade soil CBR = 1.0
Modulus of geotextile, $E = 300$ kN/m
Allowable rut depth, $r = 0.3$ m
What is the required thickness of the granular layer for an unpaved road in the presence of geotextile? (Hint: Use RFDM))

5.21 Consider:
Number of passes, $N = 2000$
Single axle load, $P = 45$ kN
Tyre inflation pressure, $p_c = 550$ kPa
Cohesive subgrade soil CBR = 1.0
Allowable rut depth, $r = 75$ mm
What is the required thickness of the granular layer for an unpaved road without geotextile, and with geotextile? (Hint: Use SFDM)

5.22  Consider:
Number of passes, $N = 10000$
Single wheel load, $P = 20000$ lb
Tyre contact area = 12 in. × 18 in.
Subgrade soil CBR = 2.0
Reinforcement ratio, as determined from the modified CBR test, $R = 1.8$
What is the required thickness of the granular layer for an unpaved road without geotextile, and with geotextile?

5.23  What are the effects of contamination of a granular layer from fine-grained subgrade soils on the performance of roadways?

5.24  Describe the steps of geotextile installations in unpaved roads, constructed in the USSR, as reported by Kazarnovsky and Brantman (1993). What improvements were observed with geotextile installations?

5.25  What is MESL? Explain its uses.

5.26  What do mean by unpaved age of a paved roadway?

5.27  Describe the concept of geosynthetic separation in paved roadways.

5.28  What are the different mechanisms of crack propagation in asphalt overlays? How can geosynthetics be beneficial in preventing such cracks?

5.29  For geotextiles used to reinforce paved roads on firm soil subgrades, the geotextile must somehow be prestressed. Can you suggest any method for prestressing the geotextiles for such an application?

5.30  What are paving fabrics? Woven geotextiles are ineffective paving fabrics. Why?

5.31  How will you determine the bitumen retention of a paving fabric for its effective application?

5.32  Regarding the proper quantity of asphalt sealant for geotextile in paved road applications, why is sealant in excess of saturation or less than saturation a problem?

5.33  What are the major causes of paving fabric-related project failures as observed in the past?

5.34  Describe the basic steps in the proper installation of a pavement overlay system with a geosynthetic interlayer.

5.35  Describe the geotextile installation steps at the subgrade level in paved roadways, as reported by Shukla *et al.* (2004). Can you suggest any other improved method of installation?

5.36  Draw a neat diagram to show the components of a railway track with the placement of the geosynthetic layer(s).

5.37  List the functions, in order of priority as per your judgment, that act when the geotextiles are placed beneath the railway track ballast in new railway track construction. Do you feel that the order of priority will change when the geotextiles are used in remediation of existing railway tracks?

5.38  The function of a geosynthetic beneath a railway track is fundamentally different from that beneath a roadway. What are the essential differences to be kept in mind while designing a railway track structure?

5.39  What are the durability criteria to be satisfied by a selected geotextile for use in railway tracks?

5.40 Describe the major manufacturing specifications for geotextiles for use in railway track rehabilitation works.

5.41 What is the minimum depth at which a geotextile should be placed beneath the bottom of a railway track tie to prevent its abrasion? How can you get this value?

5.42 Would you recommend a smooth transition from the rehabilitated railway track with geotextile to the unrehabilitated tracks? If yes, why?

5.43 What are the functions served by geosynthetics in railway track applications, as reported by Walls and Newby (1993)?

## References

AASHTO (American Association of State Highway and Transportation Officials) (1993). *Guide for Design of Pavement Structures*. American Association of State Highway and Transportation Officials (AASHTO), Washington, DC, USA.

AASHTO (2008). *Mechanistic-Empirical Pavement Design Guide – A Manual of Practice*. American Association of State Highway and Transportation Officials (AASHTO), Washington, DC, USA.

AASHTO (2013). *Geotextile Specification for Highway Applications, AASHTO Designation: M 288-06 (2011), Standard Specifications for Transportation Materials and Methods of Sampling and Testing*. Thirty-third Edition, Part 1B: Specifications M 280-R 63, American Association of State Highway and Transportation Officials, Washington, DC, USA.

AREA (American Railway Engineering Association) (1985). *American Railway Engineering Manual*, Lanham, MD, USA.

ASTM (American Society for Testing and Materials) (2013a). ASTM D4632/4632M-08, *Standard Test Method for Grab Breaking Load and Elongation of Geotextiles*. ASTM International, West Conshohocken, PA, USA.

ASTM (2013b). ASTM D3884-09, *Standard Guide for Abrasion Resistance of Textile Fabrics (Rotary Platform, Double-Head Method)*. ASTM International, West Conshohocken, PA, USA.

ASTM (2014). ASTM D6140-00, *Standard Test Method to Determine Asphalt Retention of Paving Fabrics Used in Asphalt Paving for Full-Width Applications*. ASTM International, West Conshohocken, PA, USA.

Baker, T. (1998). The most overlooked factor in paving-fabric installation. *Geotechnical Fabrics Report*, April 1998, pp. 48–52.

Barazone, M. (2000). Installing paving synthetics – overview of correct installation procedures. *Geotechnical Fabrics Report*, May 2000, pp. 17–20.

Bourdeau, P.L. and Ashmawy, A.K. (2002). Unpaved roads. Chapter 6, *Geosynthetics and Their Applications*, Shukla, S.K., Editor, Thomas Telford, London, pp. 165–183.

Cedergren, H.R. (1987). Drainage of Highway and Airfield Pavements. Krieger.

Chaddock, B.C.J. (1988). *Deformation of road foundations with geogrid reinforcement*. Research Report 140, Department of Transportation, TRRL, Crowthorne, Berkshire, U.K.

Chan, F., Barksdale, R.D. and Brown, S.F. (1989). Aggregate base reinforcement of surfaced pavements. *Geotextiles and Geomembranes*, 8, 3, pp. 165–189.

De Garidel, R. and Javor, E. (1986). Mechanical reinforcement of low-volume roads by geotextiles. *Proceedings of the 3rd International Conference on Geotextiles*. Vienna, Austria, pp. 147–152.

Fernandes, G., Palmeira, E.M. and Gomes, R.C. (2008). Performance of geosynthetic-reinforced alternative sub-ballast material in a railway track. *Geosynthetics International*, 15, 5, pp. 311–321.

Giroud, J.P., Ah-Line, A. and Bonaparte, R. (1984). Design of unpaved roads and trafficked areas with geogrids. *Proceedings of the Symposium on Polymer Grid Reinforcement*, London, pp. 116–127.

Giroud, J.P. and Han, J. (2004a). Design method for geogrid-reinforced unpaved roads. I. development of design method. *Journal of Geotechnical and Geoenvironmental Engineering, ASCE*, **130**, 8, pp. 775–786.

Giroud, J.P. and Han, J. (2004b). Design method for geogrid-reinforced unpaved roads. II. calibration and applications. *Journal of Geotechnical and Geoenvironmental Engineering, ASCE*, **130**, 8, pp. 787–797.

Giroud, J.P. and Noiray, L. (1981). Geotextile-Reinforced Unpaved Road Design. *Journal of the Geotechnical Engineering Division, ASCE*, **107**, 9, pp. 1233–1254.

Fannin, R.J. and Sigurdsson, O. (1996). Field observations on stabilization of unpaved roads with geosynthetics. *Journal of Geotechnical Engineering, ASCE*, **122**, 7, pp. 544–553.

Hirano, I., Itoh, M., Kawahara, S., Shirasawa, M. and Shimizu, H. (1990). Test on trafficability of a low embankment on soft ground reinforced with geotextiles. *Proceedings of the 4th International Conference on Geotextiles, Geomembranes and Related Products*. The Hague, The Netherlands, pp. 227–232

Holtz, R.D., Christopher, B.R. and Berg, R.R. (1997). *Geosynthetic Engineering*. BiTech Publishers Ltd., Canada.

IFAI (Industrial Fabrics Association International) (1992). A Design Primer: Geotextiles and Related Materials. Section 13 Asphalt Overlay. Industrial Fabrics Association International, St. Paul, USA.

IRC (Indian Roads Congress) (2002). IRC:SP:59, *Guidelines for Use of Geotextiles in Road Pavements and Associated Works*. Indian Roads Congress, New Delhi, India.

IRC (2011). IRC:58, *Guidelines for the Design of Plain Jointed Rigid Pavements for Highways*. Indian Roads Congress. Indian Roads Congress, New Delhi, India.

IRC (2012). IRC:37, *Tentative Guidelines for the Design of Flexible Pavements*. Indian Roads Congress, New Delhi, India.

Jay, T. (2002). Improving the performance of geosynthetic materials. *International Railway Journal*, March 2002, pp. 34–36.

Kazarnovsky, V.D. and Brantman, B.P. (1993). Geotextile reinforcement of a temporary road, Smolenskaya region, USSR. In: *Geosynthetics Case Histories* edited by G.P. Raymond and J.P. Giroud on behalf of ISSMFE Technical Committee TC9, Geotextiles and Geosynthetics, pp. 194–195.

Koerner, R.M. (1994). *Designing with Geosynthetics*. Third Edition, Prentice Hall, New Jersey, USA.

Marienfeld, M.L. and Baker, T. (1998). Paving fabric interlayer system as a pavement moisture barrier. *77th Annual Meeting, Transportation Research Board*, DC, National Academy of Sciences.

Marienfeld, M.L. and Smiley, D. (1994). Paving fabrics: the why and the how-to. *Geotechnical Fabrics Report*, June/July, pp. 24–29.

Milligan, G.W.E., Fanin, R.J. and Farrar, D.M. (1986). Model and full-scale tests of granular layers reinforced with a geogrid. *Proceedings of the 3rd International Conference on Geotextiles*. Vienna, pp. 61–66.

Perkins, S.W., Berg, R.R. and Christopher, B.R. (2002). Paved roads. Chapter 7, *Geosynthetics and Their Applications*, Shukla, S.K., Editor, Thomas Telford, London, pp. 185–201.

Rankilor, P.R. (1981). *Membranes in Ground Engineering*. John Wiley & Sons, Chichester, England, 1981.

Raymond, G.P. (1982). Geotextiles for railroad bed rehabilitation. *Proceedings of the 2nd International Conference on Geotextiles*, Las Vegas, USA, pp. 479–484.

Raymond, G.P. (1983a). Geotextile for railroad branch line upgrading. *Proceedings on Geosynthetics Case Histories, ISSMFE*, TC9. pp. 122–123.

Raymond, G.P. (1983b). Geotextiles for railway switch and grade crossing rehabilitation maritime provinces, Canada. *Proceedings on Geosynthetics Case Histories, ISSMFE*, TC9. pp. 124–125.

Raymond, G.P. (1986a). Performance assessment of a railway turnout geotextile. *Canadian Geotechnical Journal*, 23, 4, pp. 472–480.

Raymond, G.P. (1986b). Installation factors affecting performance of railroad geotextiles. *Transportation Research Record*, 1071, TRB, National Research Council of USA, pp. 64–71.

Raymond, G.P. (1999). Railway rehabilitation geotextiles. *Geotextiles and Geomembranes*, 17, 4, pp. 213–230.

Raymond, G.P. and Bathurst, R.J. (1990). Tests results on exhumed railway track geotextiles. *Proceedings of 4th International Conference on Geotextiles, Geomembranes and Related Products*. The Hague, Netherlands, pp. 197–202.

Richardson, G.N. (1997). Swamp roads and ramblings. *Geotechnical Fabrics Report*, May, pp. 17–28.

Shukla, S.K. (2006). Geosynthetic at subgrade level in paved roadways – design practice and installation. *Civil Engineering and Construction Review, New Delhi*, 19, 3, pp. 78–84.

Shukla, S.K. and Chandra, S. (1994). The effect of prestressing on the settlement characteristics of geosynthetic-reinforced soil. *Geotextiles and Geomembranes*, 13, 8, pp. 531–543.

Shukla, S.K., Chauhan, H.K. and Sharma, A.K. (2004). Engineering aspects of geotextile-reinforced roadway of the National Highway, NH-2, Varanasi zone. *Civil Engineering and Construction Review, New Delhi*, 17, 12, pp. 56–61.

Shukla, S.K. and Yin, J.H. (2004). Functions and installation of paving geosynthetics. *Proceedings of the Asian Regional Conference on Geosynthetics, GeoAsia 2004*, Seoul, Korea, pp. 314–321.

Steward, J., Williamson, R. and Nohney, J. (1977). Guidelines for use of fabrics in construction and maintenance of low-volume roads. Report No. FHWA-TS-78-205.

Tan, S.A. and Shukla, S.K. (2012). Railway tracks. Chapter 9, *Handbook of Geosynthetic Engineering*, Shukla, S.K., Editor, ICE Publishing, London, pp. 193–208.

Walls, J.C. and Newby, J.E. (1993). Geosynthetics for railroad track rehabilitation, Alabama, USA. *Proceedings on Geosynthetics Case Histories, ISSMFE*, TC9. pp. 126–127.

## Answers to selected questions

| | |
|---|---|
| 5.1 | (b) |
| 5.3 | (c) |
| 5.5 | (a) |
| 5.7 | (d) |
| 5.9 | (b) |
| 5.11 | (c) |
| 5.13 | (c) |
| 5.15 | (a) |

## Chapter 6

# Mining, agricultural and aquacultural applications of geosynthetics

## 6.1  INTRODUCTION

Several applications of geosynthetics have been presented in Chapters 3, 4 and 5. Those applications are made not only in projects related to different areas of civil engineering, but some of them are also equally applicable to mining, agricultural and aquacultural sectors as the developments in these sectors are heavily based on civil engineering infrastructure. This chapter introduces some additional applications of geosynthetics, which are made specifically in mining, agricultural and aquacultural projects in order to have either improved productivity or lower cost of operations in an environmentally friendly manner. In most of these applications, the geosynthetics get deflected into different shapes, and hence the determination of geosynthetic strain becomes an important design step. This chapter also presents the available explicit analytical expressions for the determination of geosynthetic strain for different deflected shapes as expected in civil, mining, agricultural and aquacultural fields.

## 6.2  MINING APPLICATIONS

### 6.2.1  Areas of applications and types of geosynthetics

Geosynthetics are often used in several mining and mineral processing facilities, and also in environmental mine management. In the mining-related projects, it is essentially required to design and construct structures or systems, called *containments*, to prevent the release of harmful materials into the subsoil and water environment. The major areas of application of geosynthetics in mining engineering include the following:

* Liquid (process solution, drainage water, acid mine drainage, etc.) containments.
* Treatment/process ponds.
* Evaporation ponds used for the reduction of liquid waste.
* Containments of mine tailings and other solid wastes.
* Heap leach facilities for mineral extraction and processing.
* Filtration, dewatering and disposal of mine tailings.
* Mine reclaimed fill structures.
* Surface water diversion structures.

- Environmental projects, including revegetation and erosion control.
- Stabilisation of soft/weak foundation soils for road construction.
- Stabilisation of mine tailings for soil cover placement.

Tailings ponds and the tailings impoundments/dams are the most common structures used as containments. Note that the tailings are the materials left over after the removal of the valuable fraction from the uneconomic fraction of an ore. They are distinct from the overburden, which is basically the waste rock or materials overlying an ore/mineral body that are displaced during mining without being processed. Tailing ponds are the areas of refused mine tailings where the waterborne refuse material is pumped into them to allow the sedimentation/separation of solids from the water. The pond is generally impounded with a dam, known as the tailings impoundment or tailings dam, which typically use local materials, including the tailings themselves, for their construction.

The common types of geosynthetics and their applications in the mining industry are the following (Renken *et al.*, 2006; Lupo and Morrison, 2007):

- Geotextiles: filter layers for underdrains and collection pipes, cushion layers over the geomembrane liners, erosion control, ground stabilisation.
- Geogrids: ground/foundation stabilisation for road construction, stabilisation of mine tailings for soil cover placement, remediation of other mine workings.
- Geonets: drainage layers, leak detection/collection layers.
- Geomembranes: process solution pond liner, heap leach facility liner, tailings impoundment liner systems, encapsulation of waste rock to mitigate acid mine drainage, engineered caps/covers to prevent precipitation from infiltrating into the underlying fill materials to achieve full containment or diversion of liquids.
- Geosynthetic clay liners: encapsulation or covers, barrier layers below the geomembrane liners.
- Geopipes: conveyance of runoff, drainage, process solutions and water, solution recovery, leak detection and monitoring around mine sites.
- Electrokinetic geosynthetics: dewatering of mine tailings.

The polymeric materials of the geosynthetics are the following: PP and PET for geotextiles; HDPE, PET and PP for geogrids; HDPE for geonets; HDPE, LLDPE, PVC, PP for geomembranes; and HDPE, LLDPE and PVC for geopipes.

For most mining facilities, the liners are essential components. A liner may be designed as a single composite system (Fig. 6.1(a)) or a double composite system with a leakage detection layer (Fig. 6.1(b)), depending on the site conditions, properties of the tailings material and the environmental regulations to be followed at the specific site.

A single composite liner system generally consists of a geomembrane liner placed over the compacted soil liner (Fig. 6.1(a)). It is generally used in mine facilities that experience low hydraulic head, typically less than 1–1.5 m, such as heap leach pads, lined waste rock facilities, fresh water ponds, and storm water ponds.

A double composite liner system generally consists of two geomembrane liners, an upper primary geomembrane liner and a lower secondary geomembrane liner, separated by a leak collection/drainage layer (Figure 6.1(b)). The lower secondary

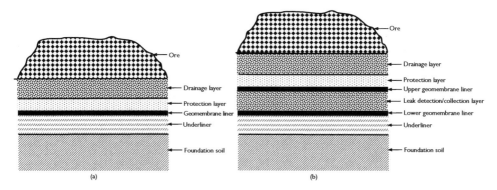

*Figure 6.1* Typical composite liners systems: (a) single composite liner system; (b) double composite liner system.

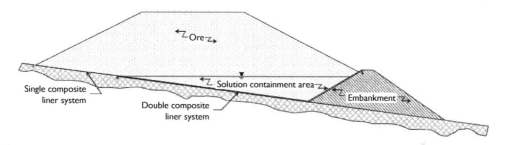

*Figure 6.2* A typical configuration of a valley leach facility (after Lupo and Morrison, 2007).

geomembrane liner rests on a compacted soil liner, called the *underliner*. This liner is generally used in mine facilities where the high hydraulic heads may occur, such as valley leach facilities (Figure 6.2), some tailings impoundments, and process solution ponds.

Note that a liner foundation/bedding soil in direct contact with the geomembrane liner reduces significantly the rate of leakage through the composite liner systems. The *overliner layer* may consist of one or more distinct layers of sand and gravel or silty soil used to protect the geomembrane liner and provide the drainage for solution. A drainage layer consisting of sand and gravel may be a part of the overliner or integrated within it.

For a better performance, the geomembrane liner should not be strained significantly. The liner is generally strained due to the total and/or differential settlements of the foundation/bedding soil. After determining the settlements analytically and/or numerically, the strain in the geomembrane liner due to deflection can easily be determined using the analytical expressions, as presented in Section 6.4. The leak detection layer consisting of sand and gravel helps control the hydraulic head on the lower geomembrane liner in a double composite liner system. The geonet geocomposite may be used as the leak detection layer under low shearing stress conditions. Solid and slotted geopipes are widely used in mining applications as solution collection pipes, leak detection pipes and underdrain pipes. These pipes are subjected to very high stresses due to their burial at greater depths even over 200 m in heap leach pads, tailings, etc.

## EXAMPLE 6.1

A geomembrane liner has to be designed for a heap leach pad/facility. If the maximum height of the ore heap above the liner can be 180 m, determine the vertical normal stress on the liner. Assume the total unit weight of the ore is 17.3 $kN/m^3$. Is this stress much higher than the load-bearing capacity of a typical shallow foundation on a dense sandy ground?

## SOLUTION

Given: Height of the ore heap, z = 180 m; and total unit weight of the ore, $\gamma = 17.3\,kN/m^3$ The vertical normal stress at the liner level is $\sigma_v = \gamma z = (17.3)(180) = 3114$ kPa or **3.1 MPa**

Yes, this stress is much higher than the load-bearing capacity of a typical shallow foundation on a dense sandy ground.

### 6.2.2　Basic concepts of design, construction and application guidelines

Special care is required for the design and construction of mining facilities with geosynthetics because of harsh loading and environmental conditions in the difficult terrains of mine sites. The design and testing methods are often non-standard, that is, the guidelines given in the standards for most civil engineering applications, as described earlier in the previous chapters, may not be applicable in their exact form. Any design, construction, and geosynthetic application procedure with a suitable deviation from the standard procedures should consider the following:

1　In most mining projects, high stresses over the geosynthetics are expected, and hence they should be estimated realistically (see Example 6.1).
2　The thickness and material of the geomembrane liner should be decided on the basis of its performance under the anticipated loads, the coarseness/angularity of the materials in contact, estimated foundation settlement and the material properties (friction, tensile strength, etc.). The stability of the geomembrane should also be checked against the flexibility, chemical resistance, exposure to harsh climatic conditions, and the cost. The theoretical analysis of geomembrane puncture may be carried out using the method suggested by Giroud *et al.* (1995). The final design should be based on suitable liner load tests (ASTM, 2010; ASTM, 2014a) and interface shear testing (ASTM, 2014b). The thickness of the geomembrane liner typically ranges from 1.5 mm to 2.5 mm.
3　Geosynthetics may be exposed to high intensities of solar radiation and exposure to UV, extreme temperatures and high winds, especially for mining projects located well above 3 km elevation, and extreme ranges of leachate properties (strongly alkaline or strongly acidic) from various ore extraction processing (Lupo and Morrison, 2007).

4  The design of mining facilities should meet both the performance requirements and the environmental demands. Most mitigation and control facilities for metal leaching (ML) and acid rock drainage (ARD) must be designed, constructed and operated to perform indefinitely under normal and extreme climatic conditions. In Canada, ML/ARD sites are required to be kept under long-term care and maintenance for at least 100 years. The environmental protection measures should be practical, robust, economical, acceptable to regulatory bodies, sustainable and environmentally sound (Renken *et al.*, 2006). The study by Lange *et al.* (2010) suggests that GCLs may be suitable for short-term containment (<4 years) in an active–passive treatment system for ARD.

5  The selection of geosynthetics must adequately assess and address the potential for chemical and thermal degradations. The impact of mining solutions/liquors may have a significant impact on geosynthetics (Hornsey *et al.*, 2010).

6  The prepared subgrade, that is, the foundation for the geomembrane liner should be firm to avoid large settlements, which may strain the geomembrane liner, thus impacting its performance.

7  The maximum allowable strains in geomembrane liners may range from 4% to 8% for HDPE geomembrane liners, and 8% to12% for LLDPE geomembrane liners (Peggs *et al.*, 2005). These recommended values should be used as a guide only for the initial design. The exact value should be based on suitable tests for the specific geomembrane liners for each project.

8  The liner foundation/bedding soil should be fine-grained, typically with the following characteristics (Lupo and Morrison, 2007):

- maximum particle size – 38 mm
- non-gap graded particle-size distribution
- high fines content (above 15% passing 75 μm)
- Moderate plasticity (plasticity index above 15%)
- saturated hydraulic conductivity of $1 \times 10^{-8}$ m/s or less

9  Fig. 6.3 shows a typical gradation envelope for the underliner soils used on several mining projects. This should be used only for the initial design. For the detailed design, the soil characterisation testing must be conducted for each project site to define the physical and hydraulic properties of the proposed underliner soil.

10  A suitable amount of bentonite, say 2% to 10% by dry weight, may be added to the local soil if required for preparing the foundation/bedding soil for the geomembrane liner.

11  The GCL may be used in place of the liner bedding soil by assessing its stability in terms of creep deformation, compressibility and internal shear strength under the anticipated normal and shear loads.

12  The thickness of the overliner as the protection layer depends significantly on the loading conditions, and is designed using a traditional point or area stress analysis. This thickness ranges from 1 m to over 5 m depending on the stress conditions.

13  The drainage material should be graded to achieve the desired hydraulic conductivity and shear strength properties, and it must minimise the damage to the geomembrane liner. The geopipes with sufficient perforations may be used to increase the drainage capacity.

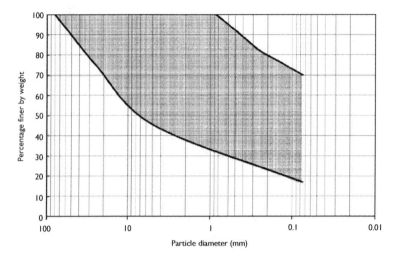

*Figure 6.3* Typical range of underliner soil gradation (adapted from Lupo and Morrison, 2007).

14  The size of the leak detection layer depends on the leakage rates through the primary liner.

15  Fig. 6.4 shows the typical gradation envelopes for the drainage layer used on several mining projects. This should be used only for the initial design. For the detailed design, the drainage material characterisation testing must be conducted for each project site to define the physical and hydraulic properties of material and its compatibility with the geomembrane liner.

16  The design of liner systems over voids or compressible fills requires a large-scale testing, preferably simulating the field conditions.

17  The overliner layer may be designed to have drainage and protection layers, solution collection pipes and air injection pipes.

18  The construction quality assurance (CQA) program, as explained in Chapter 8, should be delivered by a highly specialised team, focussing on the documentation of underliner placement, monitoring of geomembrane liner installation, and monitoring of drainage layer placement.

19  The design of pipes and/or their envelopes requires special care due to their flexibility and compressibility of the surrounding fill under very high stresses. The details of the performance of the HDPE pipe under high fill are presented by Adams *et al.* (1988). For the design, the estimation of loads over the pipes and/or pipe envelops under high fills may be based on the concept of soil arching within the fill (Terzaghi, 1943; Shukla *et al.*, 2009). The load on the pipes may be reduced significantly by including a geosynthetic (woven geotextile or geogrid) reinforcement layer above the pipes as illustrated in Figure 6.5. The load over the pipe and/or envelope may be determined by the analytical expression developed by Shukla and Sivakugan (2013). The pipe design should consider the risk and the potential consequences of the pipe failure (pipe collapse or buckling, etc.). The deflection of the top of the pipe, called the *crown deflection*, is commonly determined as the

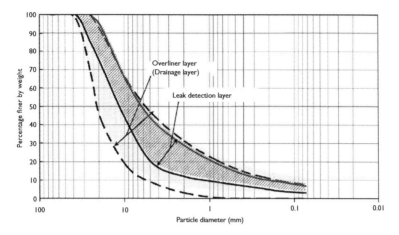

*Figure 6.4* Typical ranges for drainage layer material gradations (adapted from Lupo and Morrison, 2007).

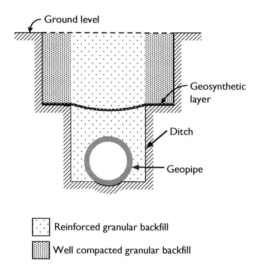

*Figure 6.5* A practical method of reducing the load from the overlying fill over a geopipe or any other conduit (after Shukla and Sivakugan, 2013).

measure of its performance. The crown deflection with respect to its diameter/vertical dimension for critical applications such as process plant pipes and pressured pipes should generally be less than 5%. In other applications, such as non-pressurised flow conditions, as seen in drainage pipes, the pipes may be operational even with crown deflections of greater than 20%, but they may exhibit severe buckling. For the initial design, the pipe envelope thickness should be sized to be two times the pipe diameter (Loop and Morrison, 2007). A large thickness allows arching within the envelope to develop and support a part of the applied load.

20 The heap leach pads utilize an engineered liner system that often is a blend of natural and geosynthetic materials in order to achieve a desired performance, such as ore heap stability, solution drainage, and efficient recovery. The design of heap leach pad liner system must consider the compatibility between geosynthetic materials (e.g. geomembrane, geopipe, geotextile, etc.), native materials, ore, and operational conditions/restrictions (Lupo, 2010).

Renken *et al.* (2006) reviewed some case studies of geosynthetic applications in mining engineering. The heap leaching process is commonly used to extract gold and/ or silver or copper from low-grade deposits. In this process, the ore is piled on a liner, typically a 1.5 mm or 2.0 mm HDPE or LLDPE geomembrane liner, and an extract solution is applied to the surface, which dissolves the metal in the ore. The metal bearing leachate (pregnant solution) is collected above the liner and pumped to a solvent extraction plant to remove and concentrate the metal. Gold is typically extracted with an alkaline extract and copper with a weak acid solution. Most of the heap leach facilities are generally single-lined (Thiel and Smith, 2004). The geomembrane liners are essentially required to be evaluated for the physical and chemical compatibility with the high stresses, and harsh environmental and operating conditions.

Neukirchner and Lord (1998) reported that a biaxial PP geogrid with a 45 cm-thick drainage layer (cobble and sandy gravel layer) was successfully used to cover a historically ponded area of the consolidated tailings pile at the Eagle Mine Superfund site, located near Minturn, Colorado. The covered tailings and wastewater treatment sludge had extremely high water contents (41–737%) and low shear strengths (undrained shear strength of 1.4 kPa–8.6 kPa).

In the present-day construction practice, the mine tailings are used as a construction material even in the construction of tailings storage facilities, including the tailings dams, which store high water content slurries. Many tailings disposal facilities contain materials that undergo extremely slow self-weight consolidation. In several mine operations it has been found that it is not adequate to rely on the self-weight consolidation to achieve the required degree of dewatering, and hence the shear strength gain. There can be a range of dewatering techniques, including installation of prefabricated vertical drains with surcharge loading and electrokinetic dewatering, as pioneered by Casagrande (1949) for accelerating the consolidation of natural soft clays. Electrokinetic dewatering of fine-grained mine wastes has been attempted by Veal *et al.* (2000) amongst others, but with relatively little success. The new generation geosynthetics that are electrically conductive, called the *Electrokinetic Geosynthetics* (EKGs), have been found to be effective in achieving the dewatering of mine tailings with additional advantages over the conventional dewatering methods.

Note that the EKGs extend the traditional functions of geosynthetic materials by incorporating electro-kinetic phenomena. They offer technical benefits over conventional electrodes in that they can be formed as strips, sheets, blankets or three-dimensional structures. They are light and easy to install and can be structured so as not to be susceptible to electro-chemical corrosion, whilst continuing to provide conventional functions of filtration, drainage, separation, reinforcement or to act as impervious membranes. Potential applications of EKG materials demonstrated in laboratory trials include: in situ decontamination of contaminated soils by electro-chemical means; electro-kinetic transport of nutrients into a soil to encourage bioremediation; treatment of industrial waste; the improvement of the performance of lime piles; and improvements in reinforced soil technology and accelerated consolidation of soil using the principle of electro-osmosis (Hamir *et al.*, 2001).

Fourie and Jones (2010) carried out experiments using the EKG, which consisted of a mesh made from a metal wire stringer coated in a conductive polymer. When used as a cathode, the tubular electrode is enveloped in a geotextile sleeve such as a nonwoven, heat bonded product. The great advantage of EKG over the conventional metallic electrodes used in soil improvement applications is that it overcomes the difficulties of corrosion, sufficient electrical contact and physical removal of water from the system (Lamont-Black, 2001). The results of the experiments carried out by Fourie and Jones demonstrate the improved estimates of power consumption during dewatering of mine tailings using EKGs. The filter tests by Hamir *et al.* (2001) show no clogging of the EKGs or loss of material through the EKGs.

The conventional geosynthetic soil reinforcing materials can also be made electrically conductive. The use of EKG reinforcement can increase the shear strength of fill and can provide a significant increase in reinforcement-soil bond strength, suggesting that formerly unsuitable wet fine-grained soils can be used to form reinforced soil structures (Hamir *et al.*, 2001).

Note that long-term field performance data for different mining applications are not yet available; and moreover, the limited warranty (20–30 years) of geosynthetics by the manufacturers does not promote their applications in mining engineering as you can see wide applications of geosynthetics in several areas of civil engineering.

## 6.3  AGRICULTURAL AND AQUACULTURAL APPLICATIONS

*Agricultural engineering* is concerned with the design and development of effective agricultural systems for agricultural production and processing, while *aquacultural engineering* deals with the design and development of effective aquacultural systems for marine and freshwater facilities.

The polymeric textile products used in agricultural applications are often called *agrotextiles*, which are similar to the geotextiles and some other types of geosynthetics. They are used in several applications, such as the following:

- Orchard netting
- Odour control floating covers
- Windbreak
- Groundcover
- Insect screen
- Ventilation screen
- Animal barrier
- Enhancement of farming practices
- Capillary barriers
- Animal waste management
- Liner for manure storage facilities

In most of the above applications, the textile products in the form of net, grid or mesh are exposed to natural weathering over a long period of time while in most applications described in Chapters 3, 4 and 5, the geosynthetics are generally not exposed to natural weather conditions, expect for a limited period during their installation before they are covered with soil and/or other similar materials. Hence it becomes important to assess the resistance to natural weathering of the textile products over a long period of time for their applications in agricultural and aquacultural applications.

In general weathering affects the tensile properties (tensile strength and strain at break) and mass of the PVC-coated materials while the thickness remains unaltered. The tensile properties of UV-protected textiles are much less affected by weathering and the mass and thickness of these materials do not change with time. It has been found that the reduction in tensile properties is the largest during the first year of exposure (Dierickx and Berghe, 2004).

In tropical and warm-temperature climates, the use of anaerobic lagoons as the first pond in waste stabilization pond systems is considered a highly cost-effective and practical way to treat municipal wastewater. While odour is a concern with anaerobic treatment, anaerobic ponds treating municipal wastewater can be designed to be relatively odour-free for sufficiently low wastewater sulphate concentrations. However, when sulphate concentrations are high, or when odour control or greenhouse gas emissions are significant issues, or when the wastewater is relatively high in organic strength resulting in commercial production of methane gas, anaerobic lagoons can be covered with the geomembrane sheet, and the biogas collected and burned both to produce energy and reduce emissions and odour. DeGarie *et al.* (2000) described the design, installation and commissioning of two 3.9-hectares floating, self-draining, geomembrane covers on the anaerobic section of two of these lagoon systems.

In general, geomembrane is used as a liner; however, nonwoven geotextile can be used as the liner in some agricultural applications. Earthen manure storage facilities often require an inexpensive but durable sealing liner. The nonwoven geotextile having a very low porosity can be used for this purpose as the manures seal media of small pore sizes (Barrington *et al.*, 1990).

Use of knitted or woven nets, commonly made from HDPE, is one of the major applications in agricultural engineering for protecting the crop from birds, crop bats, flying foxes, flies, bees, and other pests and insects. The commonly used netting systems are full canopy netting (Fig. 6.6(a)) and tunnel netting (Fig. 6.6(b)) (Rigden, 2008). In the full canopy netting system, the net is held permanently by a rigid structure of poles and tensioned cables over the entire orchard. In the tunnel netting system, a series of light frames connected by wires are erected at intervals along the row to support the net and hold it away from the tree. Appropriate netting can also protect orchards from wind and hail damage. For successful application, it is important to select the appropriate mesh shape (square, diamond or hexagonal) and size (8 to 40 mm). The choice of mesh size is governed by the types of pest(s) to be controlled, changes to orchard microclimate, and the cost. Nets with a small mesh size are generally more expensive than large mesh nets. For hail protection, small mesh nets are selected, and they need additional cost of supporting structures. The netting structure should be designed to withstand the wind loads, the weight of the net when wet and the weight of hail, if expected. The cost of netting is generally the major factor in making the decision of its application. There is always a risk of fire damage to the nets. Netting is generally regarded as an environmentally friendly and socially responsible approach to the problem of pest control. The HDPE nets may last for 10 years. Mass per unit area and thickness are good indicators of strength of a net. White nets tend to bounce the light through the mesh and reduce

(a)                                        (b)

*Figure 6.6* Orchard netting: (a) full canopy netting; (b) tunnel netting (after Rigden, 2008).

the light levels less than the black nets with equivalent mesh size. The net rolls are usually 10-15 m wide, but the may be manufactured to any width up to 50 m.

Geosynthetics are also used in aquacultural projects, such as the following:

• Enclosures (pens and cages)
• Filters for waste removal
• Lagoons and storage ponds

Enclosures (pens and cages) are often made for marine fish farming, especially in relatively less polluted coastal areas, for holding fishes within an enclosed space by maintaining a free exchange of water. A pen refers to a fixed enclosure in which the bottom is formed by the lake or sea bottom while the cage is an enclosure totally enclosed on all, or all but the top, sides by mesh or netting, whether floating at the surface or totally submerged. The pens are much larger and are stationery as their walls are fixed. The mobility of the cage is its most definite advantage over the pen. Moreover, because of the smaller size, the cages are managed easily for various purposes, including breeding and fry production of fishes.

Enclosures can be made with wood and bamboo, as done in the past, but the current practice is to use polymeric/metallic nets/meshes, which have a much longer life-span and permit better water exchange. Most cage designs currently in use are of the floating type. Cages are usually floated in rafts, and either anchored to the lake/reservoir/river/sea bottom, or alternatively connected to shore.

The periodic removal of accumulated sludge in lagoons and storage ponds in the aquaculture sector is an important task in order to increase the available volume as well as to reduce the strong odours. The sludges are effectively contained, dewatered, and stored using the geotextile filtration, which uses high-strength permeable geotextiles with uniquely designed retention properties fabricated into closed geocontainers/geotubes. Typically, the chosen fabric is inert to biological degradation and resistant to naturally encountered chemicals. The geotextile filtration operates in three basic steps: (1) confinement, (2) dewatering, and (3) consolidation (Cantrell *et al.*, 2008). The geotextile weave creates small pores that confine the fine and coarse particles of the contained material. The small pores allow excess water to escape, resulting in reduction of both water content and volume. This volume reduction allows repeated filling of the geocontainer. After the final cycle of filling and dewatering, the retained materials can continue to consolidate because residual water vapour escapes.

The bio-waste solids (fish faeces and unconverted feed) generated by the fish farming operation in the aquaculture cage site is required to be removed at some intervals for the purpose of renovating the water within the culture system. Geotextile bag systems or geotubes using flocculant aids are an efficient means for capturing and dewatering the bio-waste solids in the effluent stream from the recirculating aquaculture systems. The optimized design of the geotextile bag system depends on the following: flow rate, feed rate and solids dewatering time, and fate of the treated effluent (Guerdat et al., 2013). These factors aid in predicting the expected performance and determining the appropriate size of the geotextile bag system. Note that the recirculating aquaculture systems are used throughout the world as a means for culturing aquatic organisms in a controlled environment.

Full-scale geotextile filtration tubes approximately 5.5 m × 30.5 m treating dairy lagoon sludge were evaluated by Worley et al. (2004) and reported to have mass-based separation efficiencies of 97% of total solids. Another full-scale study using two 4.3 m × 15.2 m tubes treating dairy lagoon slurry with alum was effective in retaining 94.7% of total solids (Mukhtar et al., 2007). Note that keeping the tube full by refilling on a regular basis reduces the time to complete the dewatering process.

## 6.4 GEOSYNTHETIC DEFLECTION PROFILES AND STRAIN ANALYSIS

Geosynthetic layers in field applications may deflect under the loads and stresses due to surcharge, overburden, etc. when they are applied normal or inclined to their planes. The deflected shapes may be circular, parabolic or a combination thereof depending on the ground or hanging conditions. For a geosynthetic layer spanning over a cavity, a trench or any depression, a circular or parabolic shape as shown in Fig. 6.7 can be expected; however, the deflected shape of a geosynthetic layer resting on a foundation soil and subjected to a normal stress is not necessarily completely circular or parabolic. To simplify the mathematical derivations, it can be assumed to be a combination of both circular and parabolic curves with inflection points shown in Fig. 6.8, which were observed in the settlement profiles reported in the literature (Jarrett, 1986; Miura et al., 1990; Shukla and Chandra, 1995; Shukla, 1996).

Considering circular and parabolic concave shapes of the deflected geosynthetic with their ends at the same elevation (Fig. 6.7), Giroud (1995) presented expressions for computing the geosynthetic strain. Shukla (1996) observed that in the field applications, the two ends of the geosynthetic may be at different elevations, possibly due to inappropriate installation or as dictated by the site condition. For this field situation, an expression was presented by Shukla (1996) as an extension of the work of Giroud (1995). The expressions presented by Giroud (1995) and Shukla (1996) can only be applied to the situations where the deflected geosynthetic can be described by a concave parabola or a circular arc without any inflection point as shown in Fig. 6.7. For the other possible deflected situation in the

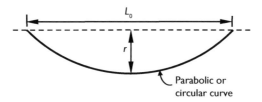

Figure 6.7 Parabolic or circular shape of a deflected geosynthetic (after Shukla and Sivakugan, 2009).

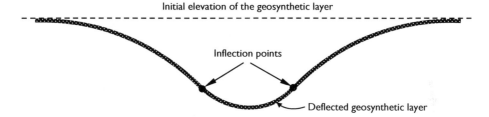

Figure 6.8 A geosynthetic layer deflected under applied normal stress in the field situation (after Shukla and Sivakugan, 2009).

field applications, such as the one shown in Fig. 6.8, where the deflected shape of the geosynthetic is described by two convex parabolas and a circular arc, Shukla and Sivakugan (2009) presented the derivation of an analytical expression for the geosynthetic strain ($\varepsilon$).

As shown in Fig. 6.9, a horizontally installed geosynthetic layer of length $L_0$ within a soil mass elongates due to deflection caused by the applied normal load. If the shape of the deflected geosynthetic is a smooth curve in a cross-section considered, the elongation of the geosynthetic in the cross-section can be determined mathematically by assuming that the shape is a curve with a known equation such as a combination of parabolas ($DGE$ and $AFB$) and a circle ($BCD$). The depth of the inflection points $B$ and $D$ on the deflected geosynthetic, measured from the initial elevation of the geosynthetic layer, is $h$. The length of the horizontal projection of the circular arc of the deflected geosynthetic is $d$, and $r$ is the rut depth or the maximum deflection of the geosynthetic, which is different from the radius $R$ of the circular arc $BCD$. It is assumed that no slippage occurs between the soil and the geosynthetic, and the geosynthetic strain is uniformly distributed. The expression for the geosynthetic strain $\varepsilon [= (L - L_0)/L_0, L$, being the length of the deflected geosynthetic], is presented as:

$$\varepsilon = \frac{h}{L_0} \left[ \sqrt{1 + \left(\frac{L_0 - d}{h}\right)^2} + \frac{1}{\left(\frac{L_0 - d}{h}\right)} \ln \left\{ \left(\frac{L_0 - d}{h}\right) + \sqrt{1 + \left(\frac{L_0 - d}{h}\right)^2} \right\} \right]$$

$$+ \frac{d}{L_0} \left[ \frac{d}{4(r-h)} + \frac{r-h}{d} \right] \sin^{-1} \left[ \frac{1}{\frac{d}{4(r-h)} + \frac{(r-h)}{d}} \right] - 1 \tag{6.1}$$

Equation (6.1) is the general expression for determining the geosynthetic strain for the deflected shape of the geosynthetic with inflection points, as shown in Fig. 6.9. This equation involves four geometric parameters: $L_0$, $d$, $r$, and $h$. Though the ratio $h/r$ has not been extensively reported in the model studies, it has been estimated to be about 0.8 from the graphical results of Jarret (1986). A suitable assumption of $h/r$ will always result in appropriate values of strain for the generalized case; the confidence in selection of $h/r$ will improve with measurement of the same in the experimental studies expected in the future.

Note that the existence of a common tangent at the inflection point has not been considered in the derivation of equation (6.1). In fact, the vertex of the parabolic curve has been assumed at the inflection point for the mathematical simplicity; therefore, the slope of the tangent to the parabola at the inflection point will never be equal to the slope of the tangent to the circular arc at this point. If this condition is considered by shifting the vertex to a suitable point, additional parameters will appear, and the equation (6.1) with new condition will have more than four parameters with replacement of $h$, which is a practically more realistic parameter than the new arbitrary parameters being introduced because of the condition of a common tangent. Since in several applications the geosynthetic remains embedded in the soil mass under the subjected loading, it can be expected that the slope of the tangents may not be exactly identical in most of the practical situations, and therefore, the general equation (6.1) proposed herein is applicable to most of the field situations.

The general equation (6.1) developed can be used to explain a few special cases discussed below.

*Case 1:* Geosynthetic layer deformed into a circular arc
For $d = L_0$, $h = 0$, the deflected shape of the curve is purely circular (Fig. 6.7), with no parabolic segment present. Here, the equation (6.1) becomes

$$\varepsilon = \left(\frac{L_0}{4r} + \frac{r}{L_0}\right)\sin^{-1}\left(\frac{1}{\dfrac{L_0}{4r} + \dfrac{r}{L_0}}\right) - 1 \qquad (6.2)$$

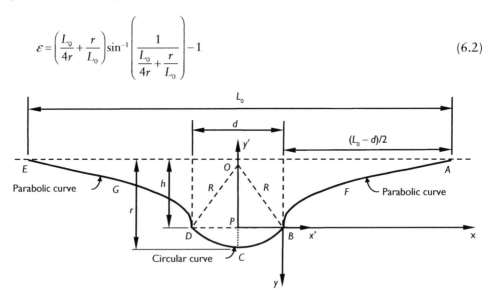

*Figure 6.9* Shape of a deflected geosynthetic layer having a combination of circular and two parabolic curves (after Shukla and Sivakugan, 2009).

Equation (6.2) is the same as the one given by Giroud (1981), and later on, considering that the relative deflection $r/L_0$ is small, its approximation was presented by Giroud (1995) as

$$\varepsilon \approx \frac{8}{3}\left(\frac{r}{L_0}\right)^2 \tag{6.3}$$

*Case 2:* Geosynthetic layer deformed into a parabolic arc
For $d = 0$, $h = r$, the deflected shape of the curve is a combination of two symmetrical convex parabolic curves (Fig. 6.10), and Equation (6.1) becomes

$$\varepsilon = \left[\sqrt{1+\left(\frac{r}{L_0}\right)^2}+\left(\frac{r}{L_0}\right)^2\ln\left\{\left(\frac{L_0}{r}\right)+\sqrt{1+\left(\frac{L_0}{r}\right)^2}\right\}\right]-1 \tag{6.4}$$

The deflected shape of the geosynthetic shown in Fig. 6.10 may be possible when a line load is applied normal to the horizontally laid geosynthetic layer resting on very weak foundation soil.

*Case 3:* Geosynthetic layer deformed into a profile consisting of circular and parabolic arcs with $h/r = 0.8$ and $d/L_0 = 0.2$.

In the field situations, $h/r$ and $d/L_0$ values may vary in some finite ranges; but for discussion of Equation (6.1), $h/r = 0.8$ and $d/L_0 = 0.2$ can be considered in the possible ranges. With these values, Equation (6.1) becomes

$$\varepsilon = \frac{4r}{5L_0}\left[\sqrt{1+\left(\frac{L_0}{r}\right)^2}+\frac{r}{L_0}\ln\left\{\left(\frac{L_0}{r}\right)+\sqrt{1+\left(\frac{L_0}{r}\right)^2}\right\}\right]+\frac{1}{5}\left(\left[\frac{L_0}{4r}+\frac{r}{L_0}\right]\right)\sin^{-1}\left(\frac{1}{\frac{L_0}{4r}+\frac{r}{L_0}}\right)-1 \tag{6.5}$$

Fig. 6.11 shows the variation of geosynthetic strain ($\varepsilon$) with relative deflection $r/L_0$ using Equations (6.2) to (6.5) and the equation for the parabolic shape of the deflected geosynthetic (Fig. 6.7) given by Giroud (1995) as

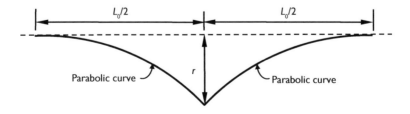

*Figure 6.10* Shape of a deflected geosynthetic layer having a combination of two parabolic curves (after Shukla and Sivakugan, 2009).

$$\varepsilon = \frac{1}{2}\sqrt{1+16\left(\frac{r}{L_0}\right)^2} + \frac{1}{8}\left(\frac{r}{L_0}\right)\ln\left[4\left(\frac{r}{L_0}\right)+\sqrt{1+16\left(\frac{r}{L_0}\right)^2}\right]-1 \qquad (6.6)$$

In Fig. 6.11, it can be seen that the geosynthetic strain ($\varepsilon$) increases with an increase in relative deflection ($r/L_0$), for all cases, irrespective of the shape of the deformed geosynthetics, which is expected intuitively. The two extreme cases corresponding to Equations (6.2) and (6.4) for the circular (Fig. 6.7) and the double convex parabolic shapes (Fig. 6.10) are the upper and lower limits, respectively for the general Equation (6.1). Equation (6.5) ($h/r = 0.8$ and $d/L_0 = 0.2$), a special case of Equation (6.1), is therefore lying between the curves for Equations (6.2) and (6.4). At low strain levels, with $r/L_0$ less than 0.25, the approximate equation (Equation (6.3)) predicts the geosynthetics strains very well, for circular and parabolic shapes as well as the combined ones shown in Figs. (6.9) and (6.10). For $r/L_0$ greater than approximately 0.25, the approximate Equation (6.3) predicts the geosynthetic strain values greater than the values calculated from any of the exact Equations (6.2), (6.4) or (6.6). Giroud (1995) noted that the approximate equation (Equation (6.3)) can be used instead of his exact Equations (6.2) and (6.6) for low values of $r/L_0$ of up to 0.15. Fig. 6.11 clearly shows that the approximation holds even for the situation where the deflected

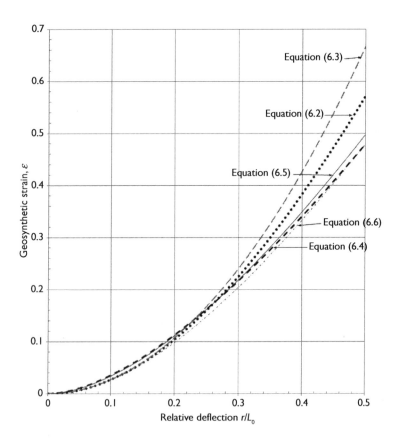

*Figure 6.11* Variation of the geosynthetic strain with relative deflection (after Shukla and Sivakugan, 2009).

geosynthetic profile is approximated by parabolic and circular arcs with inflection points. In fact, the approximation holds for $r/L_0$ values as high as 0.25.

The geomembrane or any other geosynthetic may be deflected due to wind, and in this situation, the strain in the geomembrane may be estimated using the analytical expression presented by Giroud (2009).

If the tensile strain is known, the tensile force in the geosynthetic layer may be determined using the relationship between the tensile force, $T$ (kN/m) and the tensile strain $\varepsilon$ as

$$T = E\varepsilon \tag{6.7}$$

where $E$ (kN/m) is the tensile modulus (stiffness) of the geosynthetic.

## EXAMPLE 6.2

A 20-m long geosynthetic layer deforms into a circular arc with the maximum deflection of 2 m at its centre. Determine the geosynthetic strain. If the tensile modulus (stiffness) of the geosynthetic is 100 kN/m, what will be the tensile force in the geosynthetic layer?

## SOLUTION

Given: The original length, $L_0$ = 20 m, maximum deflection, $r$ = 2 m, and tensile modulus, $E$ = 100 KN/m
The relative deflection is

$$\frac{r}{L_0} = \frac{2}{20} = 0.1$$

As $r/L$ is very small, from Equation (6.3), the geosynthetic strain is

$$\varepsilon \approx \frac{8}{3}\left(\frac{r}{L_0}\right)^2 = \frac{8}{3}(0.1)^2 = 0.027$$

From Equation (6.7), the tensile force in the geosynthetic layer is

$$T = E\varepsilon = (100)(0.027) = 2.7 \text{ kN/m}$$

## Chapter summary

1   Geosynthetics are used as essential components in mining facilities for their cost-effective and environmentally friendly performance. For most mining facilities, liners are the essential components. A liner may be designed as a single composite system or a double composite system with a leakage detection layer.

2   As the geosynthetics in mining applications are subjected to harsh climatic and load conditions, the material testing, analysis, design and field applications often require special considerations as per the site-specific conditions and environmental regulations.

3   The selection of geosynthetics for use in the mining, agricultural and aquacultural industries depends on several factors, including polymer type, puncture resistance, durability (resistance to weathering and chemical resistance), and also on the mineralogy of the bentonite in the case of geosynthetic clay liners.

4   The estimation of loads over the pipes and/or pipe envelops under high fills may be based on the concept of soil arching within the fill. The load on a pipe may be reduced significantly by including a geosynthetic (woven geotextile or geogrid) reinforcement layer above the pipe.

5   The electrokinetic geosynthetics can be effective in achieving the dewatering of mine tailings with additional advantages over the conventional dewatering methods.

6   Textile products in the form of net, grid or mesh are often used in agricultural engineering for several applications, and they are exposed to natural weathering over a long period of time. Hence it becomes important to assess the resistance to natural weathering of the textile products over a long period of time in such applications.

7   In agricultural applications, the geomembrane is normally used as the liner; however, the nonwoven geotextile can also be used as the liner in some applications.

8   The periodic removal of accumulated sludge in lagoons and storage ponds in the aquaculture sector is an important task in order to increase the available volume as well as to reduce the strong odours. The sludges are effectively contained, dewatered, and stored using the geotextile filtration, which uses high-strength permeable geotextiles with uniquely designed retention properties fabricated into closed geocontainers/geotubes.

9   Geosynthetics become deflected into different shapes in field applications. The analytical expressions presented in this chapter may be used conveniently for determining the geosynthetic strain for the design works.

## Questions for practice

(Select the most appropriate answer to the multiple-choice questions from Q 6.1 to Q 6.8.)

6.1   A double composite liner system is generally used in
   (a)   storm water ponds
   (b)   fresh water ponds
   (c)   process solution ponds
   (d)   both (a) and (c).

6.2   The thickness of the geomembrane liner in mining applications typically ranges from
   (a)   0.5 mm to 1.5 mm
   (b)   1.5 mm to 2.5 mm
   (c)   5 mm to 7.5 mm
   (d)   7.5 mm to 10.0 mm.

6.3 The dewatering of mine tailings can be achieved using
   (a) geotextiles
   (b) geosynthetic clay liners
   (c) geonets
   (d) electrokinetic geosynthetics.

6.4 The liner foundation/bedding soil should be
   (a) fine-grained soil
   (b) coarse-grained soil
   (c) sand-gravel mixture
   (d) both (b) and (c).

6.5 Geosynthetics are generally exposed to natural weather conditions over a long period of time when they are used in
   (a) geotechnical applications
   (b) transportation applications
   (c) agricultural applications
   (d) environmental applications.

6.6 In anaerobic lagoons, the odour may be reduced by covering them with
   (a) nonwoven geotextile
   (b) geomembrane
   (c) geosynthetic clay liner
   (d) geonet.

6.7 The HDPE nets used in agricultural applications may last for a maximum period of
   (a) 1 year
   (b) 5 years
   (c) 10 years
   (d) 50 years.

6.8 The bio-waste solids generated by the fish farming operation in an aquaculture cage site are removed at some intervals using
   (a) geotextile sheets
   (b) geotextile bag
   (c) geomembrane sheets
   (d) geomembrane bags.

6.9 List the major areas of application of geosynthetics in mining engineering.

6.10 What is the essential component of most mining facilities?

6.11 How does the single composite liner differ from the double composite liner? Explain with the help of neat sketches.

6.12 If the maximum height of the ore heap above the liner in a heap leach pad/facility is 120 m, determine the vertical normal stress on the liner. Assume the total unit weight of the ore is 18.5 kN/m$^3$. Do you expect the same stress over the geopipe installed within the heap leach pad at the liner level? Justify your answer.

6.13 What type of soil should be used as the liner foundation/bedding layer? List the major properties of the soil.

6.14 Enumerate the special considerations for the design of geopipes within mine fills.

6.15   Describe a practical method of reducing the load from the overlying fill over a geopipe.

6.16   What are the advantages of electrokinetic geosynthetics over the conventional metallic electrodes for the dewatering of mine tailings?

6.17   List the major application of geosynthetics in agricultural and aquacultural engineering.

6.18   Explain different netting systems used in agricultural engineering.

6.19   How are sludges produced in lagoons and storage ponds handled using the geotextiles?

6.20   Derive the general expression (Equation (6.1)) for the geosynthetic strain presented by Shukla and Sivakugan (2013), and its several special cases.

6.21   A 50-m long PP net deforms into a parabolic arc with the maximum deflection of 5 m at its centre. Determine the net strain. If the tensile modulus (stiffness) of the net is 30 kN/m, what will be the tensile force in the net?

## References

Adams, D.N., Muindi, T. and Selig, E.T. (1988). *Performance of high density polyethylene pipe under high fill*. Geotechnical Report No. ADS88-51F, Department of Civil Engineering, University MA, Amherst.

ASTM (American Society for Testing and Materials) (2010). ASTM D5617-04, *Standard test method for multi-axial tension test for geosynthetics*. ASTM International, West Conshohocken, PA, USA.

ASTM (2014a). ASTM D5514/5514M-14 (2014a). *Standard test method for large scale hydrostatic puncture testing of geosynthetics*. ASTM International, West Conshohocken, PA, USA.

ASTM (2014b). ASTM D5321/5321M-14, *Standard test method for determining the coefficient of soil and geosynthetic or geosynthetic and geosynthetic friction by the direct shear method*. ASTM International, West Conshohocken, PA, USA.

Barrington, S.F., Prasher, S.O. and Raimondo, R.J. (1990). Geotextiles as sealing liners for earthen manure reservoirs: Part 1, geotextile porosity. *Journal of Agricultural Engineering Research*, **46**, pp. 93–103.

Casagrande, L. (1949). Electro-osmosis in soils. *Geotechnique*, **1**, 3, pp. 159–177.

Cantrell, K.B., Chastain, J.P. and Mooore, K.P. (2008). Geotextiles filtration performance for lagoon sludges and liquid animal manures dewatering. *Transactions of the ASABE*, **51**, 3, pp. 1067–1076.

DeGarie, C.J., Crapper, T., Howe, B.M., Burke, B.F. and McCarthy, P.J. (2000). Floating geomembrane covers for odour control and biogas collection and utilisation in municipal lagoons. *Water Science and Technology*, **42**, 10–11, pp. 291–298.

Dierickx, W. and Berghe, P.V.D. (2004). Natural weathering of textiles used in agricultural applications. *Geotextiles and Geomembranes*, **22**, 4, pp. 255–272.

Fourie, A.B. and Jones, C.J.F.P. (2010). Improved estimates of power consumption during dewatering of mine tailings using electrokinetic geosynthetics (EKGs). *Geotextiles and Geomembranes*, **28**, 2, pp. 181–190.

Giroud, J.P. (1995). Determination of geosynthetic strain due to deflection. *Geosynthetics International*, **2**, 3, pp. 635–641.

Giroud, J.P. (2009). An explicit expression for strain in geomembrane uplifted by wind. *Geosynthetics International*, **16**, 6, pp. 500–503.

Giroud, J.P. and Noiray, L. (1981). Geotextile-reinforced unpaved road design. *Journal of the Geotechnical Division, ASCE*, **107**, 9, pp. 1233–1254.

Giroud, J.P., Badu-Tweneboah, K. and Soderman, K.L. (1995). Theoretical analysis of geomembrane puncture. *Geosynthetics International*, 2, 6, pp. 1019–1048.

Guerdat, T.C., Losordo, T.M., DeLong, D.P. and Jones, R.D. (2013). An evaluation of solid waste capture from recirculating aquaculture systems using a geotextile bag system with a flocculant-aid. *Aquacultural Engineering*, 54, 5 pp. 1–8.

Hamir, R.B., Jones, C.J.F.P. and Clarke, B.G. (2001). Electrically conductive geosynthetics for consolidation and reinforced soil. *Geotextiles and Geomembranes*, 19, 8, pp. 455–482.

Hornsey, W.P., Scheirs, J., Gates, W.P. and Bouazza, A. (2010). The impact of mining solutions/liquors on geosynthetics. *Geotextiles and Geomembranes*, 28, 2, pp. 191–198.

Jarret, P.M. (1986). Load tests on geogrid reinforced gravel fills constructed on peat subgrades. *Proceedings of the Third International Conference on Geotextiles*, Vienna, Austria, pp. 87–92

Lange, K. Rowe, R.K. and Jamieson, H. (2010). The potential role of geosynthetic clay liners in mine water treatment systems. *Geotextiles and Geomembranes*, 28, 2, pp. 199–205.

Lupo J.F. (2010). Liner system design for heap leach pads. *Geotextiles and Geomembranes*, 28, 2, pp. 163–173.

Lupo, J.F. and Morrison, K.F. (2007). Geosynthetic design and construction approaches in the mining industry. *Geotextiles and Geomembranes*, 25, 2, pp. 96–108.

Lamont-Black, J. (2001). EKG: the next generation of geosynthetics. *Ground Engineering*, pp. 22–23.

Miura, N., Sakai, A., Taesiri, Y. Yamanouchi, T. and Yasuhara, K. (1990). Polymer grid reinforced pavement on soft clay grounds. *Geotextiles and Geomembranes*, 9, 1, pp. 99–123.

Mukhtar, S., Lazenby, L.A. and Rajman, S. (2007). Evaluation of a synthetic tube dewatering system for animal waste pollution. *Applied Engineering in Agriculture*, 23, 5, pp. 669–675.

Neukirchner, R.J. and Lord, G.G. (1998). *Covering of soft mine tailings*. Tensar Earth Technologies Inc., Spring Lake, NJ.

Peggs, I.D., Schmucker, B. and Carey, P. (2005). Assessment of maximum allowable strains in polyethylene and polypropylene geomembranes. *Proceedings of the Geo-Frontiers 2005 Congress*, ASCE, Austin, TX, pp. 1–16.

Renken, K., Mchaina, D.M. and Yanful, E.K. (2006). Geosynthetics research and applications in the mining and mineral processing environment. *Proceedings of NAGS, GRI-19 Cooperative Conference*, pp. 1–20.

Rigden, P. (2008). To net or not to net, Third edition, Department of Primary Industries and Fisheries, Queensland Government, Australia.

Shukla, S.K. (1996). Discussion of "Determination of geosynthetic strain due to deflection" by J.P. Giroud', 3, 1, pp. 141–144.

Shukla, S.K. and Chandra, S. (1995). Modelling of geosynthetic-reinforced engineered granular fill on soft soil. *Geosynthetics International*, 2, 3, pp. 603–618.

Shukla, S.K. and Sivakugan, N. (2009). A general expression for geosynthetic strain due to deflection. *Geosynthetics International*, 16, 5, pp. 402–407.

Shukla, S.K., Gaurav and Sivakugan, N. (2009). A simplified extension of the conventional theory of arching in soils. *International Journal of Geotechnical Engineering*, 3, 3, pp. 353–359.

Shukla, S.K. and Sivakugan, N. (2013). Load coefficient for ditch conduits covered with geosynthetic-reinforced granular fill. *International Journal of Geomechanics, ASCE*, 13, 1, pp. 76–82.

Terzaghi, K. (1943). *Theoretical Soil Mechanics*, John Wiley and Sons, Inc., New York.

Thiel, R. and Smith, M.E. (2004). State of practice review of heap leach pad design issues. *Geotextiles and Geomembranes*, 22, 6, pp. 555–568.

Veal, C., Johnston, B. and Miller, S. (2000). The Electro-osmotic dewatering (EOD) of mine tailings. *Proceedings of the American Filtration Society Conference*, p. 10.

Worley, J.W., Bass, T.M. and Vendrell, P.F. (2004). Field test of geotextile tube for dewatering dairy lagoon sludge. ASAE Paper No. 044078, ASAE, St. Joseph, Mich.

## Answers to selected questions

6.1   (c)
6.3   (d)
6.5   (c)
6.7   (c)
6.12  2.22 MPa, No
6.21  0.053, 1.59 kN/m

# Geosynthetic applications – general guidelines and installation survivability requirements

## 7.1 INTRODUCTION

In all the field applications of geosynthetics, as explained in Chapters 3 to 6, the common objective is to install the correct geosynthetic in the correct location without having its properties impairing during the construction stage of the project. In the past, many general and project-specific guidelines have been suggested to meet this common objective (John, 1987; Ingold and Miller, 1988; Ingold, 1994; Van Santvoort, 1994, 1995; Holtz *et al.*, 1997; Pilarczyk, 2000; Koerner, 2005; Shukla, 2002, 2012; Shukla and Yin, 2006). Basically, the objectives of application guidelines are to assist the users to exercise their professional judgment and experience in developing site-specific recommendations, and to promote the use of best practices in field applications of geosynthetics.

Several guidelines applicable to the specific field applications are already presented in Chapters 3 to 6 along with their descriptions. This chapter provides many general geosynthetic application guidelines, which may be followed while working with the geosynthetics during their applications. The geosynthetics must meet some basic properties during their installation to avoid damage. Such installation survivability requirements are also presented in this chapter. Note that no two projects are identical, and hence the project-specific site conditions may dictate different requirements, techniques and guidelines. The guidelines contained in the present chapter, therefore, may not be universally applicable to all the geosynthetic applications. The project-specific guidelines will always supersede the general guidelines. Users can also find some information/guidelines on application of any specific geosynthetic from the concerned manufacturer/supplier.

## 7.2 CARE AND CONSIDERATION

In many projects, the environmental factors and the mechanical stresses during on-site storage, initial operation and construction place the most severe conditions on a geosynthetic over its projected lifetime. The successful installation of a geosynthetic is, therefore, largely dependent on the construction technique and the management of construction activities. Thus, the installation of geosynthetics in practice requires a degree of care and consideration.

In the past, most of the geosynthetic-related failures were reported to be construction-related, and a few design-related. The construction-related failures were caused mainly by the following problems:

- loss of strength due to UV exposure,
- lack of proper overlap,
- high installation stresses.

Although the general nature of installation-induced damage to geosynthetics, for example cuts, tears, splits and perforations, can be assessed by site trials, no test methods have yet been derived by which the same nature and degree of damage can be reproduced consistently in the laboratory. However, the strength reduction due to the damage during the installation might be partially or completely avoided by considering carefully the following elements, where the damage is found to be most severe:

1  firm, rock or frozen subgrades,
2  thin lift thickness using heavy equipment,
3  large soil particle size,
4  poorly graded cover soil,
5  light weight, low strength, geosynthetics.

If the type of subgrade cannot be changed, the options remain of changing the construction practice or to modifying the geosynthetic being used for a specific application. However, one may attempt to do both by recommending less severe construction practices and adopting a set of criteria on the geosynthetic strength, such as the reductions in the values of strength and strain to be taken into account when assessing the design tensile capacity of the geosynthetics.

When the geosynthetics are applied, the following aspects are also taken into account:

- temperature during placement and service life,
- possibility of leaching of UV stabilizers, resulting in the subsoil pollution,
- possibility of materials in the surroundings of the geosynthetic, which can act as catalyst in degradation process

Due care should also be taken during spreading and compaction of the fill materials on the geosynthetic layers, particularly for very soft subgrades and/or very coarse fill materials (stones, rockfills, mine tailings, etc.), in order to avoid or minimize mechanical damage to the geosynthetics.

The relationship between a geosynthetic product and its application environment should also be carefully considered.

## 7.3  GEOSYNTHETIC SELECTION

Proper geosynthetic specifications are essential for the success of any project. Due to a wide range of applications and the tremendous variety of geosynthetics available, the selection for a particular geosynthetic with specific properties is a critical decision. The selection of a geosynthetic is generally done keeping in view the general objective of its use. For example, if the selected geosynthetic is being used to function as a reinforcement, it will have to increase the stability of soil (bearing capacity, slope stability and resistance to erosion) and to reduce its deformation (settlement and lateral deformation). In order to provide stability, the geosynthetic has to have adequate strength; and to control deformation, it has to have suitable force-elongation characteristics, measured in terms of modulus (the slope of the force versus elongation curve) as explained in Section 2.3.2 (see Chapter 2). Woven geotextiles and geogrids are preferred in most reinforcement applications.

When a geosynthetic has to function for filtration/drainage applications, the most suitable product is usually a thick nonwoven needle-punched geotextile with an appropriate AOS (Apparent Opening Size). This is because of the higher permittivity and transmissivity of these nonwoven geotextiles, which is a primary requirement in such applications (Shukla, 2003).

Methods of transport, storage and placement also govern the selection of geosynthetics. The selected geosynthetic should have a certain minimum strength, thickness and stiffness so that it will be fit enough to survive the effects of placement on the ground and the loads imposed by equipment and personnel during the installation. In other words, the construction engineers should consider the field survivability/workability, transmissivity and permeability requirements of geosynthetics during their selection. These requirements can be expressed in terms of grab strength, puncture strength, burst strength, impact strength, tearing strength, permeability, transmissivity, etc. The actual values of these survivability properties of geosynthetics should be decided on the basis of the expected degree of damage (low, moderate, high or very high) on their installation in the specific field application.

In some geosynthetic applications, the colour and the surface feature of the geosynthetics should also be considered in their selection. For example, in the lining of ponds, reservoirs and canals, the white-surfaced textured HDPE geomembranes are preferred mainly because of lower heat absorption and, consequently, less expansion and contraction. White-surfaced geomembranes also allow for easier damage detection. As the geomembranes are to be left uncovered over the service life, the less dramatic temperature changes inherent in the white-surfaced geomembrane are considered to be beneficial for the containment and conveyance projects. The textured geomembrane is selected to provide a suitable working base for the installation crew (Ivy and Narejo, 2003).

Many times, cost and availability of geosynthetics may also govern their selection.

## 7.4  IDENTIFICATION AND INSPECTION

Upon receipt, each shipment of geosynthetic rolls should be inspected for conformance to product specifications and contract documents, and checked for damage. A construction quality assurance (CQA) representative should be present, whenever possible, to observe the material delivery and unloading on the site. Before storing or unrolling the geosynthetic rolls, or both, the individual roll identification should be verified and should be compared with the packing list. Irregularities should be noted and reported. Upon delivery of the rolls of geosynthetic materials, the CQA consult-ant should ensure that the conformance test samples are obtained. These samples should then be forwarded to the geosynthetic quality assurance laboratory for testing to ensure conformance with the site-specific specifications. Geosynthetic rolls not in compliance with the accepted material specifications may be rejected. The damaged, deformed or crushed geosynthetic rolls should be rejected and removed from the pro-ject site.

## 7.5 SAMPLING AND TEST METHODS

The samples of geosynthetics must be cut from the product roll supplied by the manu-facturer as per the standard sampling procedures to provide a statistically valid sam-ple for the selection of coupons and test specimens (Fig. 7.1). At least one sample is generally taken for 5000 m$^2$ or less area of the geosynthetic. Each roll selected should look undamaged and the wrapping, if any, should be intact. The first two turns of the roll should not be used for the sampling. The sample should be cut from the roll, over its full width, perpendicular to the machine direction. A mark (e.g. arrow) should be used to indicate the machine direction of the sample. When two faces of the geo-synthetic are significantly different, the sample should be marked to show which face was inside or which was outside the turn of the roll. The sample must be marked for identification purposes with the following information:

- brand/producer/supplier,
- type description,
- roll number,
- date of sampling.

The samples should be stored in a dry, dark place, free from dust at ambient tempera-ture and protected against chemical and physical damage. The sample may be rolled up but preferably not folded. Sampling may be required for three purposes: one for manufacturer's quality control (MQC) testing, one for manufacturer's quality assur-ance (MQA) testing, and a third for purchaser's specification conformance testing.

For each type of test, the required number of specimens should be cut from posi-tions evenly distributed over the full width and length of the sample, but not closer than 100 mm to the edge. The test specimens should not contain any dirt, irregular areas or other defects, and they should be conditioned as per the requirement of the test. For atmospheric conditioning, the test specimens should preferably be hung or laid flat, singly on open wire shelves allowing free access of air to all surfaces for at least 2 h. For dry conditioning, the test specimens should be placed in a desiccator until a constant mass is attained. For wet conditioning, the test specimens should be immersed in water at a temperature of $20 \pm 5$ °C for a minimum of 24 h. For most of the tests on geosynthetics, air is maintained at a relative humidity between 50% to 70% and a temperature of $21 \pm 2$ °C. If the test methods for determining the geosynthetic properties are not completely field simulated, the test values must be adjusted as dis-cussed in Section 2.6 (see Chapter 2). On any given project, the minimum average roll value (MARV) must meet or exceed the designer's specified value for the geosynthetic

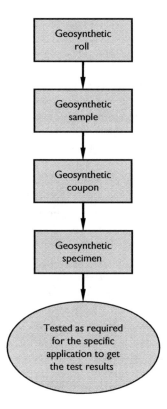

*Figure 7.1* Relationship among roll, sample, coupon, specimen and test results (Adapted from ASTM D6213–97 (ASTM, 2009)).

to be acceptable. With an understanding of the failure mechanism and the exposure environment, the user can select the appropriate test methods to best simulate the geosynthetic behaviour.

## 7.6    PROTECTION BEFORE INSTALLATION

Geosynthetics must be handled and stored properly to ensure that the specified properties are retained to perform their intended function as per the project needs. A proper choice of material and a careful handling of the geosynthetic can prevent mechanical damage during transport, storage and placement. When delivered, all the rolls of geosynthetics should be wrapped in a protective layer of plastic to avoid any damage/deterioration during their transportation.

Storage areas should be located as close as possible to the point of end use, in order to minimize subsequent handling and transportation. It is usually adequate to stack the rolls with the undamaged protective outer wrapper directly on the ground by covering with a waterproof tarpaulin or plastic sheet, provided that this is level, dry, well-drained, stable and free from sharp projections such as rock pieces, stumps of trees or bushes. The storage area must protect the geosynthetics from precipitation, standing water, UV radiation, chemicals (strong acids/bases), open flames and welding sparks, temperatures in excess of about 70 °C, vandalism, animals, and any other environmental condition that could damage the geosynthetics before end use.

Enclosed indoor storage is preferred if the geosynthetic rolls are to be stored for a long period. However, if the rolls are to be stored outside for a long period of time, some form of shading is required with an elevated base, unless the wrapper is of opaque material, to give protection against UV light attack. Acceptable limits of exposure to UV light depend upon the site environmental conditions such as temperature, latitude, time of year, wind, etc. and the assumptions used by the engineer during the design. At no time the geosynthetics should be exposed to UV light for a period exceeding two weeks. If the wrapper gets damaged and it is beyond repair, the roll should be stored by making a suitable arrangement to prevent the ingress of water. Without this, geotextiles, particularly the nonwoven geotextiles, can absorb water up to three times their weight, thereby causing handling and installation problems. In the cold regions, it is nearly impossible to unroll wet, frozen geotextile without first allowing it to thaw; therefore, geosynthetics should be protected from the freezing. Note that geosynthetics are generally hydrophobic (i.e., they repel water), so there is no wicking action in them. Where the geosynthetics are to be used as the filters, it is important to keep the wrapper intact to give a protection against the ingress of dust and mud. If a geosynthetic roll becomes wet, it must be allowed to have a few days of exposure to wind after removing the wrapper in order to dry the geosynthetic.

The geosynthetic rolls may be stacked upon one another, provided they are placed in a manner that prevents them from sliding or rolling from the stack. The height of the stack should generally be three rolls. In fact, the height of the stack should be limited to that at which safe access is provided to equipment and labourers, and at which the roll cores at the bottom of the stack are not distorted or crushed.

A good practice dictates that on site, the geosynthetic should be stored properly and handled according to the manufacturer's recommendations. In the absence of good, documented procedures, the guidelines given here may be used for general purposes.

## 7.7  SITE PREPARATION

The original ground level may be required to be graded to some predetermined formation level. During the site preparation, the sharp objects, such as boulders, stumps of trees or bushes, which might puncture or tear the geosynthetic, should be removed if they are lying on the site. All protrusions extending more than 12 mm from the subgrade surface should be removed, crushed, or pushed into the surface with a smooth-drum compactor. Disturbance of the subgrade should be minimized where soil structure, roots in the ground and light vegetation may provide additional bearing strength. All depressions and cavities must be filled with a compacted material, otherwise the geosynthetic may bridge and get torn when the fill is placed (Fig. 7.2). For critical applications, the depressions can be lined with geotextiles before filling them with the granular material. In brief, it is suggested that the subgrade should be prepared well as the geosynthetic between two stones is like grain being ground in a

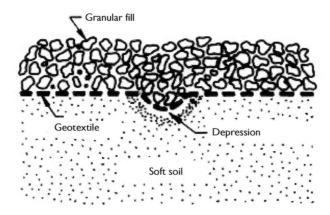

*Figure 7.2* Effect of a depression in the soft subgrade soil layer on a geotextile layer.

mill – it cannot sustain the applied load very long. If any equipment causes rutting of the subgrade, the subgrade must be restored to its originally accepted condition before the placement of the geosynthetic continues.

## 7.8   GEOSYNTHETIC INSTALLATION

The installation of a geosynthetic includes its placement and attachment to the other parts of the structure. Geosynthetic properties are only one factor in the successful installation of the geosynthetic. Proper construction and installation techniques are essential in order to ensure that the intended function of the geosynthetic is fulfilled. Placing a geosynthetic is thus the single most important step in the performance of the geosynthetic-reinforced soil system. While handling the rolls manually or by some mechanical means at any stage of installation, the load, if any, should not be taken directly by the geosynthetic. The dragging of the geosynthetic should be avoided. The entire geosynthetic should be placed and rolled out as smoothly as possible. A temporary geosynthetic subgrade covering commonly known as the *slip sheet* or *rub sheet* may be used to reduce frictional damage during the placement.

Since the geotextile opening size in some applications, such as filtration and drainage applications, is chosen with a high degree of accuracy in the design stage, it is important to observe during the installation stage that abrasion and excess straining must not enlarge the openings or even create holes before the final service state.

An overlap between the adjacent sheets must be provided while unrolling the geosynthetic into position after site preparation (Fig. 7.3). The overlap used is generally a minimum of 0.3 m, but in applications where the geosynthetic is subject to tensile stresses, the overlap must be increased or the sheets of geosynthetic sewn/bonded as explained in the following section. If possible, the overlap should not be located at the place where the transition or the edges of the cover layer may take place (Fig. 7.4).

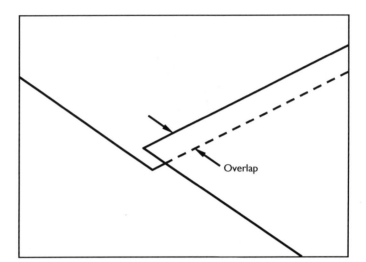

*Figure 7.3* A simple overlap.

(a)                                      (b)

*Figure 7.4* Overlap constructions: (a) wrong; (b) correct (after Pilarczyk, 2000).

It has been observed that a misunderstanding of expected imposed loads or unforeseen stresses arising from poor construction practices are the main reasons for damage, particularly mechanical damages, during the installation process. Also with a careless installation, parts of the geosynthetic may get scattered into the surroundings, resulting in a harmful influence on the environment. Therefore, the installer, that is, the party who installs, or facilitates installation of geosynthetics should consider the involved processes necessary for the perfect solution. Installation effects are often out of the designer's hands, so specifications, inspections and protective measures must be agreed upon at each individual site.

## 7.9  JOINTS/SEAMS

Geosynthetics are finite and therefore where the geosynthetic widths or lengths greater than those supplied on one roll are required, it becomes necessary to make joints or overlaps. Since the joints and the overlaps are the weakest link in the geosynthetic-reinforced soil structures they have to be limited as much as possible.

When two sheets of similar or dissimilar geosynthetics (or related materials) are attached to each other by a suitable means, the junction so formed is known as a *joint*, and when a geosynthetic sheet is physically linked to, or cast into another material (e.g. the facing panel of a retaining wall), this is known as a *connection*. When no physical attachment is involved between two geosynthetic sheets or a geosynthetic sheet and another material, this is known as an *overlap*. However, sometimes the overlap is also considered as a type of joint, called an *overlapped joint*.

There are several jointing methods, such as *overlapping, sewing, stapling, gluing, thermal bonding*, etc. Different joints, presently in use, may be classified into *prefabricated joints* and *field joints* that are basically made during the applications on the field sites. In the vast majority of cases, the geosynthetic width or length is extended simply by overlapping, which is usually found to be the easiest field method for jointing (Fig. 7.3). Overlapping by 0.3–1 m may be employed if relatively small tensile forces are developed in the geosynthetic layers to be jointed. Relatively more overlaps are required if the geosynthetic is placed under water. The overlaps involve considerable wastage of material and if not carried out with care they can be ineffective.

Geotextiles may be jointed mechanically by sewing or stapling, or chemically by means of adhesives. The term 'seam' generally refers to a series of stitches joining two or more separate pieces of geosynthetics (or related materials); however, it is also used as a synonym for 'joint'. Figure 7.5(a) shows the most suitable seam configuration known as the *prayer seams*. Another type of seam known as the lapped ('J') seam is shown in Fig. 7.5(b), which is reliably soil tight even for the fine-graded soils. Depending on the critical nature of construction, either a single or double stitch is used. Several types of threads are available (nylon, high performance polymers, etc.) depending on the type of geotextiles and type of field applications. AASHTO (2013) recommends that the threads used in joining geotextiles by sewing should consist of high strength polypropylene, or polyester. Nylon thread should not be used.

The sewn joints must be projected up so that every stitch can be inspected. High-strength geosynthetics, employed for their reinforcing potential, should normally be sewn. For jointing the geotextiles by the stapling method, corrosion resistant staples should be used. Figure 7.6 shows the stapled seam configuration. Stapling may be used with geotextiles to make temporary joints. They should never be used for structural jointing. Note that the sewn seams are most reliable and can be carried out on site using portable sewing equipment. Heat bonded or glued seams are generally not used.

For geosynthetics such as geonets and geogrids, a *bodkin joint* may be employed whereby the two overlapping sections are coupled together using a bar passed through the apertures (Fig. 7.7). Geogrids can also be sewn using a robust cord threaded through the grid apertures. Hog rings, staples, threaded loops, wires, etc. are also used for jointing geosynthetics.

The types of jointing/seaming techniques used to construct seams in plastomeric geomembranes (made from PE, PP, PET, PVC, etc.) include the following: extrusion, hot air, fusion (hot wedge/knife), ultrasonic, and electric welding methods. The *extrusion technique* encompasses extruding molten resin between two geomembranes or at the edge of two overlapped geomembranes to form a continuous bond. The *hot air technique* introduces high-temperature air or gas between two geomembrane surfaces to facilitate melting. Pressure is applied to the top or bottom geomembrane,

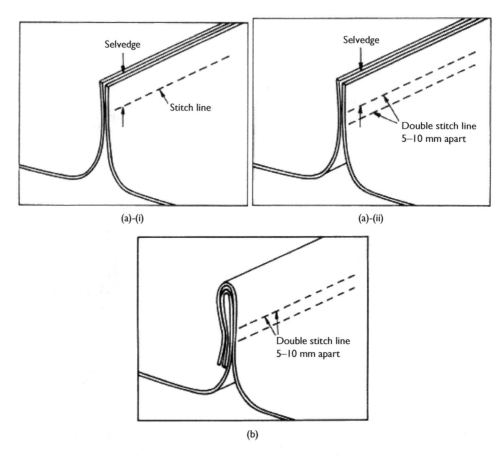

*Figure 7.5* Sewn seams: (a) face to face ('Prayer') seams – (i) single stitch line, (ii) double stitch line; (b) Lapped ('J') seam.

forcing together the two surfaces to form a continuous bond. *Hot wedge welding technique* consists of placing a heated wedge, mounted on a self-propelled vehicular unit, between two overlapped geomembrane sheets such that the surface of both sheets are heated above the melting point of the polymer. Pressure is applied to the top or bottom geomembrane, or both, to form a continuous bond. Fusion (hot wedge) welding is generally used for long seams. Extrusion welding is used for capping and patch repairs, and for joining panels at locations where fusion welding is not practical due to joint configuration. Hot air welding is also possible, but it is very sensitive to workmanship and is only used when other methods are not applicable.

Some seams are made by fusion (hot wedge) with dual bond tracks separated by a non-bonded gap, sometimes referred to as the *dual hot wedge seams* or *double track seams* (Fig. 7.8). Some typical seams in geomembranes are shown in Fig. 7.9. In general, the wedge welding with a dual-track wedge welder provides the best quality of seams and is used as a state of practice in environmental lining applications. Once finished, each air channel of the dual seam is tested by inflation. It should withstand an air pressure of 5 bars for 10 minutes without noticeable loss of pressure.

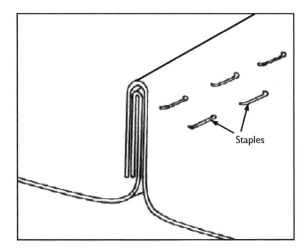

*Figure 7.6* Stapled seam.

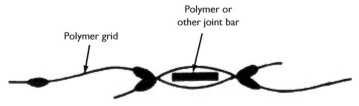

*Figure 7.7* A bodkin joint.

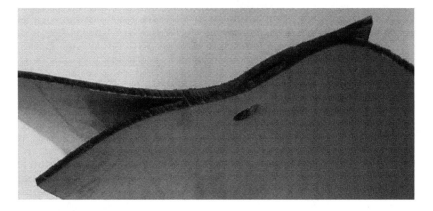

*Figure 7.8* A typical dual hot wedge seam.

Elastomeric geomembranes are made from rubbers of various types as the barrier component. They require seaming by means of solution or adhesives.

A geomembrane seam, in service, must maintain its leak-free condition. Metal hog rings should never be used when geonets are used in conjunction with geomembranes. The most important aspect of construction with geomembranes is the seam. Without proper seams, the whole concept of using the geomembrane liners as a fluid

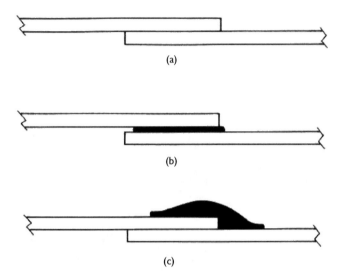

*Figure 7.9* Some typical seams in geomembranes: (a) fusion; (b) extrusion lap; (c) extrusion fillet (after Giroud, 1994).

barrier is discredited. A very important factor affecting the seaming quality is the weather condition during seaming. Normal weather conditions for seaming are as follows (Qian *et al.*, 2002):

1   temperature between 4.5 °C and 40 °C,
2   dry conditions (i.e. no precipitation or other excessive moisture, such as fog or dew),
3   winds less than of 32 km/h.

Geosynthetic clay liners are jointed by the application of bentonite at the panel joints.

It is important to understand the criterion for assessing joint performance. This is expressed in terms of the load transmission between the two pieces of the geosynthetic. In some applications, it may be essential that the load transfer capability be equal to that of the parent material. For other situations, a more important criterion may be the magnitude of the deformation of the joint under the load. The data on the tensile strength of seams/joints are necessary for all functions if the geosynthetic is to be mechanically jointed and if a load is transferred across the seams and joints.

The *seam/joint strength* is the maximum tensile resistance (that indicates the load-transfer capability), measured in kilonewtons per metre, of the junction formed by joining together two or more sheets of geosynthetics by any method (e.g. sewing or thermal bonding). The seam/joint efficiency ($E$) of a seam/joint between two geosynthetic sheets is the ratio, expressed as a percentage, of seam/joint strength to the tensile strength of unseamed/unjointed geosynthetic sheet evaluated in the same direction. Thus

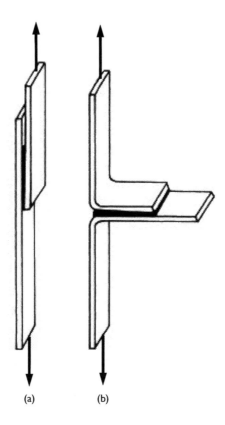

(a)                    (b)

*Figure 7.10* Geosynthetic seam specimens for (a) bonding shear strength test; (b) bonding peel strength test.

$$E = \left( \frac{T_s}{T_u} \times 100 \right) \% \tag{7.1}$$

where $T_s$ is the seam/joint strength (kN/m), and $T_u$ is the tensile strength of unseamed/unjointed geosynthetic sheet (kN/m).

Ideally, the seam/joint should be stronger than the geosynthetic being jointed and should thus never fail in tension. In practice, in the field, high efficiencies are rarely obtained. The literature generally mentions that the laboratory obtained efficiencies are usually higher than the field efficiencies. Thus, this is of little help to the field engineer trying to meet a consultant's specification. However, efforts should be made to make the seam efficiency near 100%. As the geosynthetic strength becomes higher, the seams become less efficient. Above a geosynthetic strength of 50 kN/m, even the best of seams have efficiency less than 100%, and beyond 200–250 kN/m, the best one can have approximately 50% efficiency. AASHTO (2013) recommends that when sewn seams are required, the seam strength should be equal to or greater than 90% of the specified grab strength.

Murray *et al.* (1986) undertook the research work into the seam strengths obtained from both sewn and adhesive bonded seams. Their work was comprehensive and stated that 100% efficiency could be obtained using the adhesives. With sewn joints, they described efficiencies up to 90% but they drew attention to the large deformations that were experienced. The technique of jointing geogrids by means of a bodkin joint proved to be an effective procedure whereby load carrying efficiencies of about 90% were obtained. Rankilor and Heiremans (1996) have reported that the use of adhesives can reduce the seam extension dramatically.

Since geomembranes are generally used to restrict fluid migration from one location to another in soil and rock, the integrity of their seams must be properly determined. The quality assurance (QA) engineer must analyze the seam bonding shear strength and the peel strength data obtained from the tests with tensile instrumentation on the geosynthetic seam shear and peel specimens (Fig. 7.10) to evaluate the seam quality.

If the geosynthetics are applied in reinforcement applications, overlaps and seams at right angles to the direction of the leading force are unacceptable. The termination of a geosynthetic sheet as well as its connection to another part of the structural elements needs special attention.

## 7.10    CUTTING OF GEOSYNTHETICS

The cutting of geosynthetics is a labour-intensive, time-consuming process, which in most cases can be avoided by forward planning. The maximum width of a geosynthetic roll is generally 5.3 m. The total width of an area to be covered will rarely be an exact multiple of available widths. There is less wastage of time and money if slightly larger overlaps or wraparounds are allowed to take up the excess width, than if the geosynthetic is cut on the site. In the case of walls and steep-sided embankments, the larger wraparound may enhance the compaction at the edges, and also helps to reduce erosion and may assist in the establishment of vegetation.

## 7.11    PROTECTION DURING CONSTRUCTION
##          AND SERVICE LIFE

Damage due to UV light exposure can usually be avoided by not laying more geosynthetic in a day than can be covered by fill on the same day. Unused portions of rolls must be rerolled and protected immediately. Note that when the geosynthetic is UV-stabilized, the degradation is largely reduced, but not entirely eliminated. Efforts should be made to cover the geosynthetic within 48 hours after its placement. A geosynthetic, which has not been tested for resistance to weathering, must be covered on the day of installation.

Protection of the exposed geosynthetic wall face against degradation due to UV light exposure and, to some extent, against vandalism can be provided by covering the geosynthetic with gunite (shotcrete), asphalt emulsion, asphalt products or other coatings. A wire mesh anchored to the geosynthetic may be necessary to keep the coating on the face of the wall. In the case of walls constructed from geogrids, vegetation can easily grow through (or be placed behind) the large openings, and the UV degradation of the relatively thick ribs is significantly lower. Thus the need to cover the geogrid wall face is not as compelling as with the geotextiles, and the fronts of geogrid walls are sometimes left exposed.

The chemical resistance of the geosynthetic liner to the contained liquid must be considered for the entire service life of its installation. The usually recommended minimum thickness of a geomembrane liner is 20 mils (0.50 mm) irrespective of the design calculations; however this lower limit may be 80 mils (2.0 mm) in the case of hazardous materials containment. When the secondary liner is also a geomembrane, it must be of the same thickness as the primary liner.

Geosynthetic clay liners (GCLs) are extremely sensitive to damage during and after construction owing to their small thickness and small mass of bentonite. So great care is required in applications with GCLs.

Before the placement of a granular fill on the geosynthetic, the condition of the geosynthetic should be observed by a qualified engineer to determine that no holes or rips exist in the geosynthetic. All such occurrences need a repair, and all wrinkles and folds in the geosynthetic should be removed. The following actions may result in puncturing, abrasion or excessive straining that can lead to a loss of strength or reduce the serviceability of the geosynthetic product, and therefore these actions must be avoided.

- dropping fill material from an unknown height,
- wheels passing over a relatively thin cover layer,
- compactors acting on the cover layer.

Note that the loads due to the above actions may act on the geosynthetic during installation that will never occur again.

In the road and railway constructions, the damage to the geosynthetic from the impact of dropped fill materials is usually not significant, unless the geosynthetic is very light and thin. Traffic or compaction loads cause more severe harm than fill placement (Brau, 1996).

When placing the fill on the sloped surfaces, it is always advisable to start from the base of the slope. If there is no way to avoid having stones roll downslope, their weight should not exceed 50 kg. In general, when releasing stones that weigh less than 120 kg on the geotextiles over well-prepared surfaces, the drop height should be less than 0.3 m. If the geotextile is covered by a cushion layer, the drop height may be up to 1 m. Larger stones should be placed without free fall. Compared to dumping in dry conditions, the falling height of cover materials is often much larger under water but has less impact because it falls through water, rather than air. As a rule of thumb, the impact energy when falling through water of some depth is a mere 15% of that generated by an object falling from the same height in dry conditions (Heibaum, 1998).

No equipment that could damage the geosynthetic should be allowed to travel directly on the geosynthetic. In fact, once the geosynthetic is laid, it should not be trafficked until an adequate layer of fill is placed over it, thus affording some protection; otherwise the geosynthetic may fail. One exception to this rule is where a heavyweight geosynthetic is used, which is specifically designed to be directly trafficked by vehicles, but the principle 'thicker fill is better' is valid at every site. In road, railway and embankment constructions, the first layer of fill material on the geosynthetic should have a minimum thickness of 200–300 mm, depending on aggregate size and weight of trucks/rollers. The exact answer comes only from the site tests. The maximum lift thickness may be imposed in order to control the size of the mud wave (bearing failure) ahead of the dumping due to excess fill weight. It has been observed that when more than 0.6–0.9 m of fill has been placed, the geosynthetic sustains no significant damage from trucks or vibratory rollers (USDI, 1992).

No turning of construction equipment should be allowed on the first lift. Construction vehicles should be limited in size and weight to limit initial lift rutting to 75 mm. If the rut depths exceed 75 mm, decrease the construction vehicle size and/or weight. For the initial stages of construction, low ground pressure and small dump trucks should therefore be used. For very soft formations, it is necessary to use special low-bearing pressure tracked vehicles for spreading the fill over the geosynthetic layer. During the filling operations, the blades or buckets of the construction plant must not be allowed to come into contact with the geosynthetic. A further lift may be placed after consolidation of the subgrade has increased its strength. Compaction of the first granular lift is usually achieved by tracking with the construction equipment. Smooth-drum or rubber-tyred rollers may also be considered for compaction of the first lift. A continued buildup of cover material will allow vibratory rollers to be used. If localized liquefied conditions occur, the vibratory roller should not be used. The proof rolling by a heavy rubber-tyred vehicle may provide the pretensioning of the geosynthetic by creating initial ruts, which should be subsequently refilled and leveled.

Proper care must be given during compaction of the top layers of wraparound reinforced walls and steep slopes. This is required because very high compaction results in very high stresses in the geosynthetic reinforcement due to movement of the fills in the wraparound sections as shown in Fig. 7.11, and such situations are not desirable. All vehicles and construction equipment weighing more than 1500 kg should be kept at least 2 m away from the faces of the walls or steep slopes. The fill within 2 m of the face of the wall or steep slope should be compacted using a vibro-tamper, vibrating plate compactor having a mass not exceeding 1000 kg or vibrating roller having a total mass not exceeding 1500 kg.

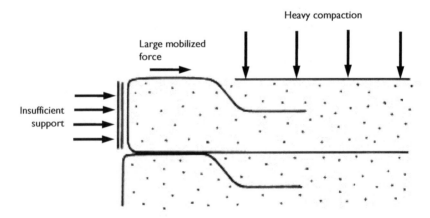

*Figure 7.11* Effects of heavy compaction (after Voskamp *et al.*, 1990).

If it is necessary to use the poorly-graded aggregate fill, and the heavy construction equipment for placement and compaction, it may be prudent to place a cushioning layer of sand above the geosynthetic.

If a geosynthetic is used in conjunction with a bituminous material, care must be taken to ensure that the temperature of the bituminous material is well below the melting point of geosynthetic. The tack-coat application quantity requires the special attention. The absence of an adequate tack coat means the loss of paving fabric system benefits and can lead to damage to the overlay. The wet geosynthetic should not be used in such applications because it creates steam, which may cause the bitumen (asphalt cement) to be stripped from the geosynthetic because of a poor bond.

In the case of liquid containment pond, to shield the geomembrane liner from ozone, UV light, temperature extremes, ice damage, wind stresses, accidental damage and vandalism, a soil cover of at least 30 cm thickness may be provided. For a proper design of the containment, a geotextile should be used beneath the geomembrane, placed directly on the prepared soil subgrade before the liner placement. The cover soil over geomembranes, installed on the sloping ground, can unravel and slump very easily, even under static conditions. To alleviate this situation somewhat, it is a common practice to taper the cover soil, laying it thicker at the bottom and gradually thinner going toward the top. Due to the water pressure caused by the groundwater, the geomembrane may be ballasted to prevent its floating.

## 7.12 DAMAGE ASSESSMENT AND CORRECTION

The ability to maintain the design function (e.g. reinforcement, separation, filtration, etc.) and/or the design properties (e.g. tensile strength, tensile modulus, chemical resistance, etc.) of a geosynthetic may be affected by the damages to the physical structure of geosynthetic during the field installation operations. Therefore, before covering the geosynthetic with soil, the engineer should examine it for defects such as holes, tears, abrasions, etc. A test section may be used to assess the worst case installation techniques (e.g. overcompaction, thin lift heights, greater drop heights, etc.) and fill materials. The damages to geosynthetics from installation operations may be quantified by evaluating the specimens cut from the samples exhumed from representative field installation sites. Damage evaluation may be performed with visual examination and/or laboratory testing of the specimens cut from the control (uninstalled/original) and exhumed samples. Laboratory tests to be performed will vary with type and function of the geosynthetic, and the project requirements. Note that the control sample should be a direct continuation of the exhumed sample so as to minimize the differences in control and exhumed specimen properties due to inherent product variability. The positions of the test specimens on the control sample, relative to the roll edge, must correspond identically with the positions of the exhumed sample. The amount, or area, of control sample to be retrieved should be equal to the area of exhumed sample. Details of the techniques for assessing the amount of damage, and documentation of installation and retrieval techniques used, may be found in the works of Bonaparte *et al.*

(1988), Bush and Swan (1988), Paulson (1990), Allen (1991), Rainey and Barksdale (1993), and Sandri *et al.* (1993).

For more critical structures, such as reinforced soil walls and embankments, it
is safest to remove the damaged section of geosynthetic, if any, entirely and replace
it with undamaged geosynthetic. In these applications, a certain degree of damage
may be acceptable, provided that this has been allowed for at the design stage. In
the low-risk applications, where the geosynthetic is not subject to significant tensile
stresses or dynamic water loading, it is permissible to patch the damaged area by
placing a new layer of geosynthetic extending beyond the defect in all directions at
a distance equal to the minimum overlap (generally 300 mm) required for adjacent
rolls.

Exposed geosynthetics, such as geomembranes used in lining of liquid containment and conveyance systems can be damaged by animals, construction equipment,
vandalism or other elements of nature. Thus, it is imperative that a meaningful
maintenance plan should be in place throughout the geomembrane service life. The
maintenance plan may include occasional seaming and anchor trench soil cover
maintenance.

## 7.13   ANCHORAGE

To maintain the position of a geosynthetic sheet before covering with soil/fill, the edges
of the sheet must be weighted or anchored in trenches (Fig. 7.12), thereby providing
the significant pullout resistance. Anchorage selection depends on site conditions. In
the case of unpaved roads, the geosynthetic should be anchored on each side of the
road. The bond length (typically around 1.0–1.5 m) can be achieved by extending
the geosynthetic beyond the required running width of the road (Fig. 7.13(a)) or by
providing an equivalent bond length by burying the geosynthetic in shallow trenches
(Fig. 7.13(b)) or by wraparound (Fig. 7.13(c)). Similar approaches can also be adopted
in other applications.

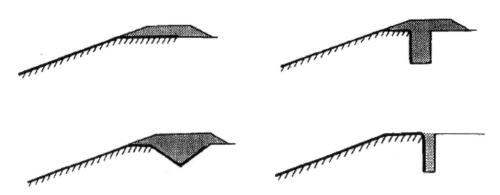

*Figure 7.12* (a) Simple run-out; and anchor trenches: (b) rectangular trench; (c) V trench; and (d) narrow trench (after Hullings and Sansone, 1997).

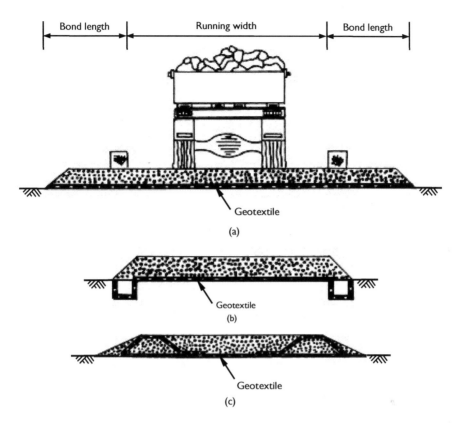

*Figure 7.13* Use of geosynthetics in unpaved road construction in different ways (after Ingold and Miller, 1988).

## 7.14  PRESTRESSING

Simple procedures such as prestressing the geosynthetic may enhance the reinforcement function in some applications. For example, to specifically add reinforcement for paved roads on firm subsoils, a geosynthetic prestressing system may be required. By prestressing the geosynthetic, the aggregate base will be placed in compression, thereby providing the lateral confinement and will effectively increase its modulus over the unreinforced case.

## 7.15  MAINTENANCE

All geosynthetic-reinforced soil structures should be subjected to a regular programme of inspection and maintenance. A habit should be developed to keep records of the inspections and any maintenance carried out, as described in Chapter 8.

## 7.16   CERTIFICATION

Certification provides a benchmark for quality, and thus it is a measure of assurance of success of geosynthetic-related projects. A quality certificate ensures that the geosynthetic delivered meets the design requirements. The initiative for certifying has to be taken by the manufacturer of the geosynthetic. It is the manufacturer's responsibility to perform thorough quality control testing for all properties requiring the certification. The same system of quality assurance should be valid for the applications of geosynthetics by the concerned contractor by issuing certificates with a complete description.

In ponds, reservoirs and any other containment of drinking water, the applied geosynthetics need a certificate, indicating that the geosynthetic concerned has been tested on the aspects of health and has been approved. In view of the increasing demand on prevention of pollution, it is recommended to inquire into the potential environmental effects. Note that there is no danger of emission of toxic materials from the geosynthetics to the environment, except from some kinds of PVC.

## 7.17   HANDLING THE REFUSE OF GEOSYNTHETICS

Geosynthetics, which become available after site clearing and demolition of a construction can be dumped on a landfill, burned or recycled. Special measures must be taken to prevent emission to the environment. Geosynthetic pieces available as waste may be converted to small fibres of desired geometrical dimensions for using them to improve the soils, for which details are presented in Chapter 9.

## 7.18   INSTALLATION SURVIVABILITY REQUIREMENTS

When a geosynthetic is used in a specific application, or in the solution of a particular engineering problem, it is for the designer to determine what properties are required. The role of the designer should be to specify the properties which the geosynthetic is required to have in order to solve a specific problem, rather than starting with a geosynthetic of predetermined properties and defining the problem which this geosynthetic might solve. However, the recommended geosynthetic should always satisfy the installation survivability requirements.

While selecting geosynthetics, particularly geotextiles, for some applications, one can follow the M 288-06 (2011) geotextile specifications laid down by the American Association of State Highway and Transportation Officials (AASHTO) to meet the requirements for the geotextile survivability from installation stresses (AASHTO, 2013). These specifications set forth a set of physical, mechanical and endurance properties that must be met or exceeded by the geotextile for its use in subsurface drainage, separation, stabilization (an interrelated group of separation, filtration and reinforcement functions), erosion control, temporary silt fence and paving fabrics. Table 7.1 provides the strength properties for three geotextile classes (Class 1, Class

2 and Class 3) that are required for survivability under typical installation conditions for different functions. Class 1 is recommended for use in more severe or harsh installation conditions where there is a great potential for geotextile damage. Class 2 can be used as a default classification in the absence of site-specific information. One can use Class 3 for mild survivability conditions.

The geotextile should conform to the properties of Table 7.1 based on the geotextile class required in Tables 7.2, 7.3, 7.4, 7.5, or 7.6 for the indicated application. Property requirements for the temporary silt fence and the paving fabrics are given in Table 7.7 and 7.8, respectively.

The specifications in Table 7.2 are applicable to placing a geotextile against a soil to allow for the long-term passage of water into a subsurface drain system retaining the in situ soil. The primary function of the geotextile in the subsurface drainage applications is filtration, which depends on the in situ soil gradation, plasticity and hydraulic conditions.

The specifications in Tables 7.3 and Table 7.4 are applicable to the use of a geotextile to function as a separator to prevent mixing of a subgrade soil and a granular cover material in situations where water seepage through the geotextile is not a primary function. The most common application of the geotextile separator is in pave-

Table 7.1 Geotextile strength property requirements (AASHTO M 288-06 (2011)) (after AASHTO, 2013).

| Property | ASTM test methods | Units | Geotextile class[1,2] | | | | | |
| | | | Class 1 | | Class 2 | | Class 3 | |
| | | | Elongation < 50%[3] | Elongation ≥ 50%[3] | Elongation < 50%[3] | Elongation ≥ 50%[3] | Elongation < 50%[3] | Elongation ≥ 50%[3] |
|---|---|---|---|---|---|---|---|---|
| Grab strength | D 4632 | N | 1400 | 900 | 1100 | 700 | 800 | 500 |
| Sewn seam strength[4] | D 4632 | N | 1260 | 810 | 990 | 630 | 720 | 450 |
| Tear strength | D 4533 | N | 500 | 350 | 400[5] | 250 | 300 | 180 |
| Puncture strength | D 6241 | N | 2750 | 1925 | 2200 | 1375 | 1650 | 990 |
| Permittivity | D 4991 | s⁻¹ | Minimum property values for permittivity, AOS and UV stability are based on the geotextile application. Refer to Table 7.2 for subsurface drainage, Table 7.3 and Table 7.4 for separation, Table 7.5 for stabilization, and Table 7.6 for permanent erosion control. | | | | | |
| Apparent opening size (AOS) | D 4751 | mm | | | | | | |
| Ultraviolet stability (retained strength) | D 4355 | % | | | | | | |

[1]Required geotextile class is designated in Tables 7.2, 7.3, 7.4, or 7.6 for the indicated application. The severity of installation conditions for the application generally dictates the required geotextile class. Class 1 is specified for more severe or harsh installation conditions where there is a greater potential for the geotextile damage, and Classes 2 and 3 are specified for less severe conditions.
[2]All numeric values represent Minimum Average Roll Values (MARV) in the weaker principal direction.
[3]As measured in accordance with ASTM D 4632.
[4]When sewn seams are required.
[5]The required MARV tear strength for woven monofilament geotextiles is 250 N.

ment structures constructed over soils with a soaked California bearing ratio (CBR) value equal to or greater than 3 (undrained shear strength ($c_u$) equal to or greater than approximately 90 kPa).

The specifications in Table 7.5 are applicable to the use of a geotextile in wet, saturated conditions, due to a high groundwater table or due to prolonged periods of wet weather, to provide the combined functions of separation, filtration and/or

*Table 7.2* Subsurface filtration (called the 'drainage' in the specification) geotextile property requirements (AASHTO M 288-06 (2011)) (after AASHTO, 2013).

| Property | ASTM test methods | Units | Requirements for Percent in situ soil passing 0.075 mm[1] | | |
| --- | --- | --- | --- | --- | --- |
| | | | < 15 | 15 to 20 | > 50 |
| Geotextile class | | | Class 2 from Table 7.1[2] | | |
| Permittivity[3,4] | D 4491 | s$^{-1}$ | 0.5 | 0.2 | 0.1 |
| Apparent opening size[3,4] | D 4751 | mm | 0.43 max. avg. roll value | 0.25 max. avg. roll value | 0.22[5] max. avg. roll value |
| Ultraviolet stability (retained strength) | D 4355 | % | 50% retained after 500 h of exposure | | |

[1]Based on the particle-size analysis of in situ soil in accordance with AASHTO T 88.
[2]Default geotextile selection. The Engineer may specify a Class 3 geotextile from Table 7.1 for trench drain applications based on one or more of the following:
  a The engineer has found Class 3 geotextiles to have sufficient survivability based on field experience.
  b The engineer has found Class 3 geotextiles to have sufficient survivability based on laboratory testing and visual inspection of a geotextile sample removed from a field test section constructed under anticipated field conditions.
  c Subsurface drain depth is less than 2 m; drain aggregate diameter is less than 30 mm; and compaction requirement is less than 95% of AASHTO T 99.
[3]These default filtration property values are based on the predominant particle sizes of in situ soil. In addition to the default permittivity value, the engineer may require geotextile permeability and/or performance testing based on engineering design for drainage systems in problematic soil environments.
[4]Site-specific geotextile design should be performed especially if one or more of the following problematic soil environments are encountered: unstable or highly erodible soils such as non-cohesive silts; gap graded soils; alternating sand/silt laminated soils; dispersive clays; and/or rock flour.
[5]For cohesive soils with a plasticity index in accordance with AASHTO T 90 greater than 7, geotextile maximum average roll value for apparent opening size is 0.30 mm.

*Table 7.3* Separation geotextile property requirements (AASHTO M 288-06 (2011)) (after AASHTO, 2013).

| Property | ASTM test methods | Units | Requirements |
| --- | --- | --- | --- |
| Geotextile class | | | see Table 7.4 |
| Permittivity | D 4491 | s$^{-1}$ | 0.02[1] |
| Apparent opening size | D 4751 | mm | 0.60 max. avg. roll value |
| Ultraviolet Stability (retained strength) | D 4355 | % | 50% retained after 500 h of exposure |

[1]Default value. Permittivity of the geotextile should be greater than that of the soil ($\Psi_g > \Psi_s$). The engineer may also require the permeability of the geotextile to be greater than that of the soil ($K_g > K_s$).

reinforcement, called stabilization. The most common stabilization application of the geotextile is in pavement structures constructed over soils with soaked CBR value between 1 and 3 ($c_u$ between approximately 30 kPa and 90 kPa). Note that the reinforcement is a site-specific design issue in most cases.

The specifications in Table 7.6 are applicable to the use of a geotextile between energy absorbing armour systems and the in situ soil to prevent the soil loss resulting in excessive scour and to prevent the hydraulic uplift pressures causing instability of the permanent erosion control system. These specifications should not be applied to other types of soil erosion control geosynthetic systems, such as turf reinforcement mats. The primary function of the geotextile in permanent erosion control systems is filtration, which depends on the hydraulic conditions and in situ soil gradation, density and plasticity.

The specifications in Table 7.7 are applicable to the use of a geotextile as a silt fence, which is a vertical, permeable interceptor designed to remove suspended soil from the overland water flow. The function of a temporary silt fence is to filter and allow the settlement of soil particles from sediment-laden water. The purpose is to prevent the eroded soil from being transported off the construction site by the water runoff.

The specifications in Table 7.8 are applicable to the use of a paving fabric, saturated with bitumen (asphalt cement), between the pavement layers. The function of the paving fabric is to act as a waterproofing and stress-relieving membrane within the pavement structure. These specifications do not describe the fabric membrane systems specifically designed for pavement joints and localized (spot) repairs.

Note that, in Tables 7.1 to 7.8, all the property values, with the exception of apparent opening size (AOS), represent the minimum average roll values (MARV) in the weakest principal direction. The values of AOS represent the maximum average roll values. The geotextile properties for each class are dependent upon the geotextile elongation. Also note that the guidelines presented here are based on the geotextile survivability from the installation stresses, and, therefore, they should be used as a base point only. The specific design and the site conditions often require individual geotextile properties and construction recommendations to be modified to ensure that the guidelines are consistent with the project needs based on engineering design and experience.

## EXAMPLE 7.1

Using the AASHTO M 288-06 (2011) specifications, recommend the properties of a geotextile (elongation, $\varepsilon < 50\%$) required for its application as a separator between the soil subgrade and the granular base course in the pavement structure. Assume that the subgrade has been cleared of all obstacles except grass, weeds, leaves and fine wood debris, and the surface is smooth and level with all depressions filled. Also assume that the maximum pressure applied by the construction equipment at the installed geotextile level is 20 kPa.

## SOLUTION

It is given that the subgrade is cleared of all obstacles except grass, weeds, leaves and fine wood debris, and the surface is smooth and level with all depressions filled. Also the maximum pressure applied by the construction equipment at the installed geotextile level is 20 kPa ($\leq$ 25 kPa). Hence from Table 7.4, the geotextile class is Class 3. For this class with elongation, $\varepsilon < 50\%$, from Tables 7.1 and 7.3, the recommended properties for the geotextile as a separator in the pavement structure are the following:

Permittivity $\geq 0.02$ s$^{-1}$
AOS $\leq 0.60$ mm
Grab strength $\geq 800$ N
Sewn seam strength $\geq 720$ N
Tear strength $\geq 300$ N
Puncture strength $\geq 1650$ N
UV stability (retained strength) $\geq$ (50% of 800 N = 400 N) after 500 h of exposure.

## EXAMPLE 7.2

Using the AASHTO M 288-06 (2011) specifications, recommend the properties of a nonwoven geotextile (elongation, $\varepsilon > 50\%$) required for its permanent erosion control application adjacent to a soil with 70% passing the 0.075 mm sieve.

## SOLUTION

For the nonwoven geotextile, from Table 7.6, the geotextile class is Class 1. For this class with elongation, $\varepsilon > 50\%$ and for soil with 70% passing the 0.075 mm sieve, from Tables 7.1 and 7.6, the recommended properties for the nonwoven geotextile as a filter in the permanent erosion control application are the following:

Permittivity $\geq 0.1$ s$^{-1}$
AOS $\leq 0.22$ mm
Grab strength $\geq 900$ N
Sewn seam strength $\geq 810$ N
Tear strength $\geq 350$ N
Puncture strength $\geq 1925$ N
UV stability (retained strength) $\geq$ (50% of 900 N = 450 N) after 500 h of exposure.

Table 7.4 Required degree of survivability as a function of subgrade conditions, construction equipment and lift thickness (Class 1, 2 and 3 properties are given in Table 7.1; Class 1+ properties are higher than Class 1, but not defined and, if used, must be specified by the purchaser)[1] (AASHTO M 288-06 (2011)) (after AASHTO, 2013).

| Subgrade conditions | Low ground-pressure equipment (≤ 25 kPa) | Medium ground-pressure equipment (>25 to ≤ 50 kPa) | High ground-pressure equipment (> 50 kPa) |
|---|---|---|---|
| Subgrade has been cleared of all obstacles except grass, weeds, leaves and fine wood debris. Surface is smooth and level so that any shallow depressions and humps do not exceed 450 mm in depth or height. All larger depressions are filled. Alternatively, a smooth working table may be placed. | Low (Class 3) | Moderate (Class 2) | High (Class 1) |
| Subgrade has been cleared of obstacles larger than small to moderate-sized tree limbs and rocks. Tree trunks and stumps should be removed or covered with a partial working table. Depressions and humps should not exceed 450 mm in depth or height. Larger depressions should be filled. | Moderate (Class 2) | High (Class 1) | Very high (Class 1+) |
| Minimal site preparation is required. Trees may be felled, delimbed, and left in place. Stumps should be cut to project not more than ± 150 mm above subgrade. Geotextile may be draped directly over the tree trunks, stumps, large depressions and humps, holes, stream channels, and large boulders. Items should be removed only if placing the geotextile and cover material over them will distort the finished road surface. | High (Class 1) | Very high (Class 1+) | Not recommended |

[1] Recommendations are for 150 to 300 mm initial lift thickness. For other initial lift thicknesses:
300–450 mm: reduce survivability requirement one level;
450 to 600 mm: reduce survivability requirement two levels;
> 600 mm: reduce survivability requirement three levels.

For special construction techniques such as prerutting, increase the geotextile survivability requirement one level. Placement of excessive initial cover material thickness may cause bearing failure of the soft subgrade.

Table 7.5 Stabilization geotextile property requirements (AASHTO M 288-06 (2011)) (after AASHTO, 2013).

| Property | ASTM test methods | Units | Requirements |
|---|---|---|---|
| Geotextile class | | | Class I from Table 7.1[1] |
| Permittivity | D 4491 | s$^{-1}$ | 0.05[2] |
| Apparent opening size | D 4751 | mm | 0.43 max. avg. roll value |
| Ultraviolet stability (retained strength) | D 4355 | % | 50% retained after 500 h of exposure |

[1]Default geotextile selection. The engineer may specify a Class 2 or 3 geotextile from Table 7.1 based on one or more of the following:

  a The engineer has found the class of geotextile to have sufficient survivability based on field experience.

  b The engineer has found the class of geotextile to have sufficient survivability based on laboratory testing and visual inspection of a geotextile sample removed from a field test section constructed under anticipated field conditions.

[2]Default value. Permittivity of the geotextile should be greater than that of the soil $\Psi_g > \Psi_s$. The engineer may also require the permeability of the geotextile to be greater than that of the soil $k_g > k_s$.

Table 7.6 Permanent erosion control geotextile property requirements (AASHTO M 288-06 (2011)) (after AASHTO, 2013).

| Property | ASTM test methods | Units | Requirements for Percent in situ soil passing 0.075 mm[1] | | |
|---|---|---|---|---|---|
| | | | <15 | 15 to 20 | >50 |
| Geotextile class | For woven monofilament geotextiles, Class 2 from Table 7.1[2] | | | | |
| | For all other geotextiles, Class 1 from Table 7.1[2,3] | | | | |
| Permittivity[1,4] | D 4491 | s$^{-1}$ | 0.7 | 0.2 | 0.1 |
| Apparent opening size[3,4,5] | D 4751 | mm | 0.43 max. avg. roll value | 0.25 max. avg. roll value | 0.22[5] max. avg. roll value |
| Ultraviolet stability (retained strength) | D 4355 | % | 50% retained after 500 h of exposure | | |

[1]Based on particle-size analysis of in situ soil in accordance with AASHTO T 88.

[2]As a general guideline, the default geotextile selection is appropriate for conditions of equal or less severity than either of the following:

a Armour layer stone weights do not exceed 100 kg, stone drop height is less than 1 m, and no aggregate bedding layer is required.

b Armour layer stone weighs more than 100 kg, stone drop height is less than 1 m, and the geotextile is protected by a 150-mm thick aggregate bedding layer designed to be compatible with the armour layer. More severe applications require an assessment of geotextile survivability based on a field trial section and may require a geotextile with higher strength properties.

[3]The engineer may specify a Class 2 geotextile from Table 7.1 based on one or more of the following:

a The engineer has found Class 2 geotextiles to have sufficient survivability based on field experience.

b The engineer has found Class 2 geotextiles to have sufficient survivability based on laboratory testing and visual inspection of a geotextile sample removed from a field test section constructed under anticipated field conditions.

c Armour layer stone weighs less than 100 kg, stone drop height is less than 1 m, and the geotextile is protected by a 150-mm thick aggregate bedding layer designed to be compatible with the armour layer.

d Armour layer stone weights do not exceed 100 kg, and stone is placed with a zero drop height.

[4]These default filtration property values are based on the predominant particle sizes of in situ soil. In addition to the default permittivity value, the engineer may require geotextile permeability and/or performance testing based on engineering design for drainage systems in problematic soil environments.

[5]See the following:

a Site-specific geotextile design should be performed especially if one or more of the following problematic soil environments are encountered: unstable or highly erodible soils such as non-cohesive silts; gap graded soils; alternating sand/silt laminated soils; dispersive clays; and/or rock flour.

b For cohesive soils with a plasticity index greater than 7, geotextile maximum average roll value for apparent opening size is 0.30 mm.

Table 7.7 Temporary silt fence geotextile property requirements (AASHTO M 288-06 (2011)) (after AASHTO, 2013).

| | | | | Requirements | |
| | | | | Unsupported silt fence | |
| Property | ASTM test methods | Units | Supported silt fence[1] | Geotextile elongation ≥ 50%[2] | Geotextile elongation < 50%[2] |
|---|---|---|---|---|---|
| Maximum post spacing | | | 1.2 m | 1.2 m | 2 m |
| Grab strength | D 4632 | N | | | |
| Machine direction | | | 400 | 550 | 550 |
| X-Machine direction | | | 400 | 450 | 450 |
| Permittivity[3] | D 4491 | s$^{-1}$ | 0.05 | 0.05 | 0.05 |
| Apparent opening size | D 4751 | mm | 0.60 max. avg. roll value | 0.60 max. avg. roll value | 0.60 max. avg. roll value |
| Ultraviolet stability (retained strength) | D 4355 | % | | 70% retained after 500 h of exposure | |

[1] Silt fence support shall consist of 14-gauge steel wire with a mesh spacing of 150 mm by 150 mm or prefabricated polymeric mesh of equivalent strength.

[2] As measured in accordance with ASTM D 4632.

[3] These default filtration property values are based on empirical evidence with a variety of sediments. For environmentally sensitive areas, a review of previous experience and/or site or regionally specific geotextile tests, such as ASTM D 5141, should be performed by the agency to confirm suitability of these requirements.

*Table 7.8* Paving fabric geotextile property requirements[1] (AASHTO M 288-06 (2011)) (after AASHTO, 2013).

| Property | ASTM test methods | Units | Requirements |
|---|---|---|---|
| Grab strength | D 4632 | N | 450 |
| Ultimate elongation | D 4632 | % | ≥50 |
| Mass per unit area | D 5261 | g/m² | 140 |
| Asphalt retention | D 6140 | l/m² | [2,3] |
| Melting point | D 276 | °C | 150 |

[1]All numeric values represent MARV in the weaker principal direction.
[2]Asphalt required to saturate paving fabric only. Asphalt retention must be provided in manufacturer's certification. Value does not indicate the asphalt application rate required for construction.
[3]Product asphalt retention property must meet the MARV value provided by the manufacturer's certification.

## Chapter summary

1  The successful installation of a geosynthetic is largely dependent on the construction technique and the management of construction activities. Hence, in general, most of the geosynthetic-related failures are reported to be construction-related.

2  The selection of a geosynthetic is generally done keeping in view the general objective of its use. Methods of transport, storage and placement also govern the selection of geosynthetics. Many times, cost and availability of geosynthetics may govern their selection.

3  Before storing or unrolling the geosynthetic rolls, or both, the individual roll identification should be verified and should be compared with the packing list.

4  Based on the quality control tests, on any given project, the minimum average roll value (MARV) must meet or exceed the designer's specified value for the geosynthetic to be acceptable.

5  Good practice dictates that on site, the geosynthetic should be stored properly and handled according to the manufacturer's recommendations.

6  During the site preparation for the geosynthetic installation, sharp objects, such as boulders, stumps of trees or bushes, which might puncture or tear the geosynthetic, should be removed if they are lying on the site.

7  When two sheets of similar or dissimilar geosynthetics (or related materials) are attached to each other by a suitable means, the junction so formed is known as the joint. There are several jointing methods, such as overlapping, sewing, stapling, gluing, thermal bonding, etc. Efforts should be made to make the seam efficiency near 100%.

8  Damage due to UV light exposure can usually be avoided by not laying more geosynthetic in a day than can be covered by fill on the same day. No equipment that could damage the geosynthetic should be allowed to travel directly on the geosynthetic.

9  To maintain the position of a geosynthetic sheet before covering with soil/fill, the edges of the sheet may be weighted or anchored in trenches in order to providing the significant pullout resistance.

10  While selecting geosynthetics, particularly geotextiles, for some applications, one can follow the M 288-06 (2011) geotextile specifications laid down by the American Association of State Highway and Transportation Officials (AASHTO) to meet the requirements for the geotextile survivability from installation stresses.

## Questions for practice

(Select the most appropriate answer to the multiple-choice e questions from Q 7.1 to Q 7.10.)

7.1  The construction-related failures of geosynthetic applications are caused mainly by
(a)  the loss of strength due to UV exposure
(b)  the lack of proper overlap
(c)  the high installation stresses
(d)  all of the above.

7.2  The minimum number of test samples for 5000 m² or less area of geosynthetic is generally taken as
(a)  1
(b)  2
(c)  3
(d)  5.

7.3  At no time the geosynthetics should generally be exposed to UV light for a period exceeding
(a)  one week
(b)  two weeks
(c)  three weeks
(d)  four weeks.

7.4  The temperature at the geosynthetic storage site should not generally exceed
(a)   21 °C
(b)   27 °C
(c)   70 °C
(d)  100 °C

7.5  The minimum geosynthetic overlap is generally
(a)   30 mm
(b)  150 mm
(c)  300 mm
(d)  600 mm.

7.6  The temperature during the geosynthetic seaming should be between
(a)  0 °C and 20 °C
(b)  4.5 °C and 20 °C
(c)  0 °C and 40 °C
(d)  4.5 °C and 40 °C.

7.7 When placing the fill material on the sloped surface, the fill material should always be kept from the
(a) base of the slope
(b) mid-point of the slope
(c) top of the slope
(d) all of the above.

7.8 When releasing the stone that weighs less than 120 kg on the geotextile layer installed over a well-prepared surface, the maximum drop height, in general, can be
(a) 0.30 m
(b) 0.45 m
(c) 0.75 m
(d) 1.00 m.

7.9 As per M 288-06 (2011) geotextile specifications, the geotextile class 2 is recommended
(a) for use in more severe or harsh installation conditions
(b) as a default classification in the absence of site-specific information
(c) for use in the mild survivability conditions
(d) all of the above.

7.10 The primary function of the geotextile in permanent erosion control systems is
(a) reinforcement
(b) separation
(c) filtration
(d) drainage.

7.11 Why does the installation of geosynthetics in practice require a degree of care and consideration?

7.12 Give reasons why the white-surfaced textured HDPE geomembranes are preferred over the other types for their use in the lining of ponds, reservoirs and canals.

7.13 What are the general guidelines regarding the proper care and handling of geotextiles during the installation process?

7.14 How can you reduce friction damage during geosynthetic installation?

7.15 What is the difference between a geosynthetic joint (seam) and a geosynthetic connection?

7.16 Describe the seaming methods for geomembrane panels.

7.17 Under what type of weather conditions should the field seaming be conducted?

7.18 List the factors influencing the seaming quality.

7.19 Give the basic description of a bodkin joint.

7.20 What type of test would you recommend to evaluate the strength of a sewn seam?

7.21 Why is a uniform compaction of subgrade soils beneath the geosynthetic layer important?

7.22 What protection is required for the geosynthetics during construction and service life?

7.23 What should be examined before covering the installed geosynthetic with soil?

7.24 What are the different approaches for providing the bond length to the installed geosynthetic?

7.25 What is the effect of prestressing the geosynthetic installed in a granular fill?

7.26 What is the purpose of certification in geosynthetic engineering?

7.27 What are the three geotextile classes that are defined for survivability under typical installation conditions for different functions, as recommended by the American Association of State Highway and Transportation Officials (AASHTO)?

7.28 Using the AASHTO M 288-06 (2011) specifications, recommend the properties of a geotextile (elongation, $\varepsilon > 50\%$) required for its application as a separator between the soil subgrade and the granular base course in the pavement structure. Assume that the subgrade has been cleared of all obstacles except grass, weeds, leaves and fine wood debris, and the surface is smooth and level with all depressions filled. Also assume that the maximum pressure applied by the construction equipment at the installed geotextile level is 40 kPa.

7.29 Using the AASHTO M 288-06 (2011) specifications, recommend the properties of a nonwoven geotextile (elongation, $\varepsilon < 50\%$) required for its application as a subsurface drain filter adjacent to a soil with 40% passing the 0.075 mm sieve under typical installation survivability conditions.

7.30 Using the AASHTO M 288-06 (2011) specifications, recommend the properties of a woven geotextile (elongation, $\varepsilon > 50\%$) required for its permanent erosion control application adjacent to a soil with 12% passing the 0.075 mm sieve.

7.31 What are the geotextile property requirements as per AASHTO M288-06 (2011) specifications for the geotextile application as a paving fabric?

## References

AASHTO (American Association of State Highway and Transportation Officials) (2013). *Geotextile Specification for Highway Applications*, AASHTO Designation: M 288-06 (2011), *Standard Specifications for Transportation Materials and Methods of Sampling and Testing*, Thirty-third Edition, Part 1B: Specifications M 280-R 63, American Association of State Highway and Transportation Officials, Washington, DC, USA.

Allen, T.M. (1991). Determination of long-term tensile strength of geosynthetics: A state-of-the-art review. *Proceedings of Geosynthetics '91*, IFAI, pp. 351–379.

ASTM (American Society for Testing and Materials) (2009). ASTM D6213-97, *Standard Practice for Tests to Evaluate the Chemical Resistance of Geogrids to Liquids*. ASTM International, West Conshohocken, PA, USA.

Bonaparte, R., Ah-Line, C., Charron, R. and Tisinger, L. (1988). Survivability and durability of a nonwoven geotextile. *Proceedings of ASCE Symposium on Soil Improvement*, pp. 68–91.

Brau, G. (1996). Damage of geosynthetics during installation – experience from real sites and research work. *Proceedings of the 1st European Conference on Geosynthetics*. Eurogeo 1, Netherland.

Bush, D.I. and Swan, D.B.G. (1988). An assessment of the resistance of TENSAR SR2 to physical damage during the construction and testing of a reinforced soil wall. *The Application of Polymeric Reinforcement in Soil Retaining Structures*, Jarrett, P.M. and McGown, A., Editors, NATO ASI Series, Kluwer Academic Publishers, Dordrecht, The Netherlands, pp. 173–180.

Giroud, J.P. (1994). Quantification of geosynthetic behaviour. *Proceedings of the 5th International Conference on Geotextiles, Geomembranes and Related Products*. Singapore, pp. 1249–1273.

Heibaum, M.H. (1998). Protecting the geotextile filter from installation harm. *Geotechnical Fabrics Report*, June–July 1998, 26–29.

Holtz, R.D., Christopher, B.R. and Berg, R.R. (1997). *Geosynthetic Engineering*. BiTech Publishers Ltd., Canada.

Hullings, D.E. and Sansone, L.J. (1997). Design concerns and performance of geomembrane anchor trenches. *Geotextiles and Geomembranes*, 15, 4–6, pp. 403–417.

Ingold, T.S. (1994). *The Geotextiles and Geomembranes Manual.* Elsevier Advanced Technology, UK.

Ingold, T.S. and Miller, K.S. (1988). *Geotextiles Handbook.* Thomas Telford Ltd., London, U.K.

Ivy, N. and Narejo, D. (2003). Canal lining with HDPE. *Geotechnical Fabrics Report*, IFAI, June/July 2003, pp. 24–26.

John, N.W.M. (1987). *Geotextiles.* Blackie, London.

Koerner, R. M. (2005). *Designing with Geosynthetics.* Fifth Edition, Prentice Hall, New Jersey, USA.

Murray, R.T., McGown, A., Andrawes, K.Z. and Swan, D. (1986). Testing joints in geotextiles and geogrids. *Proceedings of the 3rd International Conference on Geotextiles.* Vienna, Austria, pp. 731–736.

Pilarczyk, K.W. (2000). *Geosynthetics and Geosystems in Hydraulic and Coastal Engineering.* A.A. Balkema, Rotterdam, Netherlands.

Paulson, J.N. (1990). Summary and evaluation of construction related damage to geotextiles in reinforcing applications. *Proceedings of the 4th International Conference on Geotextiles, Geomembranes and Related Products*, pp. 615–619.

Qian, X, Koerner, R.M. and Gray, D.H. (2002). *Geotechnical Aspects of Landfill Design and Construction.* Prentice Hall, New Jersey.

Rainey, T. and Barksdale, B. (1993). Construction induced reduction in tensile strength of polymer geogrids. *Proceedings of Geosynthetics '93*, IFAI, pp. 729–742.

Rankilor, P.R. and Heiremans, F. (1996). Properties of sewn and adhesive bonded joints between geosynthetic sheets. *Proccedings of the 1st European Geosynthetics Conference*, Netherlands, pp. 261–269.

Sandri, D., Martin, J.S., Vann, C.W., Ferrer, M. and Zeppenfeldt, I. (1993). Installation damage testing of four polyester geogrids in three soil types. *Proceedings of Geosynthetics '93*, AFAI, pp. 743–755.

Shukla (2003). How to select geosynthetics for field applications. *Geosynthetic Pulse*, India, **1**, 5, p. 2.

Shukla, S.K. (2002). *Geosynthetics and Their Applications.* Thomas Telford, London.

Shukla, S.K. (2012). *Handbook of Geosynthetic Engineering*, Second Edition, Thomas Telford, London.

Shukla, S.K and Yin, J.H. (2006). Fundamentals of Geosynthetic Engineering, Taylor and Francis, London.

USDI (US Department of the Interior) (1992). *Design Standards # 13: Embankment Dams.* US Department of the Interior, Bureau of Reclamation , Denver.

Van Santvoort, G.) (1994). *Geotextiles and Geomembranes in Civil Engineering.* A.A. Balkema, Rotterdam, The Netherlands.

Van Santvoort, G. (1995). *Geosynthetics in Civil Engineering.* A.A. Balkema, Rotterdam, The Netherlands.

Voskamp, W., Wichern, H.A.M. and Wijk, W. van (1990). Installation problems with geotextiles, an overview of producer's experience with designers and contractors. *Proceedings of the 4th International Conference on Geotextiles, Geomembranes and Related Products*, The Hague, pp. 627–630.

## Answers to selected questions

7.1   (d)
7.3   (b)
7.5   (c)
7.7   (a)
7.9   (b)

# Chapter 8

# Quality, performance monitoring and economic evaluation

## 8.1 INTRODUCTION

The proper and intended functioning of a geosynthetic product or system in an engineered facility is strongly dependent on the quality of the material and construction, which can be checked through a well-planned field performance monitoring programme. Users of geosynthetics must, therefore, be familiar with quality evaluation and field performance monitoring techniques.

Geosynthetics are used in most applications for achieving technical benefits, environmental benefits and/or the overall cost savings. Their use may result in a lower initial cost and/or greater durability and longer life, thus reducing the maintenance costs. The cost analysis of a geosynthetic-related project needs careful handling when taking decisions for acceptance or rejection of the option of using geosynthetics in the project just only on the basis of its cost.

This chapter introduces some aspects related to the quality evaluation, the field performance monitoring, and the fundamentals of economic evaluation and related experiences from some completed geosynthetic-related projects and reported economic studies.

## 8.2 QUALITY AND ITS EVALUATION

*Quality* of a geosynthetic product or geosynthetic application is the confidence that can be placed in it, consistently meeting the numerically claimed variation limits in properties and/or functioning taken into account by the design engineer and extrapolated into the in situ conditions. Standards, test methods, testing frequencies, tolerances and corrective actions are the means by which the quality can be measured and controlled. The purpose of quality evaluation is to facilitate the continuous improvement of geosynthetic products and geosynthetic applications and to fully document the performance relative to a target.

In an ideal world, quality considerations for geosynthetics would not be needed. A design engineer would properly design and specify a material, and the contractor would install the material in accordance with the design document, occasionally calling the engineer for a design clarification. Unfortunately, we do not live in an ideal world, but in the world of the 'low bid' contractor. The low bid often means the smallest profit margin, which potentially leads to attempts to cut corners and thus sacrifice the quality. That is why a proper quality evaluation system is necessary for any construction project.

*Quality control* (QC) refers to a planned system of operational techniques and activities which sustain the quality of the geosynthetic product, or the geosynthetic application that will satisfy the needs as per the project plans, specifications, and contractual and regulatory requirements; also the use of such a system. QC is provided by the manufacturers and the installers of the various components of the geosynthetic application. Geosynthetics must be properly manufactured and installed in a manner consistent with a minimum level of quality control as determined by the testing.

To achieve QC, the manufacturer and the installer need a quality assurance system, of which the quality control is only a part. *Quality assurance* (QA) refers to a planned system of actions necessary to provide adequate confidence that a geosynthetic product or a geosynthetic application will satisfy the needs as per the project plans, specifications, and contractual and regulatory requirements. QA is provided by an organization different from the organization that provides QC. An effective QA programme must define the route by which it will achieve its stated objectives. This will require an explanation of the qualifications, roles, responsibilities, authority and interaction of all parties involved. The programme must identify and describe all the QA activities and procedures (Menoff and Eith, 1990). The direct benefits of a QA programme to the facility owner/operator are to confirm that the project was constructed in accordance with the engineering design and the specifications and to provide the technical and legal documentation and certification of the quality of the work performed. The indirect benefit is perhaps of even greater importance. This is the knowledge that the product/facility has been manufactured/constructed with as great a degree of integrity as possible, thereby minimizing the potential problems during operation and/or post-closure and ultimate liability.

Geosynthetics have the following four levels of quality management associated with them:

1   Manufacturing quality control (MQC)
2   Manufacturing quality assurance (MQA)
3   Construction quality control (CQC)
4   Construction quality assurance (CQA)

Manufacturing quality control (MQC), normally performed by the geosynthetic manufacturer, is necessary to ensure minimum (or maximum) specified values in the manufactured product (Koerner and Daniel, 1993). A factory production control scheme should be established and documented in a manual prior to a geosynthetic type being placed on the market. Subsequently, any fundamental changes in raw materials and additives, manufacturing procedures or the control scheme that affect the properties or use of a geosynthetic should be recorded in the manual. Additionally, the manufacturing quality assurance (MQA) programme provides assurance that the geosynthetics were manufactured as specified in the certification documents and contract plans and specifications. MQA includes manufacturing facility inspections, verifications, audits and evaluation of the raw materials and the geosynthetic products to assess the quality of the manufactured materials.

Construction quality control (CQC), normally performed by the geosynthetic installer, is necessary to directly monitor and control the quality of a construction project in compliance with the plans and specifications. Construction quality assurance (CQA) programme provides assurance to the owner and regulatory authority, as applicable, that the structure was constructed in accordance with plans and design specifications. The CQA includes inspections, verifications, audits, and evaluations of materials and workmanship necessary to determine and document the quality of the constructed facility. In fact, design, construction and certification reporting are three phases in the life of a project during which a CQA monitor can have the beneficial role. The CQA monitor does not design the project. An experienced CQA monitor can, however, provide a significant role in the design phase by reviewing the design based on how a contractor would consider it (Thiel and Stewart, 1993).

Quite often, MQA and CQA are performed by the same organization. On the other hand, MQC and CQC are often performed by different organizations – the manufacturer and the installer. Of course, many of the larger manufacturers have their own installation crews (Qian *et al.*, 2002). Note that although MQA/CQA and MQC/CQC are separate activities, they have similar objectives and, in a smoothly running construction project, the processes will complement one another. Conversely, an effective MQA/CQA programme can lead to identification of deficiencies in the MQC/CQC process, but an MQA/CQA programme by itself, in complete absence of an MQC/CQC programme, is unlikely to lead to acceptable quality management. Quality is thus best ensured with effective MQC/CQC and MQA/CQA programmes. A major purpose of the MQA/CQA process is to provide the documentation for those individuals who were unable to observe the entire construction process so that those individuals can make informed judgments about the quality of construction for a project. MQA/CQA procedures and results must be thoroughly documented.

The geosynthetic CQA monitors should have the following general qualifications (Thiel and Stewart, 1993):

- familiarity with construction procedures and contract issues;
- experience of reviewing the test results, the quality control data and the contractor submittals;
- familiarity with the design issues regarding the type of construction project being monitored;
- ability to effectively communicate and prepare supporting documentation;
- geosynthetic monitoring experience gained under the supervision of more experienced individuals;
- experience of reporting, communicating and resolving deficiencies, and performing remediation activities.

The CQA monitor has to have certain qualities to effectively perform his/her job. The most important quality is probably assertiveness. When the monitor observes a problem, he or she has to act quickly, clearly identify the problem, recommend the corrective action, verify that the corrective action is taken, and document the issue. The monitor does this knowing that the contractors' schedule and progress will be affected. If the monitor is working alone on a remote site, this assertiveness is essential for success and can require courage. The CQA monitors who routinely monitor the projects with geosynthetics should visit a geosynthetic manufacturing plant and a laboratory where the geosynthetic tests are performed to gain a better understanding of the materials. Note that it is better to have a representative of the design engineer performing the CQA, because of liability issues and the need for understanding the design intent. In fact, the design engineer is in the best position to evaluate the construction materials and the methods because he/she understands the minute details and knows how the design elements are interconnected.

Figure 8.1 shows the usual interaction of the various elements in a total inspection programme. Note that the flow chart includes both the geosynthetic and the natural soil materials as both require similar concern and care.

Quality assurance and quality control are recognized as the critical factors in many geosynthetic applications. Acceptance testing should be performed on a geosynthetic product to determine whether or not an individual lot of the product conforms to specified requirements. It can be done prior to the geosynthetic shipment, directly after arrival of the geosynthetic roll at the site, and/or prior to the geosynthetic installation. Irregularities should be noted and reported.

Quality control on the construction sites is generally done by index testing which has been discussed in Chapter 2. Index testing involves the use of very simple techniques, which do not give definitive design parameters for a geosynthetic, but do give reproducible results, suitable for the QC and comparison of the geosynthetic products. The users should always make at least a check for type and quantity of geosynthetics being delivered. To enable the user on site to identify the geosynthetic products as being identical to the products ordered, the following basic information should be indelibly marked on the outside wrapping of each roll so that it is clearly visible before the roll is opened:

• Name of the manufacturer and/or supplier and country of origin;
• Product name (brand name/commercial name);
• Product type (descriptive number or the code);
• Unit identification (number or other code given on each unit, for example, the roll that allows the original manufacturer to trace at a later stage the production details, including place and date of production), for example, batch number and date;
• Mass of unit, in kg;
• Unit dimensions, such as length × width or area (of material, not of package), in m;
• Mass per unit area, in g/m²;
• Thickness of material, in mm;
• Major polymer type(s) for each component;
• Geosynthetic type/classification.

The above information should also be marked inside the core so as to be visible when the roll is partly used. More general and specific details are given on the data sheet, brochures, technical sales literature, or similar documents.

A geosynthetic should be identifiable by simple means at the time of its installation, even if it is no longer in the original packaging. For the convenience of the user, the markings of the product name and type should be printed along the edge of the geosynthetic product at regular intervals of at most 5 m. A simple check on

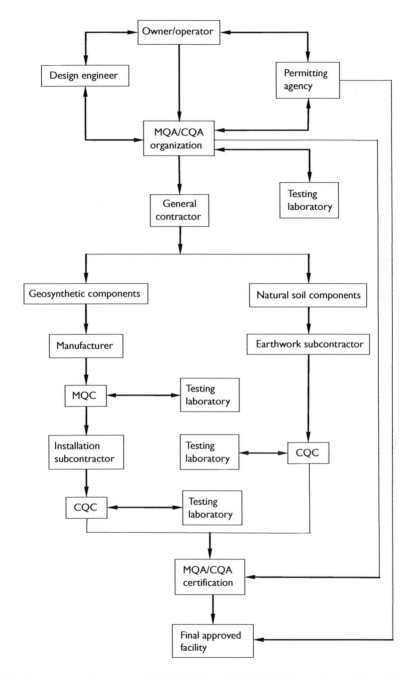

*Figure 8.1* Organizational structure of MQC/MQA and CQC/CQA inspection activities (Koerner and Daniel, 1993).

the mass per unit area may also be made using the basic equipment such as a balance and a measuring tape. In the case of high-risk applications, such as the use of geosynthetic filters in dams and geosynthetics as a soil reinforcement, testing of every roll, or at least every other roll should be performed. In such demanding applications, the most important property should be determined in addition to the basic index properties mentioned earlier. In the case of low-risk applications, such as the use of geosynthetic as a separator in unpaved roads, only the basic index tests need to be carried out for every 1 in 10 or 20 rolls. It is thus noted that the frequency and degree of QC testing are generally functions of types of application and the risk involved in that application. Also note that any test that is carried out over a long period of time and is based on a complex procedure is not considered to be a routine QC test.

For adopting the QC in reality, it is absolutely essential to have competent and professional construction inspection. The field supervisor must be properly trained to observe every phase of the construction and to ensure that:

- specified material is delivered to the project;
- rolls of geosynthetics are properly stored;
- testing requirements are verified;
- geosynthetic is not damaged during construction;
- specified sequence of construction operations is explicitly followed;
- seam/joint integrity is verified on the basis of testing and evaluation.

A trained and knowledgeable supervisor/inspector only should be allowed to perform the construction inspection. Efforts should be made to maintain good documentation of the construction activities.

## 8.3   FIELD PERFORMANCE MONITORING

As with all the geosynthetic applications, and especially with critical structures such as reinforced soil slopes and retaining walls, landfills and dams, competent and professional field inspection is absolutely essential for the successful construction. The field personnel must make sure that the specified geosynthetic is delivered to the project site, that the geosynthetic is not damaged during construction, and that the specified sequence of construction operations are explicitly followed. Other important details include the measures being adopted to minimize the geosynthetic exposure to the UV light. The field personnel should always review the checklist items for the project.

The in-situ monitoring of geosynthetics and geosynthetic-related systems usually has two goals. One addresses the integrity and safety of the system, whereas the other provides guidance and insight into the design process. It is to be noted that the purpose of the instrumentation in the geosynthetic-related projects is not only for research but also to verify the design assumptions and to control the construction.

It is important to conceive and execute a monitoring plan with clear objectives in mind. Dunnicliff (1988) provides a methodology for organizing the monitoring programme in geotechnical instrumentation. The checklist of specific steps that are recommended is given below:

1  Define the project conditions.
2  Predict the mechanism(s) that control the behaviour.
3  Define the question(s) that need answering.
4  Define the purpose of the instrumentation.
5  Select the parameter(s) to be monitored.
6  Predict the magnitude(s) of change.
7  Devise the remedial action.
8  Assign the relevant tasks.
9  Select the instruments.
10  Select the instrument locations.
11  Plan for factors influencing the measured data.
12  Establish the procedures for ensuring corrections.
13  List the purposes of each instrument.
14  Prepare a budget.
15  Write an instrument procurement specification.
16  Plan the installation.
17  Plan for regular calibration and maintenance.
18  Plan for data collection, processing, presentation, interpretation, reporting and implementation.
19  Write the contractual arrangements for field services.
20  Update the budget as the project progresses.

Such a checklist should be considered in planning for the in-situ monitoring of geosynthetic applications whenever permanent and/or critical installations are under consideration or are being otherwise challenged.

There are presently wide ranges of in situ monitoring methods/devices, which have generally resulted in reliable data. Table 8.1 provides the summary of the monitoring methods/devices as presented by Koerner (1996). In this table, the monitoring methods or devices are somewhat arbitrarily divided into recommended and optional categories. Table 8.2 gives a further description of the various methods/devices listed in Table 8.1. Since the monitoring is site-specific, its cost must be assessed on a case-by-case basis.

Geosynthetic product installation should be monitored by the engineer at a frequency appropriate to the project requirements. Excessive post-installation and other distresses should be monitored carefully by the engineer.

The experience demonstrates that the installed geomembrane liner can have potential leak paths, such as holes, tears, cuts, seam defects and burned-through zones. The damage to a geomembrane liner can be detected using the electrical leak detection systems developed in the early 1980s. Such systems have been used successfully to locate leak paths in electrically-insulating geomembranes such as polyethylene, polypropylene, polyvinyl chloride, chlorosulfonated polyethylene and bituminous geomembranes installed in ponds, reservoirs, canals, tanks and landfills. The types of potential leak paths have been related to the quality of the subgrade material, quality of the cover material, care in the cover material installation and quality of geomembrane installation.

*Table 8.1* Summary of monitoring methods/devices (after Koerner, 1996).

| Geosynthetic type | Function or application | Recommended | Optional |
|---|---|---|---|
| Geotextiles | Separation | • Water content measurements<br>• Pore water transducers | • Level surveying<br>• Earth pressure cells<br>• Inductance gauges |
| | Reinforcement | • Strain gauges<br>• Movement surveying<br>• Inclinometers<br>• Extensometers | • Earth pressure cells<br>• Inductance gauges<br>• Pore water transducers<br>• Water content measurements<br>• Settlement plates<br>• Temperature |
| | Filtration | • Water observation wells<br>• Pore water transducers | • Flow meters<br>• Turbidity meters<br>• Probes for pH, conductivity and/ or dissolved oxygen |
| | Drainage | (same as the geotextile filtration) | |
| | Barrier (e.g. reflective cracking) | • Surface deflections<br>• Level surveying<br>• Surface roughness measurements profilometry (for rut depths) | • Water content measurements |
| Geogrids | Walls | • Strain gauges<br>• Inclinometers<br>• Extensometers<br>• Monument surveying | • Earth pressure cells<br>• Piezometers<br>• Settlement plates<br>• Probes for pH<br>• Temperature |
| | Slopes | • Strain gauges<br>• Inclinometers<br>• Extensometers | • Earth pressure cells<br>• Piezometers<br>• Monument surveying |
| | Foundations | • Strain gauges<br>• Level surveying<br>• Extensometers | • Earth pressure cells<br>• Piezometers<br>• Settlement plates |
| Geonets | Drainage | • Flow meters<br>• Turbidity meters | • Probes for pH, conductivity and/or dissolved oxygen<br>• Piezometers |
| Geomembranes | Tensile stress<br>Temperature<br>Global leak monitoring | • Strain gauges<br>• Temperature measurement<br>• Flow meters<br>• Downgradient wells | • Turbidity meters<br>• Probes for pH, conductivity and/or dissolved oxygen |

(Continued)

Table 8.1  (Continued)

| Geosynthetic type | Function or application | Recommended | Optional |
|---|---|---|---|
| Geosynthetic clay liners | Global leak monitoring | • Flow meters<br>• Downgradient wells | • Turbidity meters<br>• Probes for pH, conductivity and/or dissolved oxygen |
| | Shear strength | • Extensometers<br>• Deformation telltales | • Gypsum cylinders<br>• Fiberglass wafers<br>• Strain gauges (inductance coils) |
| Geocomposites | Separation (e.g. erosion control) | • Flow meters<br>• Turbidity meters | • Level surveying |
| | Reinforcement | (same as geotextiles and geogrids) | |
| | Drainage (e.g. edge drains) | • Flow meters<br>• Turbidity meters | • Probes for pH, conductivity and/or dissolved oxygen |
| | Barrier | (same as geotextiles, geomembranes and GCLs) | |

The principle behind the electrical leak detection systems is to place a voltage across a geomembrane liner and then locate the areas where the electrical current flows through the discontinuities in the liner as shown schematically in Fig. 8.2. The liner must act as an insulator across which an electrical potential is applied. This electrical detection method of locating the potential leak paths in a geomembrane liner can be performed on exposed liners, on liners covered with water, or on liners covered by a protective soil layer. This technique can locate very smaller leak paths, smaller than 1 mm. This technique cannot be used during stormy weather when the geomembrane has been installed on a desiccated subgrade, or whenever conductive structures cannot be insulated or isolated. The details of techniques for electrical detection of leaks in geomembranes can be found in ASTM D6747-12 (ASTM, 2012).

A survey by Nosko and Touze-Foltz (2000) summarized the results of electrical damage detection systems installed at more than 300 sites and covering more than 3,250,000 m² of geomembrane liners. This survey showed that the majority of the damage (71%) was caused by stones, followed by heavy equipment (16%) (Fig. 8.3(a)). Interestingly, most of the failures (78%) were found to be located in the flat areas of the liner (bottom liner); only 9% were found at the corners and the edges of the landfills (Fig. 8.3(b)). It is also interesting to note from the reported surveys that the bulk of the defects were related to the mechanical damage caused by the placement of soil on top of the geomembrane. The readers can refer to the recommendations made by Giroud (2000) to minimize geomembrane installation and post-installation defects.

To measure directly the extent of degradation on the construction site, it may be desirable to extract the geosynthetic specimens at the following stages:

*Table 8.2* Selected description and commentary on the methods and devices listed in Table 8.2 (after Koerner, 1996).

| Category | Methods/Device | Resulting value/Information |
| --- | --- | --- |
| Surveying | Monument surveying | Lateral movement of vertical face |
| | Level surveying | Vertical movement of surface |
| | Settlement plates | Vertical movement at depth |
| Deformation | Telltales | Measures movement of fixed rods or wires, can accommodate any orientation |
| | Inclinometers | Measures vertical movement in a casing, inclined movements up to 45° |
| | Extensometers | Measures changes between two points in a borehole |
| Strain measurement | Electrical resistance gauges • Bonded foil • Weldable | Measures strain of a material over gauge length, typically, 0.25 to 150 mm |
| | Inductance gauges (coils) • Static measurements • Dynamic measurements | Measures movement between two embedded coils up to 1000 mm distance apart |
| | LVDT gauges | Measures movement between two fixed points 100 to 200 mm apart |
| Stress measurement | Earth pressure cells • Diaphragm type • Hydraulic type | Measures total stress acting on the cell, can be placed at any orientation, can also measure stress (pressure) against the walls and the structures |
| Soil moisture | Water observation wells | Measures stationary groundwater level |
| | Gypsum cylinders | Measures soil moisture content up to saturation |
| | Fiberglass wafers | Measures soil moisture content up to saturation |
| Groundwater pressure | Piezometers • Hydraulic type • Pneumatic type • Vibrating wire type • Electrical resistance type | Measures pore water pressures at any depth, can be installed as single point or in multiple point array, can be placed in any orientation |
| Temperature measurement | Bimetal thermometer | Measures temperature in adjacent area to ± 1.0°C |
| | Thermocouple | Measures temperature at a point to ± 0.5°C |
| | Thermistor | Measures temperature at a point to ± 0.1°C |
| Liquid quantity | Tipping buckets | Measures flow rates (relatively low values) |
| | Automated weirs | Measures flow rates (relatively high values) |
| | Flow meters | Measures flow rates (very high values) |
| Liquid quality | Turbidity meters | Measures suspended solids |
| | pH probes | Measures pH of liquid |
| | Conductivity probes | Measures conductivity of liquid |

1   just after installation
2   after a certain fraction of design life
3   at the end of the design life.

To monitor the condition of the extracted geosynthetic, physical and chemical analysis methods are recommended in addition to the normal index tests. In a struc-

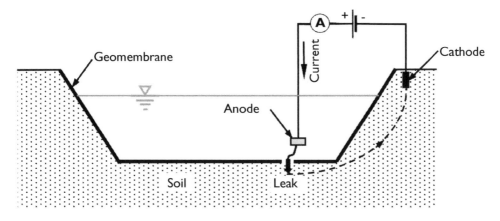

*Figure 8.2* Schematic of the electrical leak detection method.

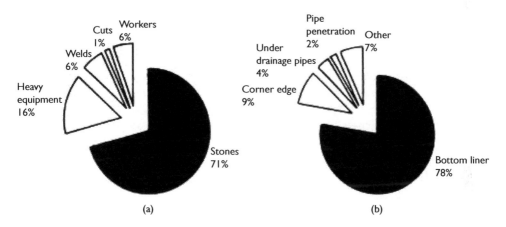

*Figure 8.3* (a) Cause of damage in geomembrane liners; (b) location of damage in geomembrane liners (modified from Nosko and Touze-Foltz, 2000; after Bouazza *et al.*, 2002).

ture whose integrity is critical, the geosynthetic specimens should be placed such that they can be extracted after a certain fraction of the design life and their condition can be compared with that predicted at the design stage. In this way, the user will obtain an advanced warning of any degradation that is occurring. For the sake of comparison with the specimens extracted, the user should retain the samples of pristine geosynthetic and a record of the original test results.

An important aspect of understanding the long-term performance of any material is to know the environmental conditions that it experiences during its service life. A key environmental condition, particularly for a geomembrane is in situ temperature. Thermocouples can be installed for the purpose of long-term temperature monitoring. The accelerated ageing test can also be performed in the laboratory (Koerner, 2001).

## 8.4   CONCEPTS OF ECONOMIC EVALUATION

The design engineer is usually confronted with an important task: whether a conventional solution or a geosynthetic-related solution should be preferred in a particular civil engineering project at a specific site. In order to give a rational decision, the data related to the following aspects should be analyzed carefully (Durukan and Tezcan, 1992):

- Relative economy
- Cost-performance efficiency (also called the cost-benefit (C/B) ratio)
- Factors of safety
- Feasibility
- Availability of materials
- Relative speed of construction

The *rate of relative economy*, $E_r$, is defined as:

$$E_r = \left( \frac{C_c - C_r}{C_r} \times 100 \right) \%$$

(8.1)

where $C_c$ is the cost of the conventional soil structure; and $C_r$ is the cost of the geosynthetic-reinforced soil structure.

The cost of a geosynthetic-related structure or application should typically be presented as an engineering estimate of the capital, operational, and maintenance costs. For having a general idea of the *cost-performance efficiency* of a geosynthetic or any other element of the geosynthetic-related structure, it can be represented as the normalized cost, $C_m$. In the case of a geosynthetic-reinforced soil retaining wall, $C_m$ can be defined as:

$$C_m = \frac{C}{T}$$

(8.2)

where $C_m$ is the normalized cost of the geosynthetic reinforcement carrying a safe tensile load of 1 kN on a 1-m-run wall; $C$ is the cost of 1 m$^2$ geosynthetic within a 1-m-run wall; and $T$ is the safe tensile resistance of one-layer geosynthetic for a 1-m-run wall. For any other reinforcing element of the structure, $C_m$ can be defined in a similar way, keeping in view the function of the element.

Note that the total cost of a geosynthetic-reinforced soil structure depends not only on the relative costs of the individual elements, but also on the geometry of the reinforced soil structure and its site location. For the purpose of determining the relative economy as well as the cost efficiency of reinforced soil structures, a comprehensive cost analysis should be performed by taking into account the costs (both direct and indirect) of various elements of the application.

The cost-benefit analysis basically consists of a comparison of the costs and benefits for a certain set of site-specific conditions. If the benefits exceed the costs for a certain set of conditions, the cost-benefit ratio is less than one. In the analysis, the preferred course of action is that which yields the smallest cost-benefit ratio (subject to being smaller than one). While making the cost-benefit analysis of environmental applications, one should consider the reduction of human health and ecological risk, and less $CO_2$ generation as the benefits of using the geosynthetics.

In general, the benefits of geosynthetic-related structures are difficult to measure like the benefits of other public projects, whereas the costs are more easily determined. For simplicity, one can only attempt to quantify the primary benefits or effectiveness with which the application goals can be met in monetary or non-monetary measure by the extent to which that alternative, if implemented, will attain the desired objective. The preferred alternative is then either the one that produces the maximum effectiveness for a given level of cost, or the minimum cost for a fixed level of effectiveness. In fact, the *cost-effectiveness method* allows us to compare alternative solutions on the basis of cost and non-monetary effectiveness measures.

The feasibility of a geosynthetic-related structure or application is strongly related to the type of problem in hand: an embankment on soft ground, a foundation, an unstable slope, a leaking dam or reservoir, a road, or a railway track. The construction time available, the speed of construction, the availability of construction materials and equipment, and the funds available are some of the factors which strongly govern the choice of using geosynthetic-related structures or applications in a particular field project.

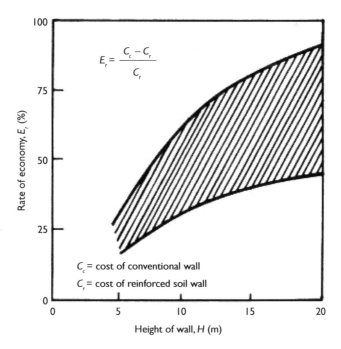

$$E_r = \frac{C_c - C_r}{C_r}$$

$C_c$ = cost of conventional wall

$C_r$ = cost of reinforced soil wall

Rate of economy, $E_r$ (%)

Height of wall, $H$ (m)

*Figure 8.4* Rate of economy in reinforced soil walls (after Anon., 1979).

## 8.5    EXPERIENCES OF COST ANALYSIS

In the case of reinforced soil walls, it is generally accepted that, under normal circumstances, and especially after a wall height of about 6 m, they become more economical, and also they are relatively easier and faster to build than their conventional counterparts (Ingold, 1982). Reinforced soil retaining walls are almost indispensable when normal slopes may not be constructed due to property line constraints, high expropriation costs, existence of important structures, or due to land being reserved for future structures. In most low to medium height retaining walls, the cost of the geosynthetic reinforcement is found to be less than 10% of the total wall cost. In these walls the greatest economy is obtained by using onsite soils for backfill and an inexpensive facing. In order to arrive at a scientific conclusion, however, a comparative cost analysis must be performed.

It has been determined by a group of researchers in the UK that the rate of relative economy of reinforced soil walls increases steadily with the height of the wall, as shown in Fig. 8.4. A similar cost effectiveness study of reinforced soil embankments by Christie (1982) in the UK showed that when space restrictions or high land-acquisition costs necessitated steep walls, it was almost unavoidable to use soil reinforcement. Murray (1982) also reported that a repair project for a cutting using the reinforced in-situ soil saved about 40% when compared with the conventional replacement techniques. It was reported by Bell *et al.* (1984) that the total cost of a series of geotextile-reinforced retaining walls varied between US $118 and US $134 per square metre of the wall surface. The average cost breakdown is shown in Table 8.3. In a blast protection embankment in London (UK), it was established by Paul (1984) that the geogrid-reinforced design was the most economical choice when compared with either the conventional reinforced concrete wall or the unreinforced soil wall. The relative costs for a 1 m run of the wall are shown in Table 8.4.

For evaluating the direct cost effect of geotextile applications in the context of India, four typical geotextile usages were examined by Ghoshal and Som (1993) for

Table 8.3 Geotextile-reinforced soil walls (after Bell et al., 1984).

| Item | Cost (US $/m²) | Share (%) |
|------|----------------|-----------|
| Geotextile | 23 | 19 |
| Labour | 7 | 6 |
| Equipment | 7 | 6 |
| Fill | 53 | 44 |
| Facing | 30 | 25 |
| Total | 120 | 100 |

Table 8.4 Cost comparisons of an embankment (after Paul, 1984).

| Wall type | Land width (m) | Cost (US $/m) |
|-----------|----------------|---------------|
| Reinforced concrete wall | 18.9 | 2625 |
| Geogrid-reinforced embankment | 13.5 | 1775 |
| Unreinforced embankment | 32.5 | 1911 |

four different regions of the country. Costs of material, labour and land were collected for the metropolitan cities Mumbai, Bangalore, Delhi and Kolkata. For identical soil data and design parameters, the variation of cost with or without geotextiles for the selected functions was determined. An examination of the economic analysis reveals that the use of geotextiles depends significantly on the unit cost of different inputs. The apparent cost-benefit derived by using the geotextiles is not uniquely determined on the basis of cost of geotextiles alone. For example, where the land cost is high, as in Mumbai, the economy of using the geotextiles becomes more predominant in the embankment slope stability function than in the separation function as required in an unpaved road. On the other hand, the separation function appears to give a greater economy in Kolkata than in Mumbai because of the higher cost of stone aggregate in Kolkata.

For estimating the cost-benefit ratio of the geosynthetic-reinforced unpaved roads, one must consider the future maintenance costs associated with both conventional and reinforced roadways. Such maintenance costs are dependent on the quality of the roadway initially constructed. In the case of geosynthetic-reinforced unpaved roads, the reduction of maintenance costs in excess of 25% has been reported (IFAI, 1992). For the subgrades having CBR values higher than 3, the geosynthetic-reinforced roadway may have an initial cost-benefit ratio near unity (that is, the additional cost of the geosynthetic is offset by a comparable savings in granular material), whereas it decreases as the strength of the existing soil subgrade decreases. This happens because for soil subgrades having CBR values lower than 3, the geosynthetic layers provide initial cost savings over alternative traditional stabilization methods (excavation and replacement, preloading and staged construction to slowly increase the subgrade strength, lime or cement stabilization) in addition to savings in the granular material (Barenberg, 1992).

Note that in the unpaved road construction, the site preparation and the type of granular material greatly controls the selection, and therefore the cost of the geosynthetic based on the survivability during its installation. Increased site preparation increases the construction costs but reduces the need for geosynthetics with higher survivability requirement. Similarly, the use of a granular soil with more angular particles produces a denser and stronger subbase/base, but increases the need and the cost for a more robust geosynthetic.

The cost effectiveness of a paving fabric interlayer system can be analyzed by comparing its cost with the cost of additional bituminous/asphaltic concrete needed to produce the same pavement rehabilitating results. For the retardation of reflective cracking, the paving interlayer system has been proven to be equivalent to approximately 30 mm of additional bituminous concrete overlay thickness. This is significant as the installed cost of a paving fabric and bituminous concrete interlayer system is generally less than one half the cost of the 25 mm of bituminous concrete. Thus, the pavement rehabilitation system with a paving fabric interlayer system is economical. In addition, the paving fabric interlayer system provides the functions of moisture barrier and stress relief for long-term fatigue resistance (IFAI, 1992).

For determining the cost-effectiveness of geosynthetic versus conventional drainage systems, simply compare the cost of the geosynthetic with the cost of a conventional granular filter layer. The use of a geosynthetic can allow a considerable reduction in the physical dimensions of the drain without a decrease in flow capacity, resulting in a reduction in the volume of excavation, the volume of granular material required, and the construction time necessary per unit length of the drain.

The total cost of a riprap-geotextile revetment system will depend on the actual application and the type of revetment selected. Additional cost for making special considerations while placing the geotextile below the water level should be considered. Cost of overlapping and pins are also required. To determine the cost-effectiveness, these points should be considered and then the cost-benefit ratios should be compared for the riprap-geotextile system versus the conventional riprap-granular filter systems or other available alternatives of equal technical feasibility and operational practicality.

For many slope stabilization projects, when all the financial aspects are considered, slope reinforcement is the most cost-effective technique to meet a change in grade. The issues include the following (Simac, 1992):

- Land acquisition costs
- Construction costs
- Soil-fill costs
- Financial return on usable area
- Future expansion

However, since the largest cost-savings component comes from reducing the land acquisition costs or optimizing the developable land from the purchased property, the slope reinforcement should be considered early in the project planning process to maximize the owner's economic benefits. It is considered that the slope reinforcement is justifiable when any of the following three site-specific factors apply to a project:

- High cost of the real estate
- Steep topography
- Expensive, unsuitable or scarce soil-fill materials

Slope reinforcement costs are offset by eliminating or minimizing the impact of these site-specific factors. Therefore, to limit the evaluation of slope reinforcement only to construction and soil-fill costs may underestimate its value to a project owner. One should evaluate the technique based on its design features and the total cost relative to other conventional solutions available. In fact, quite often, the slope reinforcement presents both the best technical and cost-efficient solution for the projects influenced by the key site-specific factors.

A reinforced slope will typically consist of three components: the soil backfill, the geosynthetic reinforcement, and a surface erosion control system. The relative cost of these components will be a function of the height of the embankment and the slope of the face. In general, the following trends have been noted (IFAI, 1992):

- The soil component will represent more than 80% of the reinforced slope cost.
- The per cent of the total cost spent on reinforcement will increase with increasing slope height.
- The per cent of the total cost spent on the erosion control will increase with increasing slope angle of the reinforced slope face.

The true savings resulting from the use of a reinforced slope will depend on the value of the real estate saved and the potential use of locally available soil fill.

*Table 8.5* Comparison of 34 canal lining test sections by the USBR (after Swihart and Haynes, 2002).

| Type of lining | Construction cost ($/ft²) | Durability (years) | Maintenance cost ($/ft²–yr) | Effectiveness at seepage reduction (%) | Benefit/cost ratio |
|---|---|---|---|---|---|
| Fluid-applied membrane | 1.40–4.33 | 10–15 | 0.010 | 90 | 0.2–1.5 |
| Concrete alone | 1.92–2.33 | 40–60 | 0.005 | 70 | 3.0–3.5 |
| Exposed geomembrane | 0.78–1.53 | 10–25 | 0.010 | 90 | 1.9–3.2 |
| Geomembrane with concrete cover | 2.43–2.54 | 40–60 | 0.005 | 95 | 3.5–3.7 |

The life cycle cost of a landfill can be categorized as follows: construction, operation (including all monitoring), closure and long-term care (LTC). In many cases, the cost of hauling is expected to be high because the landfills are located in the remote areas. The cost of road construction for borrowed materials (e.g. clayey soil) must be included in the unit cost. Thus, enough money must be made available for not only constructing but also operating, maintaining, and monitoring the landfill. A proper cost analysis must be done to ensure cash flow for performing all these tasks. Monitoring of a landfill may be required for 30–40 years after the closing of the last phase. The total cost of long-term maintenance and monitoring of a closed landfill can be higher than the cost of construction of the landfill. Methods for estimating these costs at a future date must take into account the inflation factor (Bagchi, 1994).

The United States Bureau of Reclamation (USBR) constructed 34 canal lining sections in 11 irrigation districts in four northwestern states to assess the durability, cost and effectiveness of alternate lining technologies (Swihart and Haynes, 2002; Ivy and Narejo, 2003). The lining materials include combinations of geosynthetics, shotcrete, roller compacted concrete, grout mattresses, soil, elastomeric coatings, and sprayed-in-place foam. Each test section typically covered 15,000 to 30,000 square feet. The test sections ranged in age from 1 to 10 years. The preliminary benefit/cost (B/C) ratios were calculated based on initial construction costs, maintenance costs, durability (service life), and effectiveness, determined by preconstruction and post construction ponding tests. Table 8.5 summarizes the performance of all 34-test sections dividing them into four generic categories. Exposed geomembranes were found to be 90% effective, that is, only 10% of the water was lost due to seepage. Notice also that concrete alone is only 70% effective in stopping the loss of water. Therefore, the geomembranes are almost 20% more effective in stopping loss of water as compared to the concrete. The benefit/cost ratio of geomembranes is approximately the same as that of concrete. The effectiveness of concrete at reducing seepage increases to 95% if underlined by a geomembrane. Thus, the geomembrane with a concrete cover seems to offer the best long-term performance.

The geosynthetic construction quality assurance (CQA) cost is difficult to separate from the CQA costs for the total project. For the landfill projects, it has been found that the CQA costs range between 5% and 10% of the construction costs for the elements that are being monitored. The factors that influence the cost of a CQA programme are the project duration, size, complexity and work scope, ranging from complete construction management to performing CQA only for the geosynthetic elements (Thiel and Stewart, 1993).

## EXAMPLE 8.1

Make a group of the elements comprising the total cost of a geosynthetic-reinforced soil retaining wall. Based on the past experiences available, suggest an approximate breakdown of construction costs, mainly in terms of materials, labour, plant, and others.

## SOLUTION

The elements comprising the total cost of a geosynthetic-reinforced soil retaining wall may be grouped as:

1   Foundation soil improvement, if required
2   Precast facing elements (including erection), if provided
3   Soil backfill (including haul from the borrow areas)
4   Compaction of the backfill soil
5   Laboratory and in situ testing of soil
6   Geosynthetic reinforcement and auxiliary parts
7   Granular material for use around the drainage
8   Drainage pipes and cappings
9   Transport of all materials
10  Design
11  Overhead and profit

An approximate breakdown of construction costs can be given as:

1   Materials – 60%
2   Labour – 20%
3   Plant – 15%
4   Others – 5%

---

*Chapter summary*

1   Quality of a geosynthetic product or geosynthetic application is the confidence that can be placed in it. The quality evaluation is carried out to facilitate the continuous improvement of geosynthetic products and geosynthetic applications and to fully document the performance relative to a target.
2   Geosynthetics have four levels of quality management: manufacturing quality control (MQC), manufacturing quality assurance (MQA), construction quality control (CQC) and construction quality assurance (CQA).
3   Quite often, MQA and CQA are performed by the same organization, while MQC and CQC are often performed by different organizations – the manufacturer and the installer.

4    Quality Control (QC) on the construction sites is generally done by index testing. The basic information about the geosynthetic product should be indelibly marked on the wrapping and core of each roll.

5    The in-situ monitoring of geosynthetics and geosynthetic-related systems addresses their integrity and provides the guidance and insight into the design process.

6    The cost of a geosynthetic-related structure or application should typically be presented as an engineering estimate of the capital, operational, and maintenance costs. The cost analysis should be made in terms of relative economy or cost-performance efficiency (or cost-benefit ratio).

7    The rate of relative economy of reinforced soil walls increases steadily with the height of the wall. For estimating the cost-benefit ratio of the geosynthetic-reinforced unpaved roads, one must consider the future maintenance costs associated with both conventional and reinforced roadways.

8    The life cycle cost of a landfill can be categorized as construction, operation (including all monitoring), closure and long-term care. The geomembrane with a concrete cover seems to offer the best long-term performance.

## Questions for practice

(Select the most appropriate answer to the multiple-choice questions from Q 8.1 to Q 8.13.)

8.1  Which one of the following programmes provides assurance that the geosynthetics are manufactured as specified in the certification documents and contract plans and specifications?
   (a)  CQC
   (b)  CQA
   (c)  MQC
   (d)  MQA

8.2  Manufacturing quality control (MQC) is normally performed by
   (a)  the geosynthetic manufacturer
   (b)  the geosynthetic installer
   (c)  the owner of the engineering project
   (d)  the main contractor of the project.

8.3  Construction quality control (CQC) is normally performed by
   (a)  the geosynthetic manufacturer
   (b)  the geosynthetic installer
   (c)  the owner of the engineering project
   (d)  the main contractor of the project.

8.4  'MQA and CQA are performed by the same organization'. This statement is
   (a)  never true
   (b)  always true
   (c)  sometimes true
   (d)  quite often true.

8.5 'MQC and CQC are performed by the same organization'. This statement is
(a) never true
(b) always true
(c) sometimes true
(d) quite often true.

8.6 The most important quality of a CQA monitor is
(a) ability to effectively communicate
(b) assertiveness
(c) familiarity with the design issues
(d) experience of reviewing the test results.

8.7 Quality control on construction sites is generally done by
(a) index tests
(b) personal judgment
(c) field performance tests
(d) none of the above.

8.8 The electrical detection method of locating potential leak paths in a geomembrane liner can be performed on
(a) exposed liners
(b) liners covered with water
(c) liners covered by a protective soil layer
(d) all of the above.

8.9 It is generally accepted that, under the normal circumstances, the geosynthetic-reinforced soil walls are more economical for a wall height greater than
(a) 3 m
(b) 6 m
(c) 9 m
(d) 12 m.

8.10 In most low-to-medium height geosynthetic-reinforced soil retaining walls, the cost of the geosynthetic reinforcement has been found be as high as
(a) 5% of the total wall cost
(b) 10% of the total wall cost
(c) 15% of the total wall cost
(d) 20% of the total wall cost.

8.11 In the case of geosynthetic-reinforced unpaved roads, the reduction of maintenance costs has been found to be in excess of
(a) 5%
(b) 15%
(c) 25%
(d) 45%.

8.12 Which of the following statements related to the cost of three main components of a reinforced slope is incorrect?
(a) The percent of total cost spent on surface erosion control will decrease with increasing slope angle of the reinforced slope face.

(b) The soil component generally represents more than 80% of the reinforced slope cost.

(c) The per cent of total cost spent on reinforcement will increase with increasing slope height.

(d) None of the above.

8.13 Which one of the following types of canal lining is most effective in reducing seepage?
(a) Concrete lining
(b) Geomembrane lining
(c) Geomembrane with concrete cover lining
(d) Anyone of the above

8.14 What do you mean by the term 'quality'? List the purpose of quality evaluation for geosynthetic products and applications.

8.15 What is the difference between construction quality control (CQC) and construction quality assurance (CQA)?

8.16 What technical skills are required to perform CQA?

8.17 Can the MQA organization be the same as the CQA organization

8.18 Do you agree with the statement that MQA/CQA and MQC/CQC, being separate activities, have similar objectives? Justify your answer.

8.19 Draw an organizational structure of MQC/MQA and CQC/CQA inspection activities.

8.20 What information should be marked on the outside wrapping of geosynthetic rolls for their complete identification?

8.21 How will you identify a geosynthetic by simple means at the time of its installation?

8.22 What do you mean by high-risk applications of geosynthetics? Give some practical examples. What special precautions are required in such applications?

8.23 What are the important roles of field supervisor for geosynthetic applications?

8.24 What are the goals of in-situ monitoring of geosynthetic-related systems?

8.25 What major inspections of geosynthetic-related structures are required by field personnel?

8.26 Prepare a checklist of specific steps to be included in the performance monitoring plan for a canal lining project.

8.27 Why should the cost of monitoring for the geosynthetic-related projects be assessed on a case-by-case basis?

8.28 What is the principle of an electrical leak detection system used to assess the damage to geomembrane liners?

8.29 How would you estimate the field performance of geosynthetics in severe climatic conditions?

8.30 What do you mean by the term 'Relative economy'? Explain this term by taking an example.

8.31 How will you assess the cost-performance efficiency of a geosynthetic or any other element of the geosynthetic-related structure?

8.32 How does the cost-benefit analysis of environmental applications of geosynthetics differ from their other applications?

8.33 For geosynthetic-reinforced roadways, the initial cost-benefit ratio decreases as the strength of the existing soil subgrade decreases. What are the possible reasons for this trend of variation?

8.34 How does the quality of site preparation affect the construction costs and the geosynthetic survivability requirement?

8.35 How will you analyze the cost-effectiveness of a paving fabric interlayer system?

8.36 What experience is available on the cost-effectiveness of geosynthetic versus conventional drainage systems?

8.37 In the case of slope stabilization projects, what are the important issues that must be considered to make a cost-benefit analysis?

8.38 What are the components of the life-cycle cost of a landfill project?

8.39 What are the factors that influence the cost of a CQA programme?

8.40 What difficulties do you expect in the economic evaluation of the geosynthetic-related structures?

8.41 Make a group of the elements comprising the total cost of a geosynthetic-reinforced unpaved road. Based on the past experiences available in your state/country, suggest an approximate breakdown of construction costs mainly in terms of materials, labour, plant and others.

## References

Anon. (1979). *Reinforced Earth and Other Composite Techniques*. Supplementary Report No. 457. Transportation and Roads Research Laboratory, London, UK.

ASTM (American Society for Testing and Materials) (2012). ASTM D6737–12, *Standard Guide for Selection of Techniques for Electrical Detection of Leaks in Geomembranes*. ASTM International, West Conshohocken, PA, USA.

Bagchi, A. (1994). *Design, Construction, and Monitoring of Landfills*. John Wiley & Son, Inc., New York.

Barenberg, E.J. (1992). Subgrade stabilization. In *A Design Primer: Geotextiles and Related Materials*. Industrial Fabric Association International, Minnesota, USA, pp. 10–20.

Bell, J.R., Barrett, R.K. and Ruckmann, A.C. (1984). Geotextile earth-reinforced retaining wall tests. *Transportaion Research Record*, No. 916, pp. 59–69.

Bouazza, A., Zornberg, J.G. and Adam, D. (2002). Geosynthetics in waste containment facilities. Proceedings of the 7th International Conference on Geosynthetics, France, pp. 445–514.

Christie, I.F. (1982). Economic and technical aspects of embankments reinforced with fabric. *Proceedings of the 2nd International Conference on Geotextiles*. La Vegas, USA, pp. 659–664.

Dunnicliff, J. (1988). *Geotechnical Instrumentation for Monitoring Field Performance*. J. Wiley & Sons, New York.

Durukan, Z. and Tezcan, S.S. (1992). Cost analysis of reinforced soil walls. *Geotextiles and Geomembranes*, 11, 1, pp. 29–43.

Ghoshal, A. and Som N. (1993). Geotextiles and geomembranes in India – state of usage and economic evaluation. *Geotextiles and Geomembranes*, 12, 3, pp. 193–213.

Giroud, J.P. (2000). Lessons learned from successes and failures associated with geosynthetics. *Proceedings of the 2nd European Geosynthetics Conference*, Bologna, pp. 77–118.

IFAI (Industrial Fabrics Association International) (1992). *A Design Primer: Geotextiles and Related Materials*. Section 13 Asphalt Overlay. Industrial Fabrics Association International, St. Paul, USA.

Ingold, T.S. (1982). *Reinforced Earth*. Thomas Telford Ltd., London, UK.

Ivy, N. and Narejo, D. (2003). Canal lining with HDPE. *Geotechnical Fabrics Report*, IFAI, June/July 2003, pp. 24–26.

Koerner, G. (2001). In situ temperature monitoring of geosynthetics used in a landfill. *Geotechnical Fabrics Report*, pp. 12–14.

Koerner, R.M. and Daniel, D.E. (1993). Manufacturing and construction quality control and quality assurance of geosynthetics. *Proceedings of the MQC/MQA and CQC/CQA of Geosynthetics Seminar*, Philadelphia, Pennsylvania, USA, pp. 1–14.

Koerner, R.M. (1996). The state-of-the-practice regarding in-situ monitoring of geosynthetics. *Proceedings of the 1st European geosynthetics Conference*. Eurogeo 1, Maastricht, Netherlands, pp. 71–86.

Menoff, S.D. and Eith, A.W. (1990). Quality assurance as a means to minimize long-term liability. *Geotextiles and Geomembranes*, pp. 461–465.

Murray, R. (1982). Fabric reinforcement of embankments and cuttings. *Proceedings of the 2nd International Conference on Geotextiles*. Las Vegas, USA, pp. 707–713.

Nosko, V. and Touze-Foltz, N. (2000). Geomembrane liner failure: modeling of its influence on containment transfer. *Proceedings of the 2nd European Geosynthetics Conference*, Bologna, pp. 557–560.

Paul, J. (1984). Economics and construction of blast embankments using Tensar geogrids. *Proceedings of the Conference on Polymer Grid Reinforcement*, London, UK, pp. 191–197.

Qian, X, Koerner, R.M. and Gray, D.H. (2002). *Geotechnical Aspects of Landfill Design and Construction*. Prentice Hall, New Jersey.

Simac, M.R. (1992). Reinforced slopes: a proven geotechnical innovation. *Geotechnical Fabrics Report*, pp. 13–25.

Swihart, J. and Haynes, J. (2002). *Canal-Lining Demonstration Project Year 10 Final Report*. United States Department of Interior, Bureau of Reclamation, USA.

Thiel, R.S. and Stewart, M.G. (1993). Field construction quality assurance of geotextile installations. *Proceedings of the MQC/MQA and CQC/CQA of Geosynthetics Seminar*, Philadelphia, Pennsylvania, USA, pp. 126–137.

## Answers to selected questions

| | |
|---|---|
| 8.1 | (d) |
| 8.3 | (b) |
| 8.5 | (d) |
| 8.7 | (a) |
| 8.9 | (b) |
| 8.11 | (c) |
| 8.13 | (c) |

# Chapter 9

# Fibre-reinforced soils

## 9.1 INTRODUCTION

In the previous chapters, you studied different applications of soil reinforcement in the form of continuous geosynthetic reinforcement inclusions (for example, sheets, nets, meshes, strips or bars) within a soil mass in a definite pattern. In addition to such *systematically reinforced soils*, in the past few decades several experimental and mathematical studies have been conducted to investigate the behaviour of soils reinforced randomly with different types of discrete flexible fibres, called the *randomly distributed fibre-reinforced soils*, or simply the *fibre-reinforced soils* (Shukla *et al.*, 2009).

As explained in Chapter 1, note that a *fibre* is a unit of matter characterized by flexibility, fineness, and high ratio of length to thickness/diameter. Different types of fibres are available as natural, synthetic or waste products (Fig. 9.1). As waste tyres and waste plastic materials are available in large quantities worldwide, they may be utilized in construction projects in various forms, especially in the form of fibres; otherwise they may occupy a large volume of the landfills when disposed of.

We all notice that the roots of vegetation (natural fibres) stabilize the near-surface soil that has low shear strength, mainly because of low effective stress, on both level and sloping grounds. Fibres in soil may be viewed as an artificial replication of the effects of vegetation roots. In a simple form, fibres within a soil mass mimic the role of plant roots to provide the stability of the soil structure. The concept of reinforcing soil with fibres, especially natural ones, originated in ancient times in some applications. For example, the use of natural fibres in composite construction can be seen even today in the rural areas of some countries. However, the randomly distributed fibre-reinforced soils have recently attracted increasing attention in geotechnical engineering in a scientific manner.

This chapter presents the fundamental aspects of fibre-reinforced soils, including phase concepts, engineering behaviour, reinforcement models, and possible field application areas and guidelines.

*Figure 9.1* Some natural, synthetic and waste fibres: (a) natural fibres – (i) coir fibres, (ii) jute fibres; (b) synthetic fibres – (i) polypropylene (PP) fibres, (ii) glass fibres; (c) waste fibres – (i) waste tyre fibres, (ii) waste plastic fibres.

## 9.2   BASICS OF FIBRE-REINFORCED SOIL AND PHASE CONCEPTS

The improvement of the engineering properties (strength, stiffness/modulus, etc.) of soil by the inclusion of discrete flexible fibres within the soil depends on the soil characteristics, the fibre characteristics and the test conditions as detailed below:

- *Soil characteristics:* types (cohesionless/cohesive/cohesive-frictional); gradation, particle shape and size; unit weight of soil solids; total unit weight; and water content
- *Fibre characteristics:* fibre types and materials (natural/synthetic/waste); shapes (monofilament, fibrillated, tape, mesh, etc.); fibre diameter, fibre length, aspect ratio (length to diameter ratio); specific surface; specific gravity of fibre solids; orientation; fibre content; fibre-area ratio; tensile strength, stiffness/modulus of elasticity, linear strain at failure; roughness and skin friction; melting point
- *Test conditions/variables:* confining stress, rate of loading, etc.

Note that the waste tyre fibres can be used in two different shapes, granular and fibre shaped. The properties of some fibres used by some researchers are given in Table 9.1. The fibres are also described in terms of linear mass density (kg/m), which is generally expressed in denier (grams per 9 km of the fibre) or tex (grams per 1 km of the fibre). Thus, 1 denier = 1 g/9 km, 1 tex = 1 g/km, and 1 tex = 9 denier.

The following terms are often used to describe the behaviour of fibre-reinforced soils:

Fibre content (i.e. concentration of fibres or weight ratio or weight fraction),

$$p_f = \frac{W_{sf}}{W_{ss}} \tag{9.1}$$

Aspect ratio,

$$a_r = \frac{L}{D} \tag{9.2}$$

*Table 9.1* Properties of fibres used by some researchers.

| Characteristics of fibres | Sources | | | | | |
|---|---|---|---|---|---|---|
| | Yetimoglu et al. (2005) | Consoli et al. (2009) | Ibraim et al. (2012) | Lovisa et al. (2010) | Shukla et al. (2015) | Jha et al. (2015) |
| Type | PP fibres | PP fibres | PP crimped fibres | Glass fibres | PP fibres | PE fibres |
| Specific gravity, G | 0.93 | 0.91 | | 1.7 | 0.91 | 0.99 |
| Length, L (mm) | 20 | 24 | 35 | 10–15 | 19 | 12 |
| Equivalent diameter, D (mm) | 0.05 | 0.023 | 0.1 | 0.02 | | 0.035 |
| Tensile strength, $\sigma_f$ (MPa) | 320–400 | 120 | 225 | 300 | 620–758 | 600 |
| Elongation at break/failure (%) | | 80 | 160 | | | |
| Tensile modulus, E (GPa) | 3.5–3.9 | 3 | | | | |

where $W_{ss}$ is the weight of soil solid; $W_{sf}$ is the weight of fibre solid; $L$ is the length of fibre; and $D$ is the diameter of fibre.

Fibre area ratio,

$$A_r = \frac{A_f}{A} \tag{9.3}$$

where $A_f$ is the total cross-sectional area of fibres in the shear/failure plane, and $A$ is the total area of the shear/failure plane.

**EXAMPLE 9.1**

Determine the aspect ratio of the PP fibres if their average length and diameter are 20 mm and 0.05 mm, respectively.

**SOLUTION**

Given: $L = 20$ mm and $D = 0.05$ mm
From Equation (9.2), the aspect ratio,

$$a_r = \frac{L}{D} = \frac{20}{0.05} = 400$$

Like an unreinforced soil mass, the fiber-reinforced soil mass may be represented by a three-phase system as shown in Fig. 9.2, with fibre and soil solids represented separately. The concept of phase relationships, being adopted widely in soil mechanics and geotechnical engineering for unreinforced soils, can be utilized for developing the phase relationships for fibre-reinforced soils. Some phase relationships for fibre-reinforced soils are defined below:

Void ratio of the soil mass,

$$e_s = \frac{V_{vs}}{V_{ss}} \tag{9.4}$$

Void ratio of the fibre mass,

$$e_f = \frac{V_{vf}}{V_{sf}} \tag{9.5}$$

Void ratio of the fibre-reinforced soil mass,

$$e = \frac{V_v}{V_s} \tag{9.6}$$

Volume ratio of the fibre-soil solid,

$$m = \frac{V_{sf}}{V_{ss}} \tag{9.7}$$

where $V_{vs}$ sis the volume of soil voids, $V_{ss}$ is the volume of soil solids, $V_{vf}$ is the volume of fibre voids, and $V_{sf}$ is the volume of fibre solids, $V_v$ is the volume of voids of fibre-reinforced soil, $V_s$ is the volume of solids of fibre-reinforced soil.

Shukla *et al.* (2015) presented the following phase inter-relationships:

$$e = \frac{e_s + me_f}{1 + m} \tag{9.8}$$

$$m = p_f \frac{G_s}{G_f} \tag{9.9}$$

$$e = \frac{G\gamma_w}{\gamma_d} - 1 \tag{9.10}$$

$$G = (1 + p_f)\left(\frac{1}{G_s} + \frac{p_f}{G_f}\right)^{-1} \tag{9.11}$$

where $\gamma_w$ is the unit weight of water; $G_s$ is the specific gravity of soil solids; and $G_f$ is the specific gravity of fibre solids; $G$ is the specific gravity of fibre-reinforced soil solids; and $\gamma_d$ is the dry unit weight of fibre-reinforced soil.

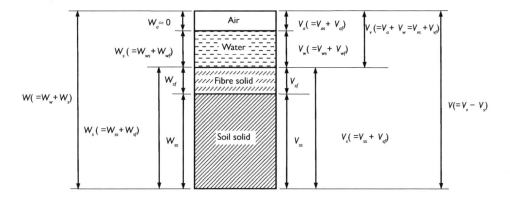

Figure 9.2 An element of fibre-reinforced soil mass represented by a three-phase system.
Note:  The symbols V and W refer to volume and weight, respectively. The subscripts a, w, v and f refer to air, water, void and fibre, respectively. The subscript s when appears first refers to solid but it refers to soil mass when appears as the second subscript.

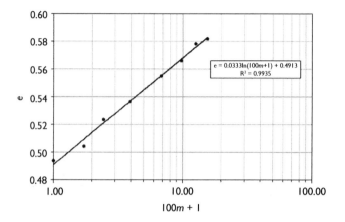

*Figure 9.3* Variation of void ratio *e* of the fibre-reinforced sand with logarithm of (100*m* + 1) (after Shukla *et al.*, 2015).

Note that the phase relationships and inter-relationships, as presented here, are fairly of general nature and can be utilized in several applications of fibre-reinforced soils in a way similar to that traditionally used in soil mechanics and geotechnical engineering for the analysis of unreinforced soils. Based on laboratory experiments, Shukla *et al.* (2015) presented the variation of void ratio, *e*, of fibre-reinforced sand with logarithm of (100*m* + 1), as shown in Fig. 9.3, with the following empirical expression:

$$e = a\ln(100m + 1) + b \tag{9.12}$$

where *a* = 0.0333 is the slope of the linear relationship and represents the type of fibre used (i.e., virgin homopolymer polypropylene), and *b* = 0.4913 is the value of *e* at *m* = 0 and represents the type of sand used (i.e., poorly graded silica sand). For other types of fibre and/or sand, *a* and *b* can be obtained from the data obtained from a minimum of four compaction tests on the fibre-reinforced soil at hand from which the measured values of *m* and the corresponding *e* can be calculated and used to determine *a* and *b*, then the equation similar to Equation (9.12) can be used for future prediction of the void ratio of that the specific fibre-reinforced soil.

It is expected that that the researchers may carry out experimental studies with several fibre types and also for cemented fibre-reinforced soils to compare their results with the developed phase relationships.

## EXAMPLE 9.2

Determine the void ratio of fibre-reinforced soil for the following two cases:

a   $m \to 0$
b   $m \to \infty$

What do you observe from the values obtained?

**SOLUTION**

a  As $m \to 0$, from Equation (9.8),

$$e = \frac{e_s + me_f}{1 + m} = \frac{e_s + (0)(e_f)}{1 + 0} = e_s$$

This indicates that the fibre content in the reinforced soil is negligible, and hence it will behave almost as the original soil mass.

b  As $m \to \infty$, from Equation (9.8),

$$e = \frac{e_s + me_f}{1 + m} = \frac{(e_s / m) + e_f}{(1/m) + 1} = \frac{0 + e_f}{0 + 1} = e_f$$

This indicates that the fibre content in the reinforced soil is almost 100%, and hence it will behave almost as the fibre mass.

## 9.3  BEHAVIOUR OF FIBRE-REINFORCED SOILS

A large number of experimental studies have been carried out to study the behaviour of fibre-reinforced soils. Most studies are based on the following tests:

- Direct shear tests
- Triaxial tests
- Unconfined compression tests
- Splitting tensile tests
- Compaction tests
- Permeability tests
- California bearing ratio (CBR) tests
- Plate load tests

The basic characteristics of fibre-reinforced soils have been reported by several researchers. The findings vary to some extent depending on the variations in the parameters, but some observations also appear to be contradictory. In general, the fibres are most effective when oriented within the soil mass in the same direction as the tensile strains caused by the applied loads. Hence, for any particular loading condition, the properties of fibre-reinforced soil depend significantly on the orientation of fibres with respect to the loading direction. The orientation of fibres in experiments as well as in the field applications depends on the method of mixing of soil with fibres, the specimen preparation method for tests, and the method of field placement.

The laboratory test specimens are prepared in two stages, namely mixing and compaction. The water, as required, is added to the dry soil, and then mixed together uniformly, followed by adding the fibres to the wet soil and mixing in order to have an even distribution of fibres. The test specimens are prepared in the desired test mould by compacting the soil-fibre mix in a specific number of layers by tamping with a rammer with a flat base; this method is called the *tamping technique*. The test specimens may also be prepared by densifying the soil-fibre mix in the desired mould by the *vibration technique*, such as vibrating the test mould filled with the soil-fibre mix on the vibration table.

It has been found that tamping and vibration techniques for preparing the fibre-reinforced specimens in moist conditions lead to a preferred near-horizontal orientation of fibres. Both techniques leave at least 80% of the fibres oriented between ± 30° of horizontal (Diambra *et al.*, 2010; Ibraim *et al.*, 2012). Hence, the consequence of an assumed isotropy as generally expected in the soil-fibre mixes may result in an overestimation or underestimation of reinforced soil design parameters, depending on the direction of practical importance. Note that these techniques generally produce a soil-fibre that resembles that of the rolled-compacted construction fills.

Several basic characteristics of fibre-reinforced soils as observed by several researchers in the experiments are summarized below:

*Observations in direct shear tests:* The relatively low modulus, fibre reinforcements (natural and synthetic fibres including metal fibres) behave as 'ideally extensible' inclusions; they do not rupture during shear. Their main role is to increase the peak shear strength and to limit the magnitude of post-peak reduction in shear resistance in dense sand. The increases of shear strength of dry sand are directly proportional to fibre area ratios, at least up to 1.7%. The shear strength increases as a result of fibre reinforcement are approximately the same for loose and dense sands. However, larger strains are required to reach the peak shear resistance in the loose case. The shear strength envelopes for fibre reinforced sand clearly show the existence of a threshold confining stress below which the fibres tend to slip or pull out. The strength envelope also indicates that the fibres do not affect the angle of internal friction of the sand. The shear strength increases are greatest for the initial fibre orientations of 60° with respect to the shear surface. This orientation coincides with the direction of maximum principal tensile strain. Increasing the length of fibre reinforcements increases the shear strength of the fibre-reinforced composite but only up to a point beyond which any further increase in fibre length has no effect (Gray and Ohashi, 1983).

The peak shear strength and initial stiffness of the sand are not affected significantly by the fibre reinforcements. The fibre reinforcements reduce the soil brittleness providing a smaller loss of post-peak strength. Hence, there appears to be an increase in residual shear strength angle of the sand by adding fibre reinforcements (Yetimoglu and Salbas, 2003).

In foundation and other applications, the soil is rarely in a dry condition. Hence Lovisa *et al.* (2010) conducted direct shear tests to investigate the effects of water content on the shear strength behaviour of sand reinforced with 0.25% randomly distributed glass fibres. The test results suggest that the peak friction angle of the fibre-reinforced sand in moist condition is approximately 3° less than that in dry condition for a relative density greater than 50%. The fibre inclusions introduce an apparent cohesion intercept to the soil in the dry state, which remains almost unchanged by an increase in water content.

*Observations in triaxial tests:* The inclusion of natural and synthetic glass fibres in sandy soil increases its strength (expressed as the major principal stress at failure) and modifies the stress-deformation behaviour with increased stiffness. The increase in strength with fibre content varies linearly up to a fibre content of 2% by weight, and thereafter approaches an asymptotic upper limit. The rate of increase is roughly proportional to the fibre aspect ratio. At the same aspect ratio, the confining pressure, and weight fraction and roughness (not stiffness) of fibres tend to be more effective in increasing the strength (Gray and Al-Refeai, 1986).

The inclusion of plastic fibres causes an increase in peak shear strength and reduction in the loss of post-peak strength. Thus, the residual strength of fibre-reinforced sand is higher as compared to that of unreinforced sand. The principal stress envelopes for fibre-reinforced sand are bilinear having a break at a confining stress, called the critical confining stress, below which the fibres tend to slip or pull out. An increase in fibre aspect ratio results in a lower critical confining stress and higher contribution to shear strength. Shear strength increases approximately linearly with increasing fibre content up to 2% by weight, beyond which the gain in strength is not appreciable (Ranjan *et al.*, 1994).

With the inclusion of fibre and increase in fibre length, the cohesive intercept of sandy soil does not change whereas the friction angle increases. The reinforcement does not affect the initial stiffness or ductility of the uncemented sand. As expected, the addition of cement to the sand significantly increases the stiffness and peak strength, and changes the soil behaviour from ductile to a noticeably brittle one. The cement content increases both the peak friction angle and the cohesive intercept. The initial stiffness of the cemented sand is not affected by fibre inclusion, since it is basically a function of cementation. The efficiency of the fibre reinforcement when applied to cemented sand is found to be dependent on the fibre length. The greatest improvements in triaxial strength, ductility, and energy absorption capacity are observed for the longer (e.g. 36 mm) fibre (Consoli *et al.*, 2002).

The friction angle is slightly affected by inclusion of polypropylene fibres in the low-plasticity soil, increasing from 30° to 31°, but the cohesion intercept increases from around 23 to 122 $kN/m^2$. (Consoli *et al.*, 2003a).

In triaxial compression, a considerable increase of shear strength is contributed by the presence of fibres; while in extension, the benefit of fibres is very limited. This behaviour confirms that the tamping technique in the moist condition generates preferential near-horizontal orientation of fibres, that is, the anisotropic distribution of fibre orientation (Diambra *et al.*, 2010).

The fibres in loose sandy soil reduce the potential for the occurrence of liquefaction in both compression and extension triaxial loadings and convert a strain softening response into a strain hardening response. When the full liquefaction of reinforced specimens is induced by strain reversal, the lateral spreading of soil seems to be prevented (Ibraim *et al.*, 2010).

Mixing of synthetic fibres increases the strength of rice husk ash (RHA), but there exists an optimum fibre content of 1.25% at which the reinforcement benefits are maximum. The stress-strain behaviour of RHA improves considerably due to an increase in fibre content. The secant modulus of reinforced RHA increases with an increase in fibre content up to 1.25% and thereafter it decreases. The shear strength parameters (cohesion and angle of internal friction) also increase with an increase in fibre content up to 1.25% and thereafter they decrease (Jha *et al.*, 2015).

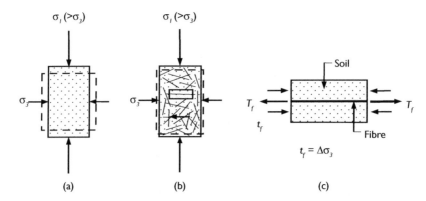

*Figure 9.4* Basic mechanism of fibre-reinforced soil in the triaxial test: (a) unreinforced cylindrical soil specimen; (b) reinforced cylindrical soil specimen; (c) magnified view of a reinforced soil element as indicated in (b).

The use of coir as random reinforcing material with expansive black cotton soil can result in improved engineering behaviour. The deviator stress at failure increases with increase in the fibre content as well as with increase in the diameter of the fibres. The maximum increase of major principal stress at failure is found to be 1.30 times over the unreinforced soil. The coir fibres help in reducing the swell potential of black cotton soil (Babu *et al.*, 2008).

Note that when the fiber-reinforced cylindrical specimen is loaded in the triaxial test with the minor principal stress (confining stress) $(\sigma_3)$ and the major principal stress $(\sigma_1)$, because of skin friction and/or adhesion between the fibres and the soil, the fibres apply the confining pressure $t_f$ $(= \Delta\sigma_3$, increase in the minor principal stress) on the soil, and in this process they get stretched with mobilization of the tensile force $T_f$ as shown in Fig. 9.4. The tensile force $T_f$ and hence $t_f$ varies with the orientation of the fibre within the soil.

*Observations in unconfined compression tests:* The inclusion of randomly oriented discrete fibres significantly improves the unconfined compressive strength (UCS) of sands. An optimum fibre length of 51 mm was identified for the reinforcement of sand specimens. A maximum performance is achieved at a fibre dosage rate between 0.6 and 1% by dry weight. The specimen performance is enhanced in both wet and dry of optimum conditions. The inclusion of up to 8% of silt does not affect the performance of the fibre reinforcement (Santoni *et al.*, 2001).

For an intermediate cement content of 5%, an increase in fibre content from 0.1 to 0.9% causes average increases in UCS of sand by about 40%. The initial stiffness of the cemented sand was not affected by fibre inclusion, since it is basically a function of cementation. The positive effect of fibre length is generally not detected in the unconfined compression tests, clearly indicating the major influence of confining pressure and the necessity of carrying out triaxial compression tests to fully observe the fibre reinforced soil behaviour (Consoli *et al.*, 2002).

The inclusion of carpet waste fibres into clay soils prepared at the same dry unit weight can significantly enhance the UCS, reduce the post-peak strength loss, and change the failure behaviour from brittle to ductile. The results also showed that the relative benefit of fibres to increase the UCS of clay soils is highly dependent on initial dry unit weight and moisture content of the soil (Mirzababaei et al., 2013).

*Observations in splitting tensile tests:* For an intermediate cement content of 5%, an increase in fibre content from 0.1 to 0.9% causes average increases in tensile strength of sand by 78%. The positive effect of fibre length is not detected in the split cylinder tests, clearly indicating the major influence of confining pressure and the necessity of carrying out triaxial tests to fully show fibre-reinforced soil behaviour (Consoli et al., 2002).

*Observations in compaction tests:* For a given compactive effort, the maximum dry unit weight of fibre-reinforced sand decreases with increasing fibre content, whereas the optimum moisture content is independent of the amount of fibres (0–0.5%) used. More compaction energy may be necessary to produce specimens with higher fibre contents at a given dry unit weight (Ibraim et al., 2010).

Kumar and Singh (2008) have observed that the addition of 0.5% by weight of fibres resulted in a decrease in optimum moisture content and maximum dry density of fly ash by 6.8% and 2.8%, respectively.

*Observations in permeability tests:* Based on very limited tests, the author has experienced that it is difficult to get the consistent results with varying fibre parameters. In general, for a given fibre content, the permeability increases with increasing length of the fibres. The effect of fibre content on the permeability depends greatly on the density of the reinforced soil as well as the confinement pressure; thus both increasing and deceasing trends can be seen. It is expected that the researchers will focus on this aspect in detail.

*Observations in California bearing ratio (CBR) tests:* The inclusion of fibres increases the CBR value of dune sand, and the improvement in the CBR can be maintained over a larger penetration range than with unreinforced sand (Al-Refeai and Al-Suhaibani, 1998).

The addition of PP fibres improves the bearing capacity of the soil by an increase in CBR values of as much as 133% (Fletcher and Humphries, 1991).

Use of fibres may improve the CBR of the sand from 6% to 34% over the unstabilized sand shoulder at similar depths. The improvement in load-bearing capacity of the fibre-stabilized sand can be attributed to the confinement of the sand particles by the discrete fibres. When the fibres are mixed into the sand, the fibres develop friction at interaction points with the particles that resists rearrangement of the particles under loading. Thus, the primary stabilization mechanism is the mechanical confinement of the sand (Tingle et al., 2002).

Fibre (rubber) inclusions with an optimum aspect ratio increase the CBR of the sand subgrade and hence may cause a substantial decrease in design thickness of the pavement. Fibre inclusions mixed with sand subgrade also provide needed tensile strength under traffic loads (Edincliler and Cagatay, 2013).

The addition of waste plastic strips to industrial wastes (fly ash, stone dust and waste recycled product) results in an appreciable increase in the CBR and the subgrade modulus. Hence these reinforced waste materials can be used in flexible pavement construction leading to safe and economical reuse of the wastes (Gill *et al.*, 2014; Jha *et al.*, 2014).

*Observations in plate load tests:* The addition of polypropylene fibres to the soil results in a noticeable stiffer response with increasing settlement, and this has been explained through a combined effect of the continuous increase in the strength of the soil at large deformations as observed in the triaxial tests, and the increase in the horizontal stresses below the test plate (Consoli *et al.*, 2003a).

The addition of polypropylene fibres to the cemented sand layer over the soil stratum improves the ultimate bearing capacity, when compared to the cemented sand layer. The fibre reinforcement significantly changes the failure mechanism by preventing the formation of tension cracks, as observed for the cemented sand layer. Instead, the fibres allow the applied load to spread through a larger area at the interface between the fibre-reinforced cemented sand layer and the underlying soil stratum (Consoli *et al.*, 2003b). The effect of fibre inclusion is found to be more pronounced for higher relative densities of sandy soil (Consoli *et al.*, 2007).

## 9.4 BASIC REINFORCEMENT MODELS

Based on experimental studies, it has been established that the strength and deformation behaviour of the fibre-reinforced soils is governed by the soil and fibre characteristics as well as by confinement and stress level. The states of stress and strain in fibre-reinforced soil during its deformation and failure are complex. However, for engineering applications, it is possible to explain the mechanism of fibre reinforcements in soils, especially the contribution of fibres to the shear strength increase, and hence the behaviour of fibre-reinforced soils by mathematical approaches/models. In the past, several attempts have been made to present such models, which are mainly based on the force-equilibrium approach, the energy dissipation approach, and the approach of superposition of the effects of soil and fibres. The force equilibrium approach is a simple one to explain the fundamentals of fibre-reinforced mechanics, and hence several researchers have presented the models based on this approach (Waldron, 1977; Gray and Ohashi, 1983; Jewel and Worth, 1987; Mahar and Gray, 1990; Ranjan *et al.*, 1996; Shukla *et al.*, 2009). Some of these models are presented here briefly.

### 9.4.1 Waldron model

Waldron (1977) proposed a force-equilibrium model to estimate the increase in strength of soil reinforced with plant roots, taking into account the tensile force developed in the reinforcement. In this model, the root-reinforced soil mass is considered as a composite material in which the roots of relatively high tensile strength are embedded in a matrix of lower tensile strength. The model consists of vertical flexible, elastic roots of uniform diameter extending an equal distance on either side of the horizontal sheer plane. This model considers only the partial mobilization of fibre tensile strength depending upon the amount of fibre elongation during the shear. The model does not place any constraint on the distribution or location of the reinforcing fibres.

## 9.4.2   Gray and Ohashi model

The concept of the Waldron model was extended by Gray and Ohashi (1983) to describe the deformation and the failure mechanism of fibre-reinforced soil and to estimate the contribution of fibre reinforcement to increasing the shear strength of soil. The model consists of a long, elastic fibre extending an equal length over either side of a potential shear plane in sand (Fig. 9.5). The fibre may be oriented initially perpendicular to the shear plane or at some arbitrary angle, $i$ with the horizontal. Shearing causes the fibre to distort, thereby mobilizing the tensile resistance in the fibre. The tensile force in the fibre can be divided into components normal and tangential to the shear plane. The normal component increases the confining stress on the failure plane thereby mobilizing additional shear resistance in the sand, whereas the tangential component directly resists the shear. The fibre is assumed to be thin enough that it offers little if any resistance to shear displacement from the bending stiffness. If many fibres are present, their cross-sectional areas are computed, and the total fibre concentration is expressed in terms of the fibre area ratio, $A_r$, defined by Equation (9.3).

The shear strength increase $\Delta S (= S_R - S_U$, $S_R$ and $S_U$, are shear strengths of reinforced soil and unreinforced soil, respectively) from the fibre reinforcement in sand thus can be estimated from the following expressions:

$$\Delta S = t_f \left( \sin\theta + \cos\theta\tan\phi_s \right) \tag{9.13}$$

for fibres oriented initially perpendicular to the shear plane (Fig. 9.5(a)), and

$$\Delta S = t_f [\sin(90^0 - \psi) + \cos(90^0 - \psi)\tan\phi] = t_f \left( \cos\psi + \sin\psi\tan\phi_s \right) \tag{9.14}$$

for fibres oriented initially at some arbitrary angle $i$ with the horizontal (Fig. 9.5(b)), with

$$\psi = \tan^{-1}\left[ \frac{1}{k + (\tan^{-1}i)^{-1}} \right] \tag{9.15}$$

where $t_f$ is the mobilized tensile strength of fibres per unit area of soil; $\phi_s$ is the angle of internal friction of the soil; $\theta$ is the angle of shear distortion; $x$ is the horizontal

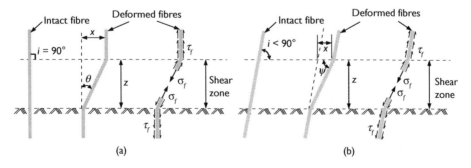

(a)                                                                   (b)

*Figure 9.5* Model for flexible, elastic fibre reinforcement extending across the shear zone of thickness z : (a) i =90°; (b) i < 90° (adapted from Gray and Ohashi, 1983; Shukla et al., 2009).

shear displacement; $z$ is the thickness of shear zone; and $k(=x/z)$ is the shear distortion ratio.

The mobilized tensile strength per unit area of soil $(t_f)$ can be estimated as

$$t_f = A_r \sigma_f = \left( \frac{A_f}{A} \right) \sigma_f \tag{9.16}$$

where $\sigma_f$ is the tensile stress developed in the fibre at the shear plane, which depends upon a number of parameters and the test variables. The fibres must be long enough and frictional enough to avoid the pullout; conversely the confining stress must be high enough so that the pullout forces do not exceed the skin friction (shear forces) along the fibre. It is also necessary to assume some sort of tensile stress distribution along the length of the fibre. Two likely or reasonable possibilities are a linear or parabolic distribution, with tensile stress a maximum at the shear plane and decreasing to zero at the fibre ends. The resulting tensile stresses at the shear plane for these two distributions are given by the following expressions (Waldron, 1977):

$$\sigma_f = \left( \frac{4E_f \tau_f}{D} \right)^{1/2} [z(\sec\theta - 1)]^{1/2} \tag{9.17}$$

for the linear distribution, and

$$\sigma_f = \left( \frac{8E_f \tau_f}{3D} \right)^{1/2} [z(\sec\theta - 1)]^{1/2} \tag{9.18}$$

for the parabolic distribution, where $E_f$ is the modulus or longitudinal stiffness of the fibre; $\tau_f$ is the skin friction stress along the fibre; and $D$ is the diameter of fibre.

### 9.4.3  Mahar and Gray model

Based on the observations made in triaxial tests and statistical analysis of randomly distributed fibre-reinforced soils, Mahar and Gray (1990) have reported that the failure surfaces in triaxial compression tests are planar and oriented in the same manner as predicted by the Mohr-Coulomb theory. This finding suggests an isotropic reinforcing action with no development of preferred planes of weakness or strength. For the fibre-reinforced soil, the principal stress envelope, that is, the plot of major principal stress at failure $(\sigma_{1f})$ versus the minor principal stress (confining stress) $(\sigma_3)$ has been found to be either curved-linear or bilinear, with the transition or the break occurring at a confining stress, called the *critical confining stress*, $\sigma_{3crit}$ (Fig. 9.6). The increase in shear strength from the fibre reinforcement can be estimated form the following expressions:

$$\Delta S = N_s \left( \frac{\pi D^2}{4} \right) (2\sigma_{3av} \tan\delta)(\sin\theta + \cos\theta \tan\phi_s)\eta \tag{9.19}$$

for $0 < \sigma_3 < \sigma_{3cri}$, and

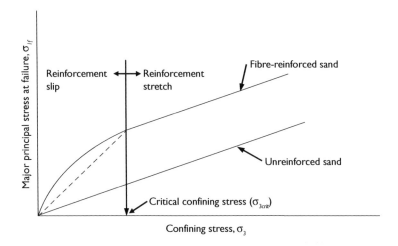

*Figure 9.6* Effect of fibre inclusion in sand on its principal stress envelope obtained from triaxial compression tests (after Mahar and Gray, 1990; Shukla and Sivakugan, 2010).

$$\Delta S = N_s \left( \frac{\pi D^2}{4} \right) (2\sigma_{3crit} \tan \delta)(\sin \theta + \cos \theta \tan \phi_s) \eta \tag{9.20}$$

for $\sigma_3 < \sigma_{3crit}$, where $N_s$ is the average number of fibres intersecting a unit area; $\delta$ is the fibre skin friction angle; and $\eta$ is an empirical coefficient depending on the soil characteristics (median particle size, $D_{50}$, particle sphericity, $S$, and coefficient of uniformity, $C_U$) and the fibre parameters (aspect ratio and skin friction). The value of $\sigma_{3crit}$ in Eqs. (9.19) and (9.20) can be determined empirically from the experimental measurements, thus depending on the soil characteristics and the fibre properties. Note that in Fig. 9.6, for $\sigma_3 < \sigma_{3crit}$, the stress envelope for fibre-reinforced sand is either linear or nonlinear depending on the types of sand and reinforcement.

The Maher and Gray model has predicted the increase in the strength of the fibre reinforced soil reasonably well. However, the width of shear zone, $z$, which significantly affects the increase in strength (Shewbridge and Sitar, 1989, 1990) has not been determined for reinforced soil. Also, the average expected orientation of fibres is statistically predicted to be perpendicular to the plane of shear failure. It is difficult to determine experimentally the exact orientation of fibres. Keeping these limitations in view, Ranjan *et al.* (1996) presented a regression analysis of a large number of triaxial tests on the fibre reinforced soil, and reported that their analysis is accurate in predicting the increase in shear strength due to inclusions of fibres in cohesionless soils.

### 9.4.4   Shukla, Sivakugan and Singh (SSS) model

Shukla et al. (2010) developed a simple analytical model for predicting the shear strength behaviour of fibre-reinforced granular soils under high confining stresses, where it can be assumed that pullout of fibres does not take place. The model presents an analytical expression for the *shear strength ratio* (SSR) as given below:

$$\text{SSR} = \frac{S_R}{S_U} = 1 + 2 \left[ \frac{p_f \left\{ \frac{G_s}{G_f(1+e_s)} \right\}}{1 + p_f \left\{ \frac{G_s}{G_f(1+e_s)} \right\}} \right] \left[ \frac{\sin i}{1 + 2\left(\frac{\sigma}{E_f}\right)\beta_1\beta_2} + \frac{\sqrt{\cos^2 i + 4\left(\frac{\sigma}{E_f}\right)\left[1 + \left(\frac{\sigma}{E_f}\right)\beta_1\beta_2\right]\beta_1\beta_2}}{\left[1 + 2\left(\frac{\sigma}{E_f}\right)\beta_1\beta_2\right]\tan\phi_s} \right] \beta_1\beta_2 \qquad (9.21)$$

with

$$\beta_1 = \frac{1 - \sin\phi_s \sin(\phi_s - 2i)}{\cos^2\phi_s} \qquad (9.22a)$$

and

$$\beta_2 = a_r \tan\delta \sin i \qquad (9.22b)$$

where $S_R$ is the shear strength of the fibre-reinforced soil, $S_U (= \sigma\tan\phi_s)$ is the shear strength of the unreinforced soil, $p_f (= W_{sf}/W_{ss})$ is the fibre content, $a_r(= L/D)$ is the aspect ratio, $E_f$ is the modulus of elasticity of fibres, $G_f$ is the specific gravity of fibre material, $\delta$ is the soil-fibre interface friction angle, $i$ is the initial orientation with respect to the shear plane, $\sigma$ is the normal confining stress, $G_s$ the specific gravity of soil particles, $\phi_s$ is the angle of shearing resistance of soil, and $e_s$ is the void ratio of soil.

## EXAMPLE 9.3

Determine the shear strength ratio (SSR) for the following three cases:

c   fibre content, $p_f = 0\%$
d   soil-fibre interface friction angle, $\delta = 0°$
e   all fibres are oriented parallel to the shear plane, $i = 0°$

Explain the physical significance of SSR obtained for each case.

## SOLUTION

a   For $p_f = 0\%$, from Equation (9.21),

$$\text{SSR} = \frac{S_R}{S_U} = 1$$

This result is well expected when there are no fibers present in the soil.

b   For $\delta = 0°$, from Equation (9.22b), $\beta_2 = 0°$, and hence from Equation (9.21),

$$SSR = \frac{S_R}{S_U} = 1$$

This suggests that if fibers do not have frictional resistance in contact with soil parti-cles, their inclusion in soil will not be useful. In this situation, there is no shear stress acting along the surface of the fibres, and therefore the fibres remain unstretched.

c   For $i = 0°$, from Equation (9.22b), $\beta_2 = 0°$, and hence from Equation (9.21),

$$SSR = \frac{S_R}{S_U} = 1$$

This suggests that if all the fibres are oriented parallel to the shear plane, there will be no increase in shear strength.

Note that Equation (9.21) is quite useful in predicting the variation of *SSR* with fibre content ($p_f$), aspect ratio ($L/D$), ratio of confining stress to modulus of fibres ($\sigma/E_f$), specific gravity of fiber material ($G_f$), and initial orientation with respect to shear plane ($i$) for any specific sets of parameters in their practical ranges. The effects of $p_f$, $a_r$ and $i$ on SSR for specific sets of parameters are shown in Figs. 9.7, 9.8 and 9.9, respectively.

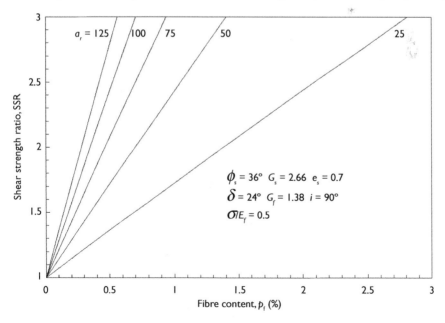

*Figure 9.7* Effect of fibre content ($p_f$) on the shear strength ratio (SSR) of fibre-reinforced soil (after Shukla et al., 2010).

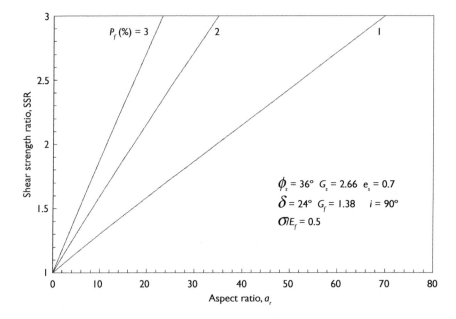

*Figure 9.8* Effect of aspect ratio ($a_r$) on the shear strength ratio (SSR) of fibre-reinforced soil (after Shukla *et al.*, 2010).

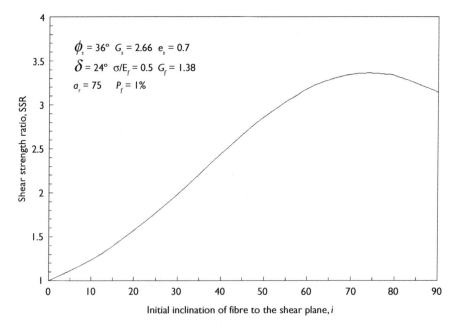

*Figure 9.9* Effect of initial inclination of fibre to the shear plane (*i*) on the shear strength ratio (SSR) of fibre-reinforced soil (after Shukla *et al.*, 2010).

Regarding the fibre-reinforced soil, it is worth noting the following:

1    The inclusion of fibres in granular soil induces cohesion, may be called the apparent cohesion, as well as an increase in the normal stress on the shear failure plane, which are proportional to the fiber content and the aspect ratio, implying that the increase in shear strength is also proportional to the fiber content and the aspect ratio.
2    The increase in shear strength of granular soil due to presence of fibres is significantly contributed by the apparent cohesion, and the contribution to the shear strength from the increase in normal confining stress is limited.
3    As the initial orientation of fibres with respect to shear plane ($i$) increases, the SSR also increases with a maximum value for a specific value of $i$ depending on the values of other governing parameters.

## 9.5    FIELD APPLICATION AREAS AND GUIDELINES

In present-day civil engineering practice, the fibre-reinforced soils have a great potential for applications in several areas. The key application areas are the following:

- Backfills behind the retaining structures
- Pavement bases and subbases, especially for low-volume roads
- Stabilization of shallow foundation soils
- Stabilization of failed slopes
- Earthwork/embankment construction over soft soils
- Lightweight fill materials
- Drainage layers
- Thermal insulator for limiting frost penetration
- Vibration damping layers
- Leachate treatment materials and cover liners in landfills

In comparison with systematically reinforced soils, the randomly distributed fibre-reinforced soils exhibit some advantages. The preparation of randomly distributed fibre-reinforced soils mimics the soil stabilization by admixtures. Discrete fibres are simply added and mixed with soil, much like cement, lime, or other additives. Randomly distributed fibres offer better strength isotropy and limit the potential planes of weakness that can develop parallel to the oriented reinforcement as included in systematically reinforced soil.

Use of randomly oriented discrete fibres for different applications as mentioned here requires investigation at a large-scale. Santoni and Webster (2001) constructed the field test sections of pavement and tested using simulated C-130 aircraft traffic with a 13,608 kg tyre load at 690 kPa tyre pressure and a 4,536 kg military cargo truck loaded to a gross weight of 18,870 kg. The test results showed that the sand-fibre stabilization over a sand subgrade supported over 1,000 passes of a C-130 tyre load with less than 51 mm of rutting. The top 102 mm of the sand-fibre layer was lightly stabilized with tree resin to provide a wearing surface. Based on the limited truck traffic tests, 203 mm thick sand-fibre layer, surfaced with a spray application of tree resin, would support substantial amounts of military truck traffic.

The influences of the engineering properties of soil and reinforcement and scale effects on the properties of the fibre-reinforced soils have not been investigated fully, and hence the actual behaviour of fibre-reinforced soil is not yet well known. Large-scale investigations of all the possible applications of fibre-reinforced soil are limited in the literature, and hence they are the subject of further study. Based on the available studies reported in the literature, some application guidelines are given below:

1   The stress-strain properties of fibre-reinforced soils are functions of fibre content, aspect ratio and skin friction along with the soil and fibre index and strength characteristics, and confining pressure. Thus, the design and construction of fibre-reinforced soil structures should properly consider all these aspects.
2   Because of the major influence of confining pressure, the design parameters should be based on the triaxial compression tests, especially on large-size specimens.
3   The method of fibre-reinforced soil placement should be similar to the method used for preparing fibre-reinforced soil specimens for tests conducted to obtain the design parameters. This is essential to keep the fibre orientations the same in both the tests and field applications.
4   In general, fibres are most influential when orientated in the same direction as the tensile strains. Therefore, for any particular loading condition, the effectiveness of fibre inclusions depends on their orientation, which in turn depends on the sample mixing and formation procedure. However, an attempt should be made to make the reasonably uniform distribution of the fibres within the soil mass.
5   More compaction energy may be necessary to produce specimens with higher fibre content at a given dry unit weight. Thus, the fibre-reinforced soil may provide an increased resistance to compaction, even in field compaction.
6   The water required to obtain the target water content should be added prior to placing the fibres on the soil to keep the fibres from sticking together during mixing.
7   The simplest method of mixing fibre and soil in a rotating drum mixer does not result in a uniform mixture due to a large difference in specific gravities between the fibre and soil. The drum mixing method usually results in the segregation of fibres even when some water is added. In fact, because of the light weight and low specific gravity of fibre (for some types it can be even less than unity) compared to soil particles, the fibre cannot be uniformly distributed in the mixture during drum rotation. An effective method of fibre inclusion can be spraying the fibres over each soil lift during field compaction, especially in pavement bases and subbases.
8   The use of natural fibres may make possible constructions which are cost-effective and environmentally friendly.

Note that reinforcing the soils with discrete flexible fibres can be a cost-effective means of improving their performance in several applications. Additionally the use of fibres from the waste materials can reduce the disposal problem in an economically and environmentally beneficial way. There is a need of further research to understand the behaviour of fibre-reinforced soils at a large scale so that the fibre-reinforcements can be routinely utilized based on a more rational analysis and design.

## Chapter summary

1  The shear strength and stiffness of soil increases with a reduced post-peak strength loss when discrete flexible fibres are mixed with the soil properly. The shear strength increase contributed by the presence of fibres is highly anisotropic because of preferred sub-horizontal fibre orientations in compacted condition.
2  The characteristics of fibre-reinforced soils are influenced by the fibre characteristics, soil properties and confining stress.
3  The orientation of fibres must be considered properly in analysis and design of fibre-reinforced soil structures.
4  An element of fibre-reinforced soil can be represented as a three-phase system. The volume ratio of fibre-soil solid is uniquely related to the void ratio of the fibre-reinforced soil.
5  The SSS model provides a more generalized expression for estimating the shear strength increase as a result of fibre inclusions in soil.
6  For field applications, the fibre content and the aspect ratio should be selected based on experimental observations with specific soil and fibres under consideration as these two parameters significantly govern the behaviour of fibre-reinforced soil, and moreover, they can be controlled easily.

## Questions for practice

(Select the most appropriate answer to the multiple-choice questions from Q 9.1 to Q 9.6.)

9.1  The specific gravity of PP fibres is
   (a)  0.91
   (b)  0.99
   (c)  1.38
   (d)  1.7.

9.2  The ratio of length of the fibre to its diameter is called
   (a)  fibre content
   (b)  aspect ratio
   (c)  area ratio
   (d)  geometrical ratio.

9.3  The tamping and vibration techniques for preparing the fibre-reinforced specimens in moist conditions lead to
   (a)  vertical orientation of fibres
   (b)  near-vertical orientation of fibres.
   (c)  horizontal orientation of fibres
   (d)  near-horizontal orientation of fibres.

9.4  With fibre inclusion in a low-plasticity soil, which of the following increases significantly?
   (a)  cohesion intercept
   (b)  angle of internal friction
   (c)  both (a) and (b)
   (d)  none of the above.

9.5   Fibre inclusion in pavement base soil
  (a)   increases the dry unit weight
  (b)   decreases the unconfined compressive strength
  (c)   increases the CBR value
  (d)   both (a) and (c).

9.6   Within the practical ranges, the shear strength of fibre-reinforced soil
  (a)   decreases with an increase in the fibre content, but increases with an increase in the aspect ratio of fibres
  (b)   increases with an increase in the fibre content, but decreases with an increase in the aspect ratio of fibres
  (c)   decreases with increase in both the fibre content and the aspect ratio of fibres
  (d)   increases with increase in both the fibre content and the aspect ratio of fibres.

9.7   Derive the following expressions in which the symbols have their usual meanings:
  (a)   $e = \dfrac{e_s + me_f}{1 + m}$
  (b)   $m = p_f \dfrac{G_s}{G_f}$
  (c)   $G = (1 + p_f)\left(\dfrac{1}{G_s} + \dfrac{p_f}{G_f}\right)^{-1}$

9.8   A sandy soil is reinforced with 1% of PP fibres. What will be the specific gravity of the fibre-reinforced soil? The specific gravity values for soil particles and fibres are 2.65 and 0.91, respectively.

9.9   List the fibre characteristics that may affect the engineering behaviour of fibre-reinforced soils.

9.10   Define the following: fibre content, aspect ratio and area ratio.

9.11   Fibres are available in the following two forms: (i) $L = 15$ mm, $D = 0.05$ mm; and (ii) $L = 25$ mm, $D = 0.1$ mm. If a sandy soil is to be reinforced, which form of PP fibres should be recommended? Justify your answer.

9.12   In the laboratory, prepare a mixture of sand and fibres by any suitable means in dry and wet conditions, and compare your observations about the fibre orientation with those reported by the researchers as presented in this chapter.

9.13   Explain the basic reinforcing mechanism of fibre-reinforced soil.

9.14   How do the compaction parameters of fibre-reinforced soil differ from those for unreinforced soil?

9.15   Under what conditions may the fibre reinforcement within the soil mass slip or rupture?

9.16   Using the SSS model, discuss the effects of the following on the shear strength of fibre-reinforced soil:
  (a)   soil-fibre interface friction angle ($\delta$)
  (b)   modulus of elasticity of fibres ($E_f$)
  (c)   angle of shearing resistance of soil ($\phi_s$)

9.17  What is the physical meaning of having a shear strength ratio (SSR) of unity for a fibre-reinforced soil?
9.18  List the potential application areas for fibre reinforcement. Visit a local construction site where the soil is being reinforced with fibres. Based on your observation, make a technical note, and identify the key aspects of the application technique.
9.19  Why should water be added prior to placing the fibres on soil in fibre-reinforced soil?
9.20  What is the most effective method of including fibre in soil? What are the difficulties in mixing fibre and soil in a rotating drum mixer in the same way as cement concrete is prepared?

## References

Al-Refeai, T. and Al-Suhaibani, A. (1998). Dynamic and static characterization of polypropylene fibre-reinforced dune sand. *Geosynthetics International*, 5, 5, pp. 443–458.

Babu, G.L.S., Vasudevan, A.K. and Sayida, M.K. (2008). Use of coir fibres for improving the engineering properties of expansive soils. *Journal of Natural Fibers*, 5, 1, pp. 61–75.

Consoli, N.C., Montardo, J.P., Prietto, P.D.M. and Pasa, G.S. (2002). Engineering behaviour of a sand reinforced with plastic waste. *Journal of Geotechnical and Geoenvironmental Engineering*, ASCE, 128, 6, pp. 462–472.

Consoli, N.C., Casagrande, M.D.T., Prietto, P.D.M. and Thome, A. (2003a). Plate load test on fibre-reinforced soil. *Journal of Geotechnical and Geoenvironmental Engineering*, ASCE, 129, 10, pp. 951–955.

Consoli, N.C., Vendruscolo, M.R.A. and Prietto, P.D.M. (2003b). Behaviour of plate load tests on soil layers improved with cement and fibre. *Journal of Geotechnical and Geoenvironmental Engineering*, ASCE, 129, 1, pp. 96–101.

Consoli, N.C., Casagrande, M.D.T., Thome, A., Rosa, F.D. and Fahey, M. (2009). Effect of relative density on plate loading tests on fibre-reinforced sand. *Geotechnique*, 59, 5, pp. 471–476.

Diambra, A., Ibraim, E., Wood, D.M. and Russell, A.R. (2010). Fibre reinforced sands: experiments and modelling. *Geotextiles and Geomembranes*, 28, 3, pp. 238–250.

Edincliler, A. and Cagatay, A. (2013). Weak subgrade improvement with rubber fibre inclusions. *Geosynthetics International*, 20, 1, pp. 39–46.

Fletcher, C.S. and Humphries, W.K. (1991). California bearing ratio improvement of remoulded soils by the addition of polypropylene fibre reinforcement. *Proceedings of the 7th Annual Meeting, Transportation Research Board*, Washington, DC, USA, pp. 80–86.

Gill, K.S., Choudhary, A.K., Jha, J.N. and Shukla, S.K. (2014). Utilization of fly ash and waste recycled product reinforced with plastic wastes as construction materials in flexible pavements. *Proceedings of the GeoCongress on Geo-Characterization and Modelling for Sustainability*, ASCE, Atlanta, Georgia, USA, GSP 234, pp. 3890–3902.

Gray, D.H. and Ohashi, H. (1983). Mechanics of fibre reinforcement in sand. *Journal of Geotechnical Engineering*, ASCE, 109, 3, pp. 335–353.

Gray, D.H. and Al-Refeai, T. (1986). Behaviour of fabric versus fiber reinforced sand. *Journal of Geotechnical Engineering*, ASCE, 112, 8, pp. 804–820.

Ibraim, E., Diambra, A., Wood, D.M. and Russell, A.R. (2010). Static liquefaction of fibre reinforced sand under monotonic loading. *Geotextiles and Geomembranes*, 28, 4, pp. 374–385.

Ibraim, E., Diambra, A., Russell, A.R. and Wood, D.M. and (2012). Assessment of laboratory sample preparation for fibre reinforced sands. *Geotextiles and Geomembranes*, 34, pp. 69–79.

Jha, J.N., Choudhary, J.N., Gill, K.S. and Shukla, S.K. (2014). Behaviour of plastic waste fibre-reinforced industrial wastes in pavement applications. *International Journal of Geotechnical Engineering*, **8**, 3, pp. 277–286.

Jha, J.N., Gill, K.S., Choudhary, A.K. and Shukla, S.K. (2015). Stress-strain Characteristics of fiber-reinforced rice husk ash. *Proceedings of the Geosynthetics 2015*, Portland, Oregon, USA, pp. 134–141.

Kumar, P. and Singh, S.P. (2008). Fibre-reinforced fly ash subbases in rural roads. *Journal of Transportation Engineering*, ASCE, **134**, 4, pp. 171–180.

Lovisa, J., Shukla, S.K. and Sivakugan, N. (2010). Shear strength of randomly distributed moist fiber-reinforced sand. *Geosynthetics International*, **17**, 2, pp. 100–106.

Maher, M.H. and Gray, D.H. (1990). Static response of sands reinforced with randomly distributed fibres. *Journal of Geotechnical Engineering, ASCE*, **116**, 11, pp. 1661–1677.

Mirzababaei, M., Miraftab, M., Mohamed, M. and McMahon, M. (2013). Unconfined compression strength of reinforced clays with carpet waste fibres. *Journal of Geotechnical and Geoenvironmental Engineering*, ASCE, **139**, 3, pp. 483–493.

Ranjan, G., Vasan, R.M. and Charan, H.D. (1994). Behaviour of plastic-fibre reinforced sand. *Geotextiles and Geomembranes*, **13**, 8, pp. 555–565.

Ranjan, G. Vasan, R.M. and Charan, H.D. (1996). Probabilistic analysis of randomly distributed fibre-reinforced soil. *Journal of Geotechnical Engineering*, ASCE, **122**, 6, pp. 419–426.

Shukla, S.K., Sivakugan, N. and Das, B.M. (2009). Fundamental concepts of soil reinforcement – an overview. *International Journal of Geotechnical Engineering*, **3**, 3, pp. 329–343.

Shukla, S.K. and Sivakugan, N. (2010). Discussion of "Fiber-reinforced fly ash subbases in rural roads' by P. Kumar and S.P. Singh". *Journal of Transportation Engineering*, ASCE, **136**, 4, pp. 400–401.

Shukla, S.K., Sivakugan, N. and Singh, A.K. (2010). Analytical model for fiber-reinforced granular soils under high confining stresses. *Journal of Materials in Civil Engineering*, ASCE, **22**, 9, pp. 935–942.

Shukla, S.K., Shahin, M.A. and Abu-Taleb, H. (2015). A note on void ratio of fibre-reinforced soils. *International Journal of Geosynthetics and Ground Engineering*, Switzerland, **1**, 3, pp. 1–5.

Tingle, J.S., Santoni, R.L. and Webster, S.L. (2002). Full-scale field tests of discrete fibre-reinforced sand. Journal of Transportation Engineering, ASCE, **128**, 1, pp. 9–16.

Santoni, R.L., Tingle, J.S. and Webster, S.L. (2001). Engineering properties of sand –fibre mixtures for road construction. *Journal of Geotechnical and Geoenvironmental Engineering*, ASCE, **127**, 3, pp. 258–268.

Santoni, R.L. and Webster, S.L. (2001). Airfields and roads construction using fibre stabilization of sands. *Journal of Transportation Engineering*, ASCE, **127**, 2, pp. 96–104.

Shewbridge, S.E. and Sitar, N. (1989). Deformation characteristics of reinforced sand in direct shear. *Journal of Geotechnical Engineering*, ASCE, **115**, 8, pp. 1134–1147.

Shewbridge, S.E. and Sitar, N. (1990). Deformation based model for reinforced sand. *Journal of Geotechnical Engineering*, ASCE, **116**, 7, pp. 1153–1170.

Yetimoglu, T. and Salbas, O. (2003). A study on shear strength of sands reinforced with randomly distributed discrete fibres. *Geotextiles and Geomembranes*, **21**, 2, pp. 103–110.

Yetimoglu, T., Inanir, M. and Inanir, O.E. (2005). A study on bearing capacity of randomly distributed fibre-reinforced sand fills overlying soft clay. *Geotextiles and Geomembranes*, **23**, 2, pp. 174–183.

Waldron, L.J. (1977). Shear resistance of root-permeated homogeneous and stratified soil. *Proceedings of the Soil Science Society of America*, **41**, 5, pp. 843–849.

## Answers to selected questions

9.1    (a)
9.3    (d)
9.5    (c)
9.8    2.60

# Subject index